BERRY PHASES IN ELECTRONIC STRUCTURE THEORY

Over the past 25 years, mathematical concepts associated with geometric phases have come to occupy a central place in our modern understanding of the physics of electrons in solids. These "Berry phases" describe the global phase acquired by a quantum state as the Hamiltonian is changed. Beginning at an elementary level, this book provides a pedagogical introduction to the important role of Berry phases and curvatures, and outlines their great influence upon many key properties of electrons in solids, including electric polarization, anomalous Hall conductivity, and the nature of the topological insulating state. It focuses on drawing connections between physical concepts and provides a solid framework for their integration, enabling researchers and students alike to explore and develop links to related fields. Computational examples and exercises throughout provide an added dimension to the book, giving readers the opportunity to explore the central concepts in a practical and engaging way.

DAVID VANDERBILT is Board of Governors Professor of Physics at Rutgers University, where he has made significant contributions to computational condensed-matter physics. He is a fellow of the American Physical Society and a member of the National Academy of Sciences, and was awarded the prestigious Rahman Prize in Computational Physics of the American Physical Society in 2006.

"This book brings together almost forty years of progress in understanding how the wavefunctions of electrons in a crystal, and in particular their continuous evolution with momentum, determine important physical properties. David Vanderbilt is one of the creators of this field, and nearly every chapter includes topics where his contributions were decisive. In addition to its scope, one way in which this book differs from others on related topics is the clear path from physical insight, through theoretical understanding, to practical methods for specific materials. This book can be read profitably by those interested in the fundamental theory of topological phases as well as those seeking to understand modern electronic structure approaches."

Joel Moore
Chern-Simons Professor of Physics, UC Berkeley

"The geometric phase and related concepts provide a unified framework for describing many fundamental properties of electrons in solids, from electric polarization to quantized effects in topological materials. Readers wishing to become familiar with these notions will find David Vanderbilt's excellent book to be an invaluable resource."

Ivo Souza
University of the Basque Country, San Sebastián

"Berry phases and associated geometric and topological concepts have transformed our understanding of electronic properties. This book provides a much needed pedagogical exposition with computational instructions which will be very valuable for students and researchers in solid state physics and materials science."

Qian Niu
University of Texas

"David Vanderbilt explicates a new exciting frontier in solid state physics and materials theory, and does so in a clear and interesting to read way. Not only does he cover every nook and cranny of this new area, but in the process clearly explains the basics of electronic structure theory, such as density functional theory (DFT) and tight-binding, that will be extremely useful and important to any student of condensed matter theory. The subject of the book is how the phases of the wave functions, neglected for decades, effects important measurable properties of materials. He covers everything from the mathematical theory of geometric phases, applications to polarization and orbital magnetism, all the way to complex applications such as three-dimensional topological insulators and beyond. To be able to write about such seemingly esoteric matters in such a clear and gripping way is the mark of a great teacher. I look forward to my second reading of the book!"

Ronald Cohen
Extreme Materials Initiative, Geophysical Laboratory, Carnegie Institution for Science

"For anyone who wants to learn about Berry phases in electronic structure and the exciting recent developments in topological insulators, I heartily recommend this book. David Vanderbilt is uniquely poised to present the concepts and practical developments in this field that has revolutionized our understanding of condensed matter. He has made some of the most important advances in electronic structure theory in the last 20 years, including the original work that has made Berry phases a central part the field, and he is known for lucid presentations. In this book Vanderbilt introduces the concepts in a way that is accessible to a nonexpert, with clear explanations and instructive examples, and yet he presents the material in the depth that it deserves. The book provides the reader with the background that is needed to understand each aspect: an excellent introduction to Berry phases and aspects of topology relevant to electrons in crystals, and the needed background in electronic structure theory. To give just one example, Wannier functions have become increasingly important in many ways, and this book provides an elegant way to understand their properties. The last three chapters are a coordinated presentation of the three most important roles of Berry phases: electric polarization, topological insulators and Weyl semimetals, and orbital magnetization. Each of these has aspects that were completely unknown or not understood only a few years ago, but now have elegant, simple explanations in terms of Berry phases. I recommend this book for anyone who wants to be a part of condensed matter theory in the twenty-first century or just to appreciate the basic ideas and phenomena of this exciting field."

Richard M. Martin
University of Illinois Urbana Champaign

BERRY PHASES IN ELECTRONIC STRUCTURE THEORY

DAVID VANDERBILT
Rutgers University

CAMBRIDGE
UNIVERSITY PRESS

University Printing House, Cambridge CB2 8BS, United Kingdom

One Liberty Plaza, 20th Floor, New York, NY 10006, USA

477 Williamstown Road, Port Melbourne, VIC 3207, Australia

314-321, 3rd Floor, Plot 3, Splendor Forum, Jasola District Centre, New Delhi - 110025, India

79 Anson Road, #06-04/06, Singapore 079906

Cambridge University Press is part of the University of Cambridge.

It furthers the University's mission by disseminating knowledge in the pursuit of education, learning and research at the highest international levels of excellence.

www.cambridge.org
Information on this title: www.cambridge.org/9781107157651
DOI: 10.1017/9781316662205

First published 2018

A catalogue record for this publication is available from the British Library

Library of Congress Cataloging in Publication data
Names: Vanderbilt, David, author.
Title: Berry phases in electronic structure theory /
David Vanderbilt (Rutgers University, New Jersey).
Description: Cambridge : Cambridge University Press, 2018. |
Includes bibliographical references and index.
Identifiers: LCCN 2018018455 | ISBN 9781107157651 (Hardback)
Subjects: LCSH: Electronic structure. | Geometric quantum phases. |
Quantum electronics. | Electronics.
Classification: LCC QC176.8.E4 V36 2018 | DDC 530.4/11–dc23
LC record available at https://lccn.loc.gov/2018018455

ISBN 978-1-107-15765-1 Hardback

Additional resources for this publication at http://minisites.cambridgecore.org/berryphases

Contents

Preface

This book introduces a class of mathematical methods connected with geometric phases, also known as Berry phases, and the important role that these have come to play in recent years in the band theory of electrons in crystals. The "Berry phase" terminology derives from the seminal papers of Michael Berry in the mid-1980s, in which he highlighted the importance of geometric phases in a wide range of physical contexts and advocated for a recognition of geometric phases as representing a new paradigm in theoretical physics. These concepts rapidly came to play a central role in aspects of atomic, molecular, nuclear, and optical physics, but were initially rather slow to enter into the theory of the electron bands in crystalline materials, commonly known as electronic structure theory.

This began to change in the 1990s with the introduction of the "modern theory of polarization," and then in the early 2000s with developments in the theory of Wannier functions and the emergence of a proper theory of orbital magnetization. However, the pace accelerated greatly with the exciting developments in the theory of topological insulators in the mid- to late 2000s, and by 2010 it was clear that the role of Berry phases in the electronic structure of crystalline materials had become very topical. In the years since then, the field has been advancing rapidly, with the interest in topological states of crystalline materials developing in many directions.

These advances convinced me of the need for a pedagogical text that could capture and synthesize the core of these developments. With this in mind, I began the process of turning some informal lecture notes, prepared earlier for a graduate-level special-topics course, into a proper text. The opportunity to teach this course again in 2016 provided a further impetus to continue the writing, and also to develop an accompanying set of computational examples and exercises. The scientific understanding of topological phases continued to develop rapidly during the course of the writing, with topological semimetals and other novel topological states attracting much attention. Where possible, I have expanded the later chapters to reflect some of this recent progress as well.

This book is the result.

I owe a debt of gratitude to a large number of individuals who provided invaluable comments and suggestions as this book developed. First, I warmly acknowledge the feedback offered by the students in the special-topics courses I taught on this subject, as well as other students and postdocs who were at Rutgers during those years. Among those who notably contributed useful ideas, comments, and corrections are Victor Aleksandrov, Sinisa Coh, Philipp Eck, Qiang Han, Wenshuo Liu, Jisoo Moon, Alexey Soluyanov, Maryam Taherinejad, Nicodemos Varnava, Wenhan Wu, and Tahir Yusufaly. Others who gave a careful reading to some sections and provided me with especially valuable feedback and corrections include Heung-Sik Kim, Se Young Park, and Shuchen Zhu. I also wish to thank Barry Bradlyn, Charlie Kane, Joel Moore, and Qian Niu for assistance in clarifying some subtle points of physics, and for helping me to understand how best to present certain topics that lie somewhat outside the scope of my own expertise.

Special thanks are due to Sinisa Coh for his role as the prime developer and maintainer of the PYTHTB code package, which plays such an important part in the scheme of this book.

I wish to thank the Simons Foundation for providing a Simons Fellowship in Theoretical Physics in 2014, which funded a sabbatical extension during which the first part of this book was written. In addition, I thank the Department of Chemistry at Princeton University for providing a visiting appointment that allowed me to devote some quiet days to writing away from the distractions of my office at Rutgers. I would also like to acknowledge the importance of research funding from the National Science Foundation and the Office of Naval Research, which provided resources in support of the research conducted in my group on the topics topics discussed here.

I would especially like to thank my colleagues in the Department of Physics and Astronomy at Rutgers University for providing a consistently supportive intellectual environment over the course of my career at Rutgers.

Three other individuals deserve special mention. Richard Martin, who provided valuable comments on pedagogical approaches to the material presented here, has been a constant source of inspiration and support over the course of my professional career. Raffaele Resta, who likewise provided important feedback on the structure of the book, is perhaps the person whose scientific career most closely parallels the topics addressed here. Our interactions over the years have been deeper and more pervasive than the occasional formal collaborations might suggest, and his scientific impact on me and on the field has been profound. Finally I must acknowledge Ivo Souza, with whom I have collaborated frequently over the years on topics connected with the themes of this book. Ivo has been the most conscientious reader of my draft chapters as they have developed. He has frequently suggested ways of improving or extending a line of argument, and has been relentless in finding errors and inconsistencies in the text. His assistance has been inestimable, and I thank him profoundly.

I am deeply grateful to my wife Roslyn, whose love and encouragement have been a great source of support in the course of writing this book.

Lastly, I would like to thank all of the graduate students and postdocs who have worked with me over the years. We have shared the excitement and joy of developing some of the concepts discussed in this book, designing computational schemes to implement them, and exploring their physical consequences. It has been a great journey, and a pleasure to have all of them along for the ride.

Acronyms

1D	one-dimensional
2D	two-dimensional
3D	three-dimensional
4D	four-dimensional
AHC	anomalous Hall conductivity
ARPES	angle-resolved photoemission spectroscopy
bcc	body-centered cubic
BZ	Brillouin zone
CS	Chern–Simons
DFT	density-functional theory
fcc	face-centered cubic
GGA	generalized gradient approximation
HWF	hybrid Wannier function
IC	itinerant circulation
LC	local circulation
LCAO	linear combination of atomic orbitals
LDA	local-density approximation
LSDA	local spin-density approximation
ME	magnetoelectric
MLWF	maximally localized Wannier function
QAH	quantum anomalous Hall
QSH	quantum spin Hall
SI	international system of units
SOC	spin-orbit coupling
SVD	singular value decomposition
TB	tight binding
TI	topological insulator
TR	time reversal
TRIM	time-reversal invariant momentum (or momenta)

Introduction

Two linked concepts have been central to the theory of quantum mechanics since its development nearly a century ago. One states that the probability of a measurement is given by the squared norm of the corresponding quantum amplitude, so that the *phase* of that amplitude has no importance to the final measurement. The other asserts that the interference phenomena associated with phase *differences* can modulate the amplitudes, thereby giving rise to measurable effects. One famous example of the latter is the double slit experiment. Another is the Aharonov–Bohm effect, in which the interference that occurs when a beam of charged particles is split and recombined is found to depend on the magnetic flux enclosed between the two legs of the path. Nevertheless, we tend to regard these as exceptional cases, and for most practical purposes we often think it is safe to ignore quantum phases.

Until recently, this has been the dominant point of view in solid state physics. In the theory of the electronic structure of crystalline systems, almost all quantities of interest can be expressed as Bloch-state expectation values, including the band energies, charge densities, spin densities, local densities of states, total energies, forces, stresses, spin magnetic moments, and so on. In computing such quantities, the overall phase in front of any given Bloch function plays no role.

In a paradigm shift that has occurred over the last 25 years, however, it has become clear that the phases of the electronic states are essential for the description of some properties of crystalline materials, including electric polarization, orbital magnetization, and anomalous Hall conductivity. In fact, the overall phase of a given Bloch function is still immaterial, but the *relative phase* between states at nearby times, or for nearby Hamiltonians, or for nearby wavevectors of Bloch functions, can have crucial physical consequences. In these cases, the appropriate formalism is that of a geometric phase, or Berry phase, which is defined in terms of a global phase difference that accumulates when tracing the evolution of a quantum state in time, or along a path in the parameter space of the Hamiltonian, or along a path in the space of Bloch wavevectors (i.e., the Brillouin zone) of a crystalline system.

In the last 15 years, a related paradigm shift has occurred with the discovery and development of the theory of topological insulators. Previously, the distinction between different phases of matter had almost always rested on Landau-theory concepts of symmetry and order parameters. By contrast, the topological insulator Bi_2Se_3 is known to be identical with its trivial analogue Sb_2Se_3 in all respects as far as symmetry is concerned, and no order parameter can be defined to distinguish them. The difference is that the Bloch functions in Bi_2Se_3 are twisted in a manner connected with their phase evolution across the Brillouin zone, while those of Sb_2Se_3 are not. Again, Berry phases and curvatures are at the heart of these developments.

The purpose of this book is to give a pedagogical introduction to the mathematics of Berry phases and the closely related concepts of Berry connections, Berry curvatures, and Chern numbers, and then explain how these concepts have come to play a central role in several aspects of electronic structure theory. In particular, they have come to underlie the modern theories of electric polarization, anomalous Hall conductivity, orbital magnetization, magnetoelectric coupling, and topological insulators and semimetals.

The intended audience is that of advanced graduate students and researchers in the fields of condensed matter and electronic structure theory. A solid background in quantum mechanics and solid state physics, preferably at the graduate level, is assumed. Virtually all of the concepts described in this book rely only on a description at the single-particle level, so that the formalisms and techniques of many-body physics are not essential here. A prior familiarity with electronic structure calculations in the density-functional context, or the simpler tight-binding formulation, will be useful but not essential. The book is intended to be accessible to both experimental and theoretical students, although the former may wish to skim over a few sections that have a more formal mathematical flavor.

The book is organized as follows. The first chapter provides a pedagogical introduction to the *physics* of the concepts and phenomena to be discussed in the book. That is, I explain why, on physical grounds, we might expect the electric polarization to be defined only modulo a quantum, or why the anomalous Hall conductivity of an insulator should be quantized. Since the proper introduction of the mathematical formalism needed to discuss these topics has been deferred to later chapters, this has entailed some compromises. Readers who find this chapter to be hard going are encouraged to skim or skip some sections, or the entire chapter, and then return to it later after working through some of the subsequent material.

The second chapter gives a brief review of the theory of electrons in crystalline solids. Most of this coverage should be familiar to the typical reader, but the chapter also provides an introduction to tight-binding theory and its implementation in the open-source PYTHTB package, which will be used for examples and exercises throughout the later chapters, and to linear response theory. The mathematical framework of the theory of Berry phases (also known as geometrical phases) and curvatures is then carefully introduced in Chapter 3. Initially this is done in the context of the ground state of a finite system whose Hamiltonian is undergoing an adiabatic deformation described by one or more external parameters. The same theory is then applied to crystalline solids in which the wavevector of the Bloch eigenstates serves as an internal parameter, possibly together with external parameters. The treatment here is primarily mathematical, but some hints will appear as to the physical significance of the Berry-related quantities.

The remainder of the book is devoted to developing those connections between the mathematics and physics. Chapter 4 is devoted to the Berry-phase theory of electric polarization, quantized adiabatic charge transport, and surface charge. The ability to compute the Berry-phase polarization is now a standard feature of almost all well-known electronic-structure code packages, but subtleties often arise in the interpretation of the results, and these are discussed at some length. Chapter 5 then introduces the theory of topological insulators and semimetals, starting with the 2D quantum anomalous Hall state (which we may regard as the "mother of all topological insulators"), and then discussing 2D quantum spin Hall insulators, 3D strong topological insulators, topological crystalline insulators, and Weyl semimetals. Our story then concludes in Chapter 6 with a discussion of orbital magnetization and magnetoelectric coupling, and especially the role of the Chern–Simons "axion" coupling, which is closely related to the theory of topological insulators.

The book is structured so that it can form the basis of an advanced special-topics course at the graduate level. It should also serve well for a motivated individual interested in learning the material in an independent context. Exercises are provided throughout, and in formulating these I have tried to strike a balance between theory and practice. Many of the exercises are of the traditional form where students are asked to complete a derivation, apply a theory to a special case, or extend some argument given in the text. The remainder are computational in nature and require that students familiarize themselves with the use of the PYTHTB code package, which is detailed in Appendix D. This open-source package is written in the PYTHON programming language, and can be downloaded from standard PYTHON repositories or from the Cambridge University Press website at http://minisites.cambridgecore.org/berryphases. An elementary knowledge of

PYTHON programming is required, but this is not hard to pick up with the aid of the examples given and the various PYTHON tutorials that can be found online. None of the calculations are computationally demanding, so any laptop or desktop computer with PYTHON installed can be used to complete these exercises. While it would be possible to teach this material without making use of the computational exercises, I believe that the ability to illustrate the formal theory by practical computations adds an important element to the learning experience.

1

Invariance and Quantization of Charges and Currents

The theoretical framework of solid state physics is built on two principal pillars, those of electromagnetism and quantum mechanics. The former is needed to specify the microscopic forces acting between electrons and nuclei, while the latter provides the framework for describing states of the system. The close marriage of electromagnetism and quantum mechanics is taken for granted at this fundamental level.

However, there is another, more macroscopic level at which electromagnetism and quantum mechanics become intimately intertwined in the physics of materials. Here, there are subtleties that are not often emphasized in the standard texts. For example, virtually all elementary electrodynamics texts discuss the electrical and magnetic properties of material media and introduce the concepts of macroscopic electric polarization and magnetization. However, this discussion is usually carried out in the context of a simplified model in which solids are assembled out of nonoverlapping classical charge or current distributions representing "atoms" or "molecules," with little real attention paid to the quantum nature of the sources of polarization and magnetization.

Indeed, in real polarized solids, as, for example, in ferroelectric $BaTiO_3$, the electrons are distributed throughout the crystal according to the Bloch solutions of the Schrödinger equation. The resulting electron charge distribution peaks around the atomic positions, of course, but it does not fall to zero between atoms. As a result, there is no natural way to decompose the electron charge into "polarized units." Similar problems exist for the orbital currents that give rise to macroscopic magnetization. The puzzle of how properly to define polarization and magnetization, and how to compute their responses to external fields and strains, has been a recurrent theme of theoretical and computational condensed-matter theory, with important new developments continuing to occur even into the present decade.

But there is yet another sense in which electromagnetism and quantum mechanics can combine to produce effects – sometimes surprising and beautiful

effects – at the macroscopic scale, especially for insulating systems. The quintess-
ential example is the quantum Hall effect, discovered in 1980, in which a two-
dimensional electron gas is subjected to a perpendicular magnetic field at low
temperature. If the Fermi level lies in a mobility gap between a pair of adjacent
Landau levels, then a transverse conductivity appears that is quantized to an integer
multiple of e^2/h, where e is the charge quantum and h is Planck's constant.

This brings us to the theme of the present chapter. The discovery of the
quantum Hall effect provided one of the first hints that concepts of quantization,
invariance, and topology may play an important role in condensed-matter theory.
There were, of course, many other hints. The most obvious was the phenomenology
of superconductivity, where quantization appears prominently in such properties as
the magnetic flux in vortices and the Josephson-junction response. The discovery
of the fractional quantum Hall effect in 1982 demonstrated that such quantization
concepts could also play a central role in strongly correlated systems. More
recently, the blossoming of interest in "topological insulators," which will be
discussed in Chapter 5, has led to a renewed interest in, and understanding of, the
role of topology in classifying phases of matter.

The purpose of this chapter is to introduce and describe a class of phenomena
in which concepts of topology and quantization play an important role, beginning
with "ordinary" electric and magnetic responses of conventional materials, but
especially including the anomalous Hall conductivity, and eventually connecting
with the physics of topological insulators.

The emphasis in this chapter will be on physical arguments; the introduction
of the formalism needed for a proper discussion of these phenomena will be
deferred to later chapters. Nevertheless, the physical arguments to be presented here
constrain the form of any candidate theory. For example, it will become clear in
Section 1.1 that any proposed definition of electric polarization that does not allow
for an ambiguity modulo a quantum must necessarily be incorrect, while the same
statement does not apply to orbital magnetization. Similarly, strong arguments will
be given as to why the anomalous Hall conductivity of a two-dimensional insulator
must be quantized, suggesting a connection to some kind of topological index. As
we shall see, these expectations are indeed satisfied by the detailed developments
that will be presented in the later chapters.

The concept of adiabatic evolution of a crystalline insulating system as a function
of one or more external parameters will play an important role in many of the
arguments to be given in this chapter. Moreover, in most cases we will argue
that the conclusions can be extended from noninteracting to interacting systems,
as long as an adiabatic connection exists between them. On the one hand, by a
noninteracting system we mean one whose electronic Hamiltonian is built entirely
out of one-particle operators. Typically these include the electronic kinetic energy

$\sum_i \mathbf{p}_i^2/2m$ and potential energy $\sum_i V(\mathbf{r}_i)$, where $\mathbf{r}_i$ and $\mathbf{p}_i$ are the coordinates and momenta of the i'th electron and $V(\mathbf{r})$ is the crystal potential, but the system may also be augmented to account for spinor electrons, external fields, or relativistic corrections. The eigenstates of such a system can be regarded as a single Slater determinant constructed from the occupied one-particle orbitals. On the other hand, *interacting* systems are those that also have two-particle or higher interactions in the Hamiltonian, such as the all-important Coulomb interaction $\sum_{ij} e^2/|\mathbf{r}_i - \mathbf{r}_j|$ involving the inverse of the distance between pairs of electrons. While the physical arguments to follow will mostly be formulated in the noninteracting context, they usually apply equally well to any interacting system that can be reduced to a noninteracting one along an adiabatic path that takes the interaction strength to zero, provided the system remains insulating everywhere along the path. In such cases we shall say that the properties in question are *robust against interactions*.

Of course, in real insulating materials such as Si or MgO or $BaTiO_3$, the two-particle Coulomb interaction is enormously important, and a theory in which it is neglected would be a disaster. Instead, we almost invariably start from a mean-field approximation in which each electron feels an electrostatic potential coming from the nuclei and the *average* charge cloud of the other electrons. In most cases, this noninteracting mean-field Hamiltonian is sufficiently close (in some physical sense) to the true interacting one that an adiabatic path can, in principle, connect them. The mean-field theory may be Hartree–Fock theory, or more commonly, density-functional theory, in which the mean-field potential contains Hartree, exchange, and correlation terms that approximate the effects of the true two-body Coulomb interactions (see Section 2.1.1). However, it should be kept in mind that for some strongly interacting systems, such as those displaying superconductivity or the fractional quantum Hall effect, no such path exists and the conclusions may have to be modified. In such cases, if one imagines turning up the interaction strength from zero, there is typically a critical interaction strength at which some kind of phase transition occurs; beyond this point the system has changed qualitatively, so that arguments based on an adiabatic connection no longer apply.

The concept of "robustness against interactions" described previously has a counterpart in *robustness against disorder*. Here one imagines starting with a perfect crystal and smoothly tuning the strength of the disorder up from zero. If the Fermi energy stays in a mobility gap, one may be able to argue that some claimed properties persist, at least up to some critical disorder strength at which the behavior changes qualitatively. We shall see examples of such arguments later in this chapter.

Finally, let me add a word of encouragement for those who may have difficulty following some of the arguments in this chapter on a first reading. You may encounter concepts that have not yet been adequately introduced, or arguments that you cannot absorb in detail. You should feel free to skim these portions of the text, just catching the flavor of the arguments and getting a feel for the motivations, or even to skip sections entirely. The picture should become clearer after we get into the details in the subsequent chapters, at which time it may be useful to return to this chapter and give some parts of it a second reading.

1.1 Polarization, Adiabatic Currents, and Surface Charge

The macroscopic electric polarization $\mathbf{P}$, which carries the physical meaning of dipole moment per unit volume, is one of the central concepts of the electrodynamics of material media. However, the proper definition of polarization for a crystalline insulator becomes far from obvious when the electrons in the solid are treated quantum mechanically.

To illustrate the problem, consider the sketch in Fig. 1.1(a), which typifies the discussion in most electromagnetism textbooks. The solid is considered to be decomposable into polarized entities, perhaps atoms or molecules, which are well separated by vacuum regions in which the charge density falls to zero. In this case the obvious choice is to define $\mathbf{P} = \mathbf{d}/V_{\text{cell}}$, where $\mathbf{d}$ is the dipole moment of the entity and V_{cell} is the unit cell volume. By comparison, Fig. 1.1(b) shows what a charge-density contour plot might look like in a real polarized crystal. The charge does not come subdivided into convenient packages; on the contrary, the quantum charge cloud extends throughout the unit cell.

In this case, how are we to define a dipole per unit cell? An approach that may seem reasonable at first sight is to make some choice of unit cell and let

$$\mathbf{P} = \frac{1}{V_{\text{cell}}} \int_{\text{cell}} \mathbf{r}\,\rho(\mathbf{r})\,d^3r, \tag{1.1}$$

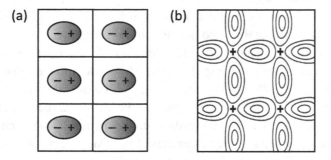

Figure 1.1 (a) Sketch of charge distribution in an idealized classical insulating crystal. Charges are decomposed into isolated packages well separated by charge-free interstitial regions. (b) The charge distribution in a realistic insulator in which the electrons are treated quantum mechanically. No such interstitial regions exist.

that is, we compute the dipole moment of the (nuclear and electronic) charge density $\rho(\mathbf{r})$ inside this cell and divide by the cell volume. Unfortunately, as we shall see later in Section 4.1, this does not lead to a unique answer; different choices of unit cell boundaries lead to different results for **P**. Of course, for a centrosymmetric crystal it is natural to choose a cell centered on an inversion center, which will give zero as it should, but what should we do for a noncentrosymmetric crystal? This is, after all, precisely the case where **P** matters.

It turns out that the problem goes quite deep. According to the "modern theory of polarization," developed in the 1990s, we now understand that even if the periodic charge density $\rho(\mathbf{r})$ is perfectly known, this information *is not enough, even in principle,* to determine **P**. Some other ingredients are needed. We seek a *bulk* definition of **P**, so the answer is not to use information about the surfaces; instead, we must find some deeper way of characterizing the ground state in the interior of the crystal that goes beyond the information carried in $\rho(\mathbf{r})$.

In Section 4.2 we shall derive a proper bulk expression for **P**, but the focus of our present interest is to ask a different question. Here we inquire: based on physical principles, what properties must any hypothetical bulk definition of electric polarization exhibit to have a chance of being correct? For example, we expect any valid formula for the polarization to produce results that respect the symmetries when symmetries are present. Are there other principles that constrain the form of a possible definition of polarization?

First, consider a crystalline insulator that is changing slowly in time, but in such a way that it remains unit-cell periodic. This might, for example, result from the application of a weak external electric or magnetic field, or from the slow motion of a zone-center optical phonon. If this causes **P** to change, then we expect there to be a macroscopic current density **J** flowing in the interior of the crystal[1] and satisfying

$$\mathbf{J} = \frac{d\mathbf{P}}{dt}. \tag{1.2}$$

In general, a part of this current will be given by the displacements of the positively charged nuclei, but certainly the electrons must contribute as well.

Second, consider a static situation in which there is a slow *spatial* variation of the crystal Hamiltonian, such that the crystal Hamiltonian H and the resulting **P** vary slowly with a macroscopic spatial coordinate **r** that is regarded as being defined only on length scales much larger than the atomic one. Then we expect a macroscopic charge density

$$\rho(\mathbf{r}) = -\nabla \cdot \mathbf{P}(\mathbf{r}) \tag{1.3}$$

---

[1] The macroscopic **J** is by definition the unit-cell average of the microscopic current density $\mathbf{J}(\mathbf{r})$ deep in the bulk.

to appear. This is what is referred to as a "bound" charge density in elementary electrostatics texts. Note that Eqs. (1.2) and (1.3) are consistent with each other in the case that the system varies slowly in both space and time, since then the charge conservation condition $\dot{\rho} = -\nabla \cdot \mathbf{J}$ is evidently satisfied.

Third, consider a static situation in which a crystalline insulator with polarization $\mathbf{P}$ has a surface normal to $\hat{\mathbf{n}}$. We expect bound charge to accumulate at the surface, and if the surface is insulating such that free charge is absent, we expect a relation of the form

$$\sigma_{\text{surf}} = \mathbf{P} \cdot \hat{\mathbf{n}}, \tag{1.4}$$

where σ_{surf} is the macroscopic surface charge. As we shall see shortly, Eq. (1.4) is not quite correct as written, but we expect that some relation like Eq. (1.4) must hold.

Eqs. (1.2–1.4) provide powerful constraints on any bulk theory of electric polarization. For example, they provide strong hints that the polarization $\mathbf{P}$ cannot be uniquely defined; instead, it is only well defined modulo a "quantum," in a sense that will be made precise shortly. This remarkable result is at the heart of the modern theory of polarization. Let us see how this comes about.

1.1.1 Surface Charge

Figure 1.2(a) shows a toy model of an ionic crystal such as NaCl; we can take the filled and open circles to represent cations and anions, respectively, which form an ideal checkerboard lattice in the bulk. However, atoms typically relax at the surface, as, for example, in Fig. 1.2(b), where the cations are shown relaxing outward relative to the anions. We imagine computing the fully quantum-mechanical electron charge densities for cases (a) and (b), and then extracting the macroscopic surface charge σ_{surf} in both cases. We now ask: what is the change of σ_{surf} as the atoms relax, taking the system from (a) to (b)?

The answer is exactly zero!

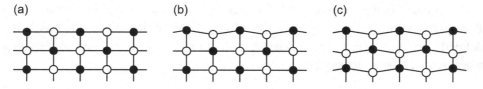

Figure 1.2 (a) Initial configuration of the surface of an ionic crystal composed of cations and anions indicated by filled and open circles, respectively. (b) Relaxed surface configuration, but with no change of bulk structure. (c) Surface structure after a bulk polar distortion has been applied.

Before defending this remarkable statement, we have to add some conditions –
namely, that the surface and bulk are both insulating, with the Fermi level falling in
a gap common to both the bulk and the surface. Also, this "experiment" has to be
done in complete vacuum, so that no charge can arrive at the surface from above.

The argument runs as follows. Imagine slowly moving the surface atoms from
configuration (a) to configuration (b), and ask what current flows into the surface
region. None can come from the vacuum, and none can travel laterally along the
surface, because of the conditions specified in the previous paragraph. Moreover,
the atoms do not move deep in the bulk during this surface distortion, so the
current $\mathbf{J}$, and in particular its component J_z normal to the surface, remain exactly
zero there. So if no current flows into the surface region, the net surface charge
cannot change; that is, the surface charge density σ_{surf} is exactly invariant.[2]

This argument is only valid so long as the surface remains insulating, since
otherwise the current flowing into the surface region is uncontrolled. There is also
another, very important, implicit assumption: that the charge $\rho(\mathbf{r})$ and adiabatic
current $\mathbf{J}(\mathbf{r})$ in an insulator depend only on the local Hamiltonian and its (slow)
time variation. This is often referred to as the "nearsightedness principle," a phrase
coined by Kohn (1996) and elaborated by Prodan and Kohn (2005), who have
argued that it is a universal property of all insulators, even with interactions
and disorder. The basic concept is that if one makes a local perturbation, such
as a Ge substitution in a Si crystal, the resulting change in charge density
decays exponentially away from the site of the perturbation with a decay length
ξ characteristic of the insulator.[3] Strictly speaking, this statement is true only
when charge self-consistency is neglected; clearly a charged impurity in Si would
generate a long-range electric field that polarizes the medium far away, so that the
change in the microscopic $\rho(\mathbf{r})$ would have a longer power-law decay. In many
cases, however, the conclusions can be shown to remain unchanged when charge
self-consistency is carefully taken into account.[4]

Returning to our main line of argument, we have seen that the surface charge
density σ_{surf} of an insulator is exactly unchanged by a variation of the surface
Hamiltonian, provided that the surface remains insulating along the path of
variation. But this is also what we should expect based on Eq. (1.4), since the

---

[2] This argument can be tightened up by introducing a "pillbox" with in-plane radius R extending from a height H
above the surface to a depth D below it, where R, H, and D are all much larger than the atomic length scale a,
and relating the macroscopic currents flowing into the pillbox with the change of surface charge density inside.

[3] Roughly speaking, ξ is determined by the exponential decay of the Green's function $G(\mathbf{r}, \mathbf{r}'; E)$ with $|\mathbf{r} - \mathbf{r}'|$
for E deep in the gap.

[4] For example, to argue that the change in the surface charge $\Delta\sigma$ between Figs. 1.2(a–b) vanishes, one can use a
2D Fourier analysis of the Poisson equation to show that if $\Delta\sigma = 0$, the microscopic charge variations produce
a Coulomb perturbation that decays exponentially into the bulk; moreover, if, beyond this decay length scale,
the bulk is unperturbed, then no current can flow to the surface. Having found a self-consistent solution with
$\Delta\sigma = 0$, one then invokes uniqueness to conclude that this is *the* solution.

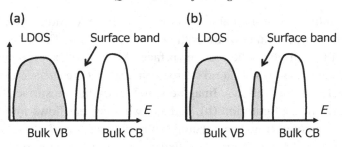

Figure 1.3 (a) Sketch of the local density of states (LDOS) at the surface of an insulating crystal, where there is a surface band (center) that lies entirely within the bulk gap between the valence band (VB) and conduction band (CB). An insulating surface results when the surface band is either (a) entirely empty or (b) completely filled.

polarization $\mathbf{P}$ deep in the bulk is unchanged by the adiabatic variation at the surface. Furthermore, we can consider making some slow variation of the *bulk* Hamiltonian instead, as shown in Fig. 1.2(c) for the case in which the bulk atoms have been displaced according to the pattern of a polar zone-center phonon. In this case, a macroscopic current $\mathbf{J} \cdot \hat{\mathbf{n}}$ flows into the surface region, so that $d\sigma_{\mathrm{surf}}/dt = \mathbf{J} \cdot \hat{\mathbf{n}}$ or, using Eq. (1.2), $d\sigma_{\mathrm{surf}}/dt = \hat{\mathbf{n}} \cdot d\mathbf{P}/dt$. Integrating with respect to time, we again conclude that a relation like Eq. (1.4) should be correct.

However, Eq. (1.4) cannot quite be correct as written! That is, if we assume that Eq. (1.4) is correct for any insulating surface of a crystalline insulator, and if we assume that $\mathbf{P}$ is a unique local property of the bulk Hamiltonian, we arrive at a contradiction. For example, consider a case like that shown in Fig. 1.3. The sketch shows the local density of states at the surface of an insulating crystal, with the shaded region at left and the unshaded region at right corresponding to the occupied bulk valence band and empty bulk conduction band, respectively. We also assume that there is a surface-state band whose bandwidth is small enough that it fits entirely inside the bulk gap, as shown in the figure. In this case we can fulfill the insulating-surface condition in two ways: we can either put the surface Fermi level E_F in the lower gap, leaving the surface band empty, as in Fig. 1.3(a), or we can raise E_F until the surface band is filled as in Fig. 1.3(b). The surface charge density clearly differs by $-e/A_{\mathrm{surf}}$ (or by $-2e/A_{\mathrm{surf}}$ for spin-paired electrons), where $e > 0$ is the charge quantum and A_{surf} is the surface unit-cell area for a surface normal to $\hat{\mathbf{n}}$.

How can two different values of σ_{surf} be consistent with a unique $\mathbf{P}$ in Eq. (1.4)? It seems we either have to give up on the notion of a bulk definition of electric polarization entirely, or else have to broaden the notion such that different "branch choices," differing from one another by an integer multiple of some "quantum," are equally valid. While the latter point of view may seem surprising and unnatural at first sight, in what follows we give several further arguments designed to show that it is, in fact, the correct approach.

The conclusion to be drawn from the preceding argument is that we have to modify Eq. (1.4) to become

$$\sigma_{\text{surf}} = \mathbf{P} \cdot \hat{\mathbf{n}} \quad \text{mod} \quad \frac{e}{A_{\text{surf}}}, \tag{1.5}$$

where we admit to an uncertainty as to which is the correct "branch choice" for $\mathbf{P} \cdot \hat{\mathbf{n}}$, as indicated by the modulo at the end of the equation. Even this way of writing things runs a danger of being misleading, since it could be read as suggesting that $\mathbf{P} \cdot \hat{\mathbf{n}}$ has a definite value, as σ_{surf} really does, and that these differ by the modulo. To be more precise, we can adopt a notation similar to that introduced by Vanderbilt and Resta (2006) and Resta and Vanderbilt (2007), in which Eq. (1.5) is written as

$$\sigma_{\text{surf}} := \mathbf{P} \cdot \hat{\mathbf{n}}. \tag{1.6}$$

Here the special notation $:=$ means that the object on the left-hand side is equal to *one of the values* on the right-hand side when $\mathbf{P} \cdot \hat{\mathbf{n}}$ is regarded as a multivalued object whose values are separated by the lattice of values e/A_{surf}. Clearly $\hat{\mathbf{n}}$ is uniquely defined, so this implies that $\mathbf{P}$ carries the branch-choice uncertainty. For this argument to work on any surface facet, we must have that $\mathbf{P}$ is only defined modulo a 3D lattice of values separated by

$$\Delta\mathbf{P} = \frac{e\mathbf{R}}{V_{\text{cell}}}, \tag{1.7}$$

where $\mathbf{R}$ is a lattice vector, since then $\Delta\mathbf{P} \cdot \hat{\mathbf{n}} = e(\hat{\mathbf{n}} \cdot \mathbf{R})/V_{\text{cell}} = me/A_{\text{cell}}$ (m an integer) as required. An equation such as Eq. (1.6) does not pretend to specify *which* branch choice of $\mathbf{P}$ gives the correct value, only that *one of them* does.

Before proceeding, it is useful to generalize the previous 3D discussion to 2D and 1D. The current density $\mathbf{J}$ that has units of C/m²s in 3D acquires units of C/ms in 2D (sometimes denoted as a sheet current $\mathbf{K}$), and becomes a simple current I with units of C/s in 1D. Correspondingly, the polarization $\mathbf{P}$ has units of C/m², C/m, and C in 3D, 2D, and 1D, respectively. The "quantum of surface charge" $\Delta\sigma_{\text{surf}} = e/A_{\text{surf}}$ in 3D becomes a quantum of edge charge $\Delta\lambda_{\text{edge}} = e/b_{\text{edge}}$ in 2D, where b_{edge} is the edge lattice constant, and a quantum of end charge $\Delta Q_{\text{end}} = e$ for a 1D chain. In 3D or 2D, the quantum of surface charge is consistent with an uncertainty of the polarization

$$\Delta\mathbf{P} = e\mathbf{R}/\Omega_{\text{cell}}, \tag{1.8}$$

where $\mathbf{R}$ is a lattice vector and Ω_{cell} is the cell volume V_{cell} in 3D or the cell area A_{cell} in 2D. In 1D, we have $R = \Omega_{\text{cell}} = a$, the lattice constant, so that the quantum of polarization is just the charge quantum e. The arguments given in the previous paragraphs can then be generalized to 2D or 1D in an obvious way.[5]

As a precaution, it is useful to point out that the distinction between "bound" and "free" charge, while intuitive in certain simplified contexts, is not always clean in practice. For example, either Fig. 1.3(a) or 1.3(b) could be used as a reference for the absence of free charge at the surface, and in the case of a one-third-filled surface band, we could say with equal justification that the free charge is $-e/3A_{\mathrm{surf}}$ or $+2e/3A_{\mathrm{surf}}$. Some additional sources of charge, such as those arising from charged defects or stoichiometric variations, are not easily classified as "bound" or "free." Thus, these terms should be regarded as useful and intuitive in certain situations, but not as fundamental concepts that can be put on a rigorous footing in general.

1.1.2 Adiabatic Loop and Charge Pumping

From Eq. (1.2) it follows that an adiabatic evolution of a bulk crystalline insulator leads to a change of polarization

$$\Delta \mathbf{P} = \int_{\mathrm{i}}^{\mathrm{f}} \mathbf{J}(t)\, dt. \tag{1.9}$$

If $\mathbf{P}$ were uniquely defined, it would follow that $\Delta\mathbf{P} = 0$ for any cyclic evolution, meaning along a loop in parameter space that returns the crystalline Hamiltonian to its starting point at the end of the loop – provided, as always, that the system remains insulating along the path. But it is not hard to find counterexamples to show that this is not always the case. For example, Fig. 1.4 shows a sketch of a sliding charge-density wave in an insulating 1D crystal. In Fig. 1.4(a–c) this is modeled in extreme simplicity by the 1D electronic Hamiltonian

$$H = \frac{p^2}{2m} - V_0 \cos(2\pi x/a - \lambda), \tag{1.10}$$

where a is the lattice constant and λ is a cyclic parameter that returns the Hamiltonian to itself after running from 0 to 2π. For a sufficiently large value of V_0 it seems clear that the electron of charge $-e$ is semiclassically localized around the minimum of the potential at $x = a\lambda/2\pi$, so that a charge of $-e$ is pumped by $+a\hat{x}$ during one cycle. There is no tunneling involved here; the physics is just that of a ground-state wave function adiabatically following a gradual change in the crystal Hamiltonian.

---

[5] Actually, the essential physics described here is 1D physics, and the generalization to 2D or 3D can be regarded as being accomplished by introducing one or two orthogonal dimensions over which averages are taken in a standard way.

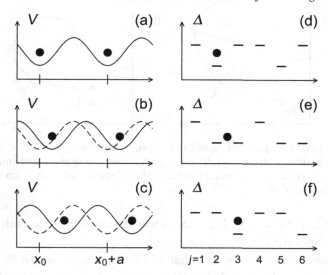

Figure 1.4 (a–c) Evolution of the model described by Eq. (1.10) at three increas-ing values of the parameter λ. (d–f) Same but for the model of Eq. (1.11), where the site energy Δ is $V_0 \cos(2\pi j/3 - \lambda)$. Both cases result in the transport of electrons, indicated semiclassically as black dots, to the right.

Figure 1.4(d–f) shows a slightly more sophisticated model of the same phe-nomenon, in which a simple 1D tight-binding model of (spinless) s orbitals at one-third filling is modulated by a sliding cosine potential,

$$H = -t_0 \sum_j (c_j^\dagger c_{j+1} + \text{h.c.}) - V_0 \sum_j \cos(2\pi j/3 - \lambda)\, c_j^\dagger c_j , \qquad (1.11)$$

so that the actual period is a if the spacing between sites is $a/3$. Here $c_j^\dagger$ and c_j are creation and annihilation operators, respectively, for an electron on site j, and the uniform nearest-neighbor hopping $-t_0$ plays the role of a kinetic energy. Again, if V_0 is large, our intuition tells us that the electron will be confined primarily to one or two sites at any given λ, as shown in Fig. 1.4(d–f), such that it would be transported by a over the course of one cycle. We can write this as

$$\Delta P_{\text{cyc}} \equiv \oint J(t)\, dt = Ne, \qquad (1.12)$$

where N is an integer (-1 in our example) and $\oint$ indicates a line integral around a closed cycle.

This result – a pumping of one electron by one lattice vector during one cycle – is thus inconsistent with the notion of a completely well-defined electric polarization P! At the same time, it is fully consistent with Eq. (1.7) (recall that the quantum of

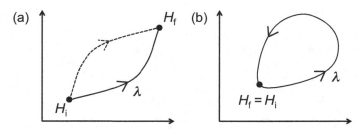

Figure 1.5 (a) In a space of insulating crystal Hamiltonians H, the system is carried adiabatically along two different paths connecting the same initial H_i at λ_i to the same H_f at λ_f. (b) Same but for a closed path, such that λ_i and λ_f label the same Hamiltonian.

polarization is just $\Delta P = e$ in 1D), which was obtained from an entirely different surface-charge argument!

Consider now what happens to either of the models shown in Fig. 1.4 as the magnitude of V_0 is gradually reduced. We might expect that some "leakage" would occur as V_0 becomes comparable to the scale of the kinetic energy, such that the total transport of charge as given by Eq. (1.9) would fall below e. However, if $\mathbf{P}$ is to have any meaning at all, this cannot be the case! In fact, it turns out that the pumped charge remains exactly quantized and does not fall as V_0 is reduced.[6] This fundamentally quantized nature of adiabatic charge transport, which was derived in a seminal paper of Thouless (1983) and subsequently shown to be robust against interactions and disorder (Niu, 1986, 1991; Niu and Thouless, 1984), provides the essential physical underpinning for the modern theory of polarization.[7]

An important aspect of Eq. (1.12) is that time is a passive variable in the following sense. Let us return to 3D, and for the moment consider an *open* path such that the system is carried from λ_i with Hamiltonian H_i at initial time t_i to λ_f with Hamiltonian H_f at final time t_f, as shown by the solid path in Fig. 1.5(a). We interpret $\mathbf{J} = d\mathbf{P}/dt$ in Eq. (1.2) by noting that even if $\mathbf{P}$ is ill defined modulo the quantum $e\mathbf{R}/V_{\text{cell}}$, its derivative with respect to the parameter λ describing the Hamiltonian is fully well defined since $e\mathbf{R}/V_{\text{cell}}$ is just a constant,[8] so that it doesn't matter which branch choice of $\mathbf{P}$ is used. Now the temporal variation of $\mathbf{P}$ really comes about via λ, so that $\mathbf{P}(t) = \mathbf{P}(\lambda(t))$, and the analog of Eq. (1.9) becomes

$$\Delta\mathbf{P}_{i\rightarrow f} \equiv \int_i^f \mathbf{J}(t)\,dt = \int_i^f \frac{d\mathbf{P}}{d\lambda}\frac{d\lambda}{dt}\,dt = \int_i^f \left(\frac{d\mathbf{P}}{d\lambda}\right) d\lambda \qquad (1.13)$$

---

[6] This requires that the adiabatic condition must remain satisfied, which requires that the rate of traversal of the loop has to become slower and slower as V_0 is reduced and the gap becomes small. Nevertheless, for any given V_0, the adiabatic charge transport goes to the quantized value in the limit of a sufficiently slow evolution around the loop.

[7] While the discussion in these papers was not framed in terms of "polarization," the essential physics of the modern theory of polarization was already contained there.

[8] Strictly speaking, we require that elastic strains are not occurring as a function of time, so that $\mathbf{R}$ is constant; otherwise, subtleties associated with the distinction between "proper" and "improper" piezoelectric responses enter in (see, for example, Vanderbilt, 2000).

upon invoking the chain rule. We see that time has dropped out! For a full cycle Eq. (1.12) becomes, in 3D,

$$\Delta \mathbf{P}_{\text{cyc}} = \oint \left(\frac{d\mathbf{P}}{d\lambda} \right) d\lambda = \frac{e\mathbf{R}}{V_{\text{cell}}}, \tag{1.14}$$

which is again our quantum of polarization. If $\mathbf{P}$ were uniquely defined, the usual rules of calculus would lead us to rewrite the open-path expression in Eq. (1.13) as $\Delta \mathbf{P}_{i \to f} = \mathbf{P}_f - \mathbf{P}_i$, but this is inappropriate here.[9] It is safer to adopt the notation of Eq. (1.6) and write that

$$\Delta \mathbf{P}_{i \to f} := \mathbf{P}_f - \mathbf{P}_i. \tag{1.15}$$

Here the left-hand side has a definite value for a definite path, and we claim that this is equal to *one* of the multiple values given by regarding $\mathbf{P}$ as a lattice-valued quantity on the right-hand side.

This logic also implies that if we consider following two different paths connecting H_i to H_f, such as those indicated by the solid and dashed lines in Fig. 1.5(a), we should get the same result modulo the quantum. This is really just the same as saying that the change in polarization around a closed loop, as shown in Fig. 1.5(b), must be zero or a multiple of the quantum. After all, the difference between the two paths is the same as following one path and then the other in reverse, which is itself a closed path.

To summarize, at this point we have two strong arguments for the notion that electric polarization $\mathbf{P}$ has to be understood as being well defined only modulo the quantum of Eq. (1.7) in crystalline insulators. We shall now give two more arguments supporting this hypothesis.

1.1.3 Slow Spatial Variation in a Supercell

In the previous section, we considered a slow *temporal* variation of a periodic bulk crystal and arrived at Eq. (1.14), expressing the notion of quantized adiabatic charge transport. We now consider a slow *spatial* variation of the Hamiltonian instead, and show that we can arrive at the same conclusion.

The argument goes as follows. Consider a crystal with lattice constants a, b, and c along $\hat{\mathbf{x}}$, $\hat{\mathbf{y}}$, and $\hat{\mathbf{z}}$, respectively, and construct an $N \times 1 \times 1$ supercell, corresponding to a large unit cell of size $L = Na$ along $\hat{\mathbf{x}}$ and of primitive dimensions along $\hat{\mathbf{y}}$ and $\hat{\mathbf{z}}$. Furthermore, arrange that the local Hamiltonian at position x varies slowly along a closed cycle as x runs from 0 to its periodic image at Na. This time we shall let the closed loop be parametrized by λ going from 0 to 1, so that $H_{\lambda=0} = H_{\lambda=1}$, which is a convention we shall frequently adopt later. The situation is sketched in

---

[9] For example, it would seem to imply that $\Delta \mathbf{P}$ must vanish for any closed cycle.

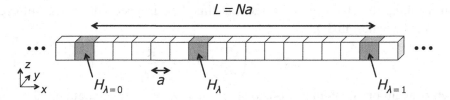

Figure 1.6 Supercell with long repeat distance along $L = Na$ arranged such that the local Hamiltonian H_λ undergoes a cyclic variation as x cycles from $x = 0$ to $x = L$.

Fig. 1.6. Now since the primitive cells are neutral for any given value of λ, the total charge inside the supercell is given by integrating the "bound" charge $-\nabla \cdot \mathbf{P}$, yielding

$$Q = bc \int_0^{Na} (-\nabla \cdot \mathbf{P})\, dx = -bc \int_0^{Na} \frac{dP_x(\lambda(x))}{dx}\, dx = -\frac{V_{\text{cell}}}{a} \int_0^1 \left(\frac{dP_x}{d\lambda}\right) d\lambda,$$
(1.16)

where x is now seen to be a passive variable, just as time was in Eq. (1.13). It is a simple consequence of Bloch's theorem that filling one band in the band structure contributes exactly $-e$ to the charge in the unit cell. Applying this to our supercell, which after all is just a crystal with a strangely shaped unit cell, we argue that $Q = -Ne$ on the left-hand side of Eq. (1.16) for some integer N. This leads directly to

$$\Delta P_{x,\text{cyc}} = \oint \left(\frac{dP_x}{d\lambda}\right) d\lambda = \frac{Nea}{V_{\text{cell}}},$$
(1.17)

which is just Eq. (1.14) specialized to the x component of $\mathbf{P}$.

Thus we see that an argument based on a static crystal Hamiltonian with slow spatial variation yields the same result as one based on a spatially uniform crystal Hamiltonian that varies slowly in time: the polarization $\mathbf{P}$, if it is well defined at all, is only well defined modulo a quantum of polarization $e\mathbf{R}/V_{\text{cell}}$.

1.1.4 Fictitious Physics of Classical Point Charges

In the previous three subsections, arguments based on quantum-mechanical electronic charge densities and currents were invoked to argue that $\mathbf{P}$ is only well defined modulo a quantum. But do we really need to invoke quantum mechanics? Here we see that purely classical considerations can lead to the same conclusion!

In our real world it is a good approximation to treat crystals as composed of classical positively charged nuclei and distributed electron charge clouds, as

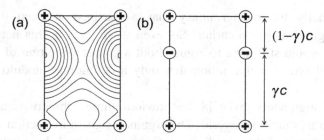

Figure 1.7 (a) A realistic crystal in which a quantum-delocalized electron charge cloud (shown by density contours) coexists with positive classical point charges. (b) An imaginary crystal composed of positive and negative classical point charges.

sketched in Fig. 1.7(a). But imagine that we live in a fictitious world in which matter is composed entirely of classical point particles of charge Ze (Z an integer) connected by springs or similar two- or three-body forces.[10] Figure 1.7(b) shows a typical crystal built out of two particles of charge $\pm e$, where the symmetry is such that we expect a nonzero polarization along the $\hat{z}$ direction. One might expect that all the subtleties discussed previously should disappear in such a simple world, but this is not at all the case. For how shall we decide what value of electric polarization P_z to assign to the crystal shown in Fig. 1.7(b)?

The problem is that we have to decide upon a "basis" to describe the contents of the unit cell, meaning a list of the charges eZ_j and their coordinates τ_j, chosen such that a periodic repetition of this basis "tiles" the space and reproduces the crystal in question. Figure 1.7(b) is sketched for the case of a 3D tetragonal crystal with charges $Z_1 = +1$ and $Z_2 = -1$. We can choose their locations to be $\tau_1 = 0$ and $\tau_2 = \gamma c \hat{z}$ ('$-$' above '$+$' by γc), but an equally valid choice of atomic basis is $\tau_1 = c\hat{z}$ and $\tau_2 = \gamma c\hat{z}$ ['$+$' above '$-$' by $(1 - \gamma)c$]. The polarization is the dipole per unit cell

$$\mathbf{P} = \frac{e}{V_{\text{cell}}} \sum_j Z_j \tau_j, \tag{1.18}$$

which gives $-\gamma e\hat{z}/A$ or $(1 - \gamma)e\hat{z}/A$, respectively, for the two choices of basis, where A is the basal cell area. In fact, an infinite number of choices is possible, as, for example, with $\tau_1 = 0$ and $\tau_2 = \mathbf{R} + \gamma c\hat{z}$ for any lattice vector $\mathbf{R}$; this correctly tiles the space to give the specified crystal, regardless of the choice of $\mathbf{R}$, but the inferred polarization is then only well defined modulo $e\mathbf{R}/V_{\text{cell}}$.

---

[10] These need to be stronger than the Coulomb force at short distances to prevent the collapse of opposite charges onto one another.

This is exactly the indeterminacy that we arrived at from the quantum-mechanical arguments given earlier! So, even if we lived in this fictitious world, our physicists would still have to argue about a proper definition of polarization, and would still arrive at the notion that only a definition "modulo a quantum" makes sense.

In fact, the arguments given in the previous three subsections all have their counterparts in this fictitious world. The arguments given in Section 1.1.1, that an adiabatic adjustment of the coordinates of surface "atoms" (here "charges") does not change the macroscopic surface charge, are equally valid here. We found there that we could change the macroscopic surface charge by Ne/A_{surf} (integer N), which we can do here by simply adding or removing a layer of charges at the surface. We would again conclude that Eq. (1.6) is valid: $\sigma_{surf} = \mathbf{P} \cdot \hat{\mathbf{n}}$ for one of the many possible values of $\mathbf{P}$. Similarly, the kind of adiabatic loop discussed in Section 1.1.2 would correspond here just to the motion of the charges. Figure 1.8 shows two possible cycles, in which the positive charges remain fixed but the negative charges migrate as a function of the loop parameter. The first results in $\Delta\mathbf{P} = \mathbf{0}$, while the second results in a change by our now-familiar quantum $\Delta\mathbf{P} = e\mathbf{R}/V_{cell}$, here with $\mathbf{R} = -c\hat{\mathbf{z}}$. Finally, we can consider a supercell with a long repeat period along $\hat{\mathbf{z}}$, with a spatial variation chosen according to the same loop; if built from N primitive cells, this will result in the presence of N positive and $N-1$ negative charges in the supercell, and arguments parallel to those in Section 1.1.3 apply.

To summarize, all of the preceding arguments are consistent with a viewpoint in which the polarization $\mathbf{P}$ should be regarded as well defined only modulo $e\mathbf{R}/V_{cell}$ or, equivalently, as a *multivalued* quantity whose equally valid values are separated by $e\mathbf{R}/V_{cell}$. We shall refer to such a mathematical object as *lattice-valued*. For physical statements that can be expressed in differential form, such as $\rho = -\nabla \cdot \mathbf{P}$ or $\mathbf{J} = d\mathbf{P}/dt$, this uncertainty modulo the quantum introduces no difficulty, as

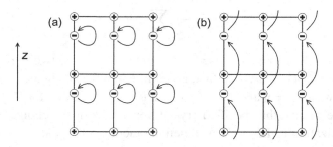

Figure 1.8 Fictitious classical crystal undergoing a cyclic evolution. (a) Negative charges return to themselves such that $\Delta\mathbf{P} = \mathbf{0}$. (b) Negative charges replace partners in the next cell such that $\Delta\mathbf{P} = -e\hat{\mathbf{z}}/A$. Figure adapted from Resta and Vanderbilt (2007) with permission of Springer.

long as we insist on staying on the same branch while carrying out the derivative. In contrast, some questions such as "What is the macroscopic surface charge σ_{surf} on *this* facet of the crystal?" or "How much net current $\int \mathbf{J}\, dt$ flowed during *this* adiabatic evolution?" can be only *partially* answered by a knowledge of the (lattice-valued) bulk polarization. We can say that the answers are, respectively, $\sigma_{\text{surf}} = \mathbf{P} \cdot \hat{\mathbf{n}}$ or $\int \mathbf{J}\, dt = \mathbf{P}_{\text{f}} - \mathbf{P}_{\text{i}}$ for *one of the valid values* of $\mathbf{P}$ or $\mathbf{P}_{\text{f}} - \mathbf{P}_{\text{i}}$, but without knowing more about the particular surface or the particular path, we cannot say which one is correct.

1.1.5 Robustness against Weak Interactions

The preceding arguments were framed in the context of a noninteracting description of the electrons. The interacting Hamiltonian is of the form

$$H_{\text{full}} = \sum_i \left[\frac{p_i^2}{2m} + V_{\text{bare}}(\mathbf{r}_i) \right] + \sum_{\langle ij \rangle} \frac{e^2}{|\mathbf{r}_i - \mathbf{r}_j|}, \qquad (1.19)$$

where V_{bare} is the unscreened attraction of the electrons to the nuclei, but we typically approximate Eq. (1.19) with a mean-field (MF) Hamiltonian

$$H_{\text{MF}} = \sum_i \left[\frac{p_i^2}{2m} + V_{\text{MF}}(\mathbf{r}_i) \right], \qquad (1.20)$$

where V_{MF} includes terms that approximately describe the interaction of an individual electron with the collective electron cloud. Because only one-electron operators appear in Eq. (1.20), its overall ground state is just a single Slater determinant of all of the occupied one-particle eigenfunctions of $p^2/2m + V_{\text{MF}}(\mathbf{r})$ – that is, the Bloch states $\psi_{n\mathbf{k}}(\mathbf{r})$. Typically H_{MF} is constructed using some variety of density-functional theory, which we shall encounter in Section 2.1.1, and this works quite well for ordinary insulators such as Si, GaAs, and BaTiO_3.

However, we have made various exact claims in the previous subsections: that the surface charge is exactly unchanged by a surface perturbation, that the charge pumped by an adiabatic cycle is exactly e times a lattice vector, and so on. It is then natural to ask: do these statements remain exactly true in the context of the full interacting Hamiltonian of Eq. (1.19)? If so, we say that these statements are *robust against interactions*. This is not to say that the polarization $\mathbf{P}$ of a certain crystal is the same according to Eqs. (1.19) and (1.20); rather, we only mean to say that the quantum of polarization $e\mathbf{R}/V_{\text{cell}}$ plays exactly the same role, and has the same value, in both cases.

In some situations, the answer to our question is a simple "yes." For example, when discussing Fig. 1.2(a–b), we argued that the change in the macroscopic surface charge $\Delta\sigma$ resulting from an adiabatic modification of the surface Hamiltonian must vanish exactly if the surface gap does not close. The arguments given there apply equally well for the case of the full many-body Hamiltonian. Similarly, the argument given in Section 1.1.3 works equally well if we just accept that the number of electrons per cell remains quantized to an integer even in the many-body case. But is this true? Here the answer is a qualified "yes." One can show that it *is* true for any many-body system that can be adiabatically connected to a noninteracting mean-field one. To clarify what is meant by this, consider the Hamiltonian

$$H = (1 - \mu)\, H_{\mathrm{MF}} + \mu\, H_{\mathrm{full}}, \qquad (1.21)$$

where μ running from 0 to 1 plays the role of an adiabatic parameter that gradually replaces the mean-field approximation to the interaction with the true many-body one.[11] Assuming the crystal remains insulating along this path, which is very likely except perhaps for certain types of strongly correlated materials, it can be shown that the number of electrons per unit cell remains quantized to an integer, so that the macroscopic charge density remains exactly constant as μ is varied.[12] The argument of Section 1.1.3 then again leads to the quantization of adiabatic charge transport expressed by Eq. (1.14), as was demonstrated by other methods by Niu and Thouless (1984) (see also Niu, 1986, 1991).

We know, however, that there are some strongly interacting gapped phases that *cannot* be adiabatically connected to a single-particle insulator. An example is the 1/3 fractional quantum Hall state, discovered by Tsui et al. (1982), where the quantized adiabatic charge transport comes in units of $e/3$ because of the existence of quasiparticles with charge $\pm e/3$. Something similar happens for certain spin liquids and other topological phases (see Section 5.5.3). Typically these phases are characterized by multiple degenerate ground states, so that arguments based on a unique ground state do not necessarily apply.

This is a fascinating subject, but we shall have little more to say about it here. We shall content ourselves with observing that most "not too strongly" interacting systems, such as those that can be adiabatically connected to a mean-field single-particle description, obey the same rules as described in Sections 1.1.1–1.1.4. In this sense, the conclusions in those sections are "robust against interactions." Some

---

[11] It would be a mistake to try a similar procedure for $H = H_0 + \mu H_{\mathrm{int}}$, where H_0 and H_{int} are the two terms in Eq. (1.19); in this case the starting system would be so different that metallic phases would likely be encountered along the path.

[12] This is essentially Luttinger's theorem, which states that the noninteger part of the number of electrons per cell in a metal is proportional to the k-space volume under the Fermi surface, specialized to the case of an insulator for which this volume vanishes. See also Ex. 1.2.

of them are also "robust against disorder" when this is carefully defined, but a discussion of this interesting topic falls outside the scope of this book.

Exercises

1.1 Regarding the argument leading to Eq. (1.17), why is it necessary to assume that the crystal is insulating everywhere along the path – that is, for all values of λ? Is this condition sufficient, as well as necessary, to ensure an insulating supercell?

1.2 Give an argument to show that the charge per unit cell does not change at all when the parameter μ controlling the strength of interactions in Eq. (1.21) is varied. *Hint*: consider varying μ in a localized region in the interior of the sample, and consider the current flowing into this region during the variation.

1.3 Above 435°C $KNbO_3$ adopts a simple cubic perovskite structure with lattice vectors $\mathbf{a}_1 = (a, 0, 0)$, $\mathbf{a}_2 = (0, a, 0)$, and $\mathbf{a}_3 = (0, 0, a)$. Two valid choices of basis for describing the atomic positions are: $\tau_K = (0, 0, 0)$, $\tau_{Nb} = (a/2, a/2, a/2)$, $\tau_{O1} = (0, a/2, a/2)$, $\tau_{O2} = (a/2, 0, a/2)$, and $\tau_{O3} = (a/2, a/2, 0)$; or $\tau_{Nb} = (0, 0, 0)$, $\tau_K = (a/2, a/2, a/2)$, $\tau_{O1} = (a/2, 0, 0)$, $\tau_{O2} = (0, a/2, 0)$, and $\tau_{O3} = (0, 0, a/2)$. First explain why these describe the same structure. Then compute $\mathbf{P}$ via Eq. (1.18) with $Z_K = 1$, $Z_{Nb} = 5$, and $Z_O = -2$, and show that these are consistent with each other modulo the quantum $e\mathbf{R}/V_{cell}$.

1.2 Orbital Magnetization and Surface Currents

In elementary electromagnetism we learn that the polarization $\mathbf{P}$ and magnetization $\mathbf{M}$ are the corresponding electric and magnetic order parameters that couple to external electric and magnetic fields. Our focus here is on the spontaneous magnetization $\mathbf{M}$ in magnetic materials, in analogy to the spontaneous polarization $\mathbf{P}$ discussed in the previous section. $\mathbf{M}$ has two components, $\mathbf{M} = \mathbf{M}_{spin} + \mathbf{M}_{orb}$. In most ferromagnetic materials $\mathbf{M}_{spin}$ is the dominant component. It describes the excess population of spin-up over spin-down electrons, and is straightforward to describe theoretically.[13] More subtle is the orbital component $\mathbf{M}_{orb}$, which we can think of as arising from circulating currents on the atoms that make up the crystal as roughly sketched in Fig. 1.9. This is typically induced by the spin-orbit coupling (SOC), which takes the effective form $H_{SOC} = \xi \mathbf{L} \cdot \mathbf{S}$ on a given atomic site, where $\mathbf{S}$ and $\mathbf{L}$ are the spin and orbital angular momentum operators on this site,

---

[13] See the discussion of the local spin-density approximation on p. 43.

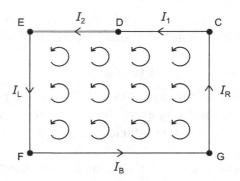

Figure 1.9 Sketch of a rectangular sample cut from a 2D insulating ferromagnet. The edge currents I are argued to be identical (see text) despite different surface cuts and terminations.

respectively. Since the SOC is a relativistic effect, its strength ξ tends to be largest for heavy atoms at the bottom of the Periodic Table.

From elementary electromagnetism we know that just as the electric polarization $\mathbf{P}$ is related to surface charges through Eq. (1.4), an orbital magnetization $\mathbf{M}_{\text{orb}}$ manifests itself in the form of a surface current

$$\mathbf{K}_{\text{surf}} = c\,\mathbf{M}_{\text{orb}} \times \hat{\mathbf{n}}, \tag{1.22}$$

where $\hat{\mathbf{n}}$ is the surface normal. (The speed of light c appears because we adopt Gaussian units throughout.) However, we argued earlier that $\mathbf{P}$ can be well defined only as a bulk quantity modulo some quantum, leading to the replacement of Eq. (1.4) by Eq. (1.6). We may now ask: is $\mathbf{M}_{\text{orb}}$ uniquely defined as a bulk quantity, or does it, too, suffer from a similar kind of ambiguity?

This time the answer is comforting: The orbital magnetization is indeed well defined in the conventional sense. To see this we can invoke Eq. (1.22), which says that if $\mathbf{M}_{\text{orb}}$ is well defined, then all surface currents are determined by this equation. Turning this around, we argue that if it is impossible to modify the surface current by any local surface treatment, then the surface current can be regarded as the manifestation of a unique bulk property.

Consider a finite sample, such as the rectangular one cut from a 2D insulating[14] magnet as shown in Fig. 1.9. In general, the manner of truncation may be different on the left (L), bottom (B), and right (R) sides. Just to elaborate the situation, we also imagine that the top surface is broken into two surface patches with different surface terminations labeled by 1 and 2. If we can argue that the edge currents are

---

[14] The argument also applies to a metallic sample, although with some elaboration to rule out the role of bulk currents.

all identical, $I_L = I_B = I_R = I_1 = I_2$, then it follows that these are all reflections of a unique bulk M_{orb}. But this immediately follows from conservation of charge. For example, if $I_R \neq I_1$, then there would be a buildup of charge $dQ_C/dt = I_R - I_1$ at point C, and similarly for the other junctions at D, E, F, and G. But we are trying to describe a "stationary state," specifically the ground state of this rectangular sample, and it is a basic postulate of quantum mechanics that observables such as these junction charges are independent of time in any stationary state. In other words, such a charge accumulation would have nowhere to go, and is therefore impossible. So all surface currents are the same, and equal to c times a unique bulk magnetization M_{orb}. The argument is easily generalized to the surface currents $\mathbf{K}_{surf}$ of a 3D sample.

Note the asymmetry between the physics of $\mathbf{P} \leftrightarrow \sigma_{surf}$ and $\mathbf{M}_{orb} \leftrightarrow \mathbf{K}_{surf}$. In the former case, we argued that we could change the surface charge of an insulator by a quantum (one electron per surface cell) via some local change in the surface treatment, so that $\mathbf{P}$ is only well defined modulo a quantum. Here, we find that there is no way to modify $\mathbf{K}_{surf}$ by any local change of the surface, so that we can regard $\mathbf{M}_{orb}$ as being absolutely well defined.[15]

The surface currents $\mathbf{K}_{surf}$ are examples of *dissipationless currents*. Ohm's law plays no role; the currents run forever without slowing down. Moreover, they occur even when the bulk material and its surfaces are all insulating. The existence of dissipationless currents is taken for granted in elementary quantum mechanics, as for example in the $|nlm\rangle = |211\rangle$ state of the H atom,[16] but they seem a little more "creepy" when they occur at the macroscopic scale, and in an insulating system. These surface currents are really just "bound currents," which from another point of view can be regarded as a book-keeping fiction. That is, even if each individual current loop is entirely confined to its own unit cell, so that no "free current" flows at the surface, we claim the existence of a surface current $\mathbf{K} = \mathbf{K}_{bound}$. This current is "real" in the sense that it generates macroscopic magnetic fields, contributing, for example, to the discontinuity in the surface-parallel component of the macroscopic $\mathbf{B}$ at the surface of a ferromagnet. Dissipationless currents will play a role in many of the phenomena to be discussed later.

Finally, we point out that the arguments we have given here are generally robust against interactions and disorder. For example, in Fig. 1.9 we can imagine that the many-body terms, controlled by μ in Eq. (1.21), are turned off in surface region 1 but not in surface region 2, or that disorder is introduced to one region but not

---

[15] This argument that $\mathbf{M}_{orb}$, and therefore $\mathbf{M}$, is absolutely well defined rests on an implicit assumption that magnetic monopoles do not exist in nature. If they did, then $\mathbf{M}$ would have to join $\mathbf{P}$ as being multivalued in principle (although not in practice).

[16] A better example might be the $1s^2 2s^2 2p^1$ configuration of the boron atom with $m = 1$, which is a ground, not an excited, state.

the other. We still must have $dQ_B/dt = 0$ and so forth, so all surface currents are identical, and the concept of a bulk $\mathbf{M}_{orb}$ is still allowed. It may seem especially surprising that the presence of surface disorder does not reduce the surface current at all, but this is typical of dissipationless currents.

1.3 Edge Channels and Anomalous Hall Conductivity

1.3.1 Edge Channels

It is not too surprising that we cannot modify the edge currents I in Fig. 1.9 by changing the surface properties if the surface is insulating, but it is less obvious if the surface is metallic. To analyze this situation it is useful to introduce the concept of an *edge band structure*.

Consider again a 2D insulating crystal like that shown in Fig. 1.9, and assume it is built out of rectangular $a \times b$ unit cells in the x and y directions. We focus on the top edge, and assume it shares the same periodicity a parallel to the edge as in the bulk. Then Bloch's theorem applies in the x direction and k_x is a good quantum number. We adopt the convention that all edge currents are defined in the sense of positive (i.e., counterclockwise) circulation, and apply the same for the edge wavevector k, so we have $k = -k_x$ for the states on the top edge, with k running from 0 to $2\pi/a$.

To construct the edge band structure, one plots the allowed energies of both bulk-like and edge-localized states as illustrated in Fig. 1.10. Bulk states can be labeled by wavevectors (k_x, k_y) and band index n; allowing k_y and n to run over all possible values for a given k_x generates the shaded regions in the figure, representing the projected bulk valence bands (lower region) and conduction bands (upper region). In addition to the bulk bands there may be edge states that appear in the gap, as shown in the figure by the line that emerges out of the valence band in (a) and the

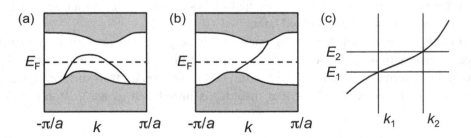

Figure 1.10 (a–b) Possible edge band structures of a 2D insulating ferromagnet. Shaded regions are projected bulk bands; isolated curves are edge states. The Fermi energy is at E_F. (c) Close-up of an edge state in the vicinity of E_F.

one that crosses the gap in (b). Since there are no bulk states at these energies, the wave functions of the edge states must be exponentially localized near the edge.[17]

Both of the surface states in Fig. 1.10 are shown crossing at least once through the Fermi energy E_F. In the vicinity of such a crossing, the energy position of E_F will clearly influence the contribution of that edge state to the total edge current I. Doesn't this suggest that a local modification at the edge – as, for example, by raising E_F slightly at the edge – will allow us to modify the edge current, in contradiction to what was claimed in Section 1.2?

Let's see how I depends on E_F. Suppose we raise the Fermi energy from $E_F = E_1$ to E_2 in a rigid-band picture (i.e., the Hamiltonian remains unchanged). Then the additional current along $-\hat{x}$ at the top edge, coming from an upward-crossing band like that shown in Fig. 1.10(c), is

$$\Delta I = \int_{k_1}^{k_2} \left(\frac{dk}{2\pi} \right) (-ev_g)$$

$$= \frac{-e}{2\pi\hbar} \int_{k_1}^{k_2} \frac{dE}{dk} dk = \frac{-e}{h} (E_2 - E_1). \tag{1.23}$$

Here, $1/2\pi$ is the fundamental density of states (number per unit length per unit wavevector), $-e$ is the charge associated with each newly occupied state, and $v_g = \hbar^{-1}dE/dk$ is the group velocity of the state. Equation (1.23) is a remarkable result, indicating that the additional current carried by the edge channel is only a function of the energy shift, independent of the shape of the edge band structure curve. Thus, Eq. (1.23) leads to

$$\frac{\partial I}{\partial \phi} = e \frac{\partial I}{\partial E_F} = -\frac{e^2}{h}, \tag{1.24}$$

where we have used the fact that I depends on $E_F + e\phi$ in a rigid-band picture to convert the derivative with respect to electrostatic potential ϕ (at fixed E_F) into one with respect to E_F (at fixed ϕ). The result in Eq. (1.24) is evidently universal, being composed only of constants of nature. To summarize the behavior, it is useful to define a quantity with units of conductance,

$$G = \frac{\partial I}{\partial \phi}. \tag{1.25}$$

This "circulatory conductance" characterizes the change in counterclockwise edge current as the electrostatic potential of the sample is uniformly raised relative to a fixed Fermi level. Since electrons carry negative charge, a rising ϕ lowers the

---

[17] Figure 1.10 is drawn for the case of a ferromagnetic bulk 2D material, since if time-reversal symmetry were present we would expect a left–right symmetry in the bands' structures. The figure also assumes the absence of M_x mirror symmetry.

energies of the surface states relative to E_F, causing them to become more populated and contribute more negative current in the $-\hat{x}$ (counterclockwise) direction, for a negative I. Thus $G = -e^2/h$ for the case of a surface state like that in Fig. 1.10(b) with the Fermi velocity pointing in the positive sense of circulation around the boundary.[18]

A similar argument shows that the contribution to G from a down-going crossing along $-\hat{x}$ (Fermi velocity in the negative sense of circulation), like the one shown toward the right in Fig. 1.10(a), is e^2/h. When both contributions are taken into account in Fig. 1.10(a), they cancel such that $G=0$. That is, the contributions from edge channels with positive and negative sense of circulation cancel each other, and we again conclude that a modification of the edge band structure has no influence on the edge current, consistent with a unique definition of orbital magnetization.[19]

We are then left with the perplexing question of whether a situation like that shown in Fig. 1.10(b) is possible. This case has one edge channel in the direction of counterclockwise circulation, but none in the clockwise direction. Such a situation is clearly impossible if the system has time-reversal (TR) symmetry, but recall that we are interested here in magnetic systems in which TR symmetry has been broken.

One thing we can say right away is that if the sample has an excess up-crossing edge state on any one edge, then the same must be true on *every* edge, regardless of edge orientation or termination. For example, suppose E_F is uniformly raised throughout the bulk and on all surfaces; we can argue, as we did in Section 1.2, that the surface current must remain the same on all surfaces, as there would otherwise be a violation of charge conservation at the corners. But we have seen that the change in surface current is related to the count of surface states crossing the gap, thus proving the assertion.

For this reason we refer to the edge state as *chiral*, circulating all the way around the boundary in a unique sense, either clockwise or counterclockwise.[20] Of course, one edge segment could have two up-going and one down-going gap crossings; only the excess number figures into the argument.

Since all edges must have the same net chirality, we are led to the conclusion that this number must really be a property of the bulk insulator itself. This argument is much like the one given earlier in Section 1.2 for orbital magnetization to be a bulk

---

[18] Because of the way e^2 appears in all the formulas, the results would be the same for a fictitious physics with positively charged electrons. That is, an edge state with group velocity in the counterclockwise sense of circulation still gives a negative G, just as for real electrons.

[19] If E_F had different values at the two crossings, this would indicate that the system is out of equilibrium, and is therefore not in a stationary ground state. In practice, charge flow and scattering between edge channels would eventually reestablish a globally uniform E_F.

[20] In a slightly different language, we can say that the low-energy excitations, or quasiparticles, travel in only one sense of circulation around the boundary.

quantity: if all edges act alike, that behavior must reflect a bulk property. Here we conclude that each bulk insulator is characterized by a number C, such that every edge obeys

$$n_{\text{up}} - n_{\text{down}} = -C, \tag{1.26}$$

where $n_{\text{up}} - n_{\text{down}}$ is the excess number of up-crossings versus down-crossings in the edge band structure as viewed in the counterclockwise direction. The minus sign on the right side of Eq. (1.26) is a convention chosen such that the circulatory conductance defined in Eq. (1.25) is just

$$G = C \frac{e^2}{h}. \tag{1.27}$$

Needless to say, the number of up-going or down-going edge channels is intrinsically an integer; there can be no such thing as 1.43 up-going edge channels. So, if crystals with this behavior are allowed in nature, the scenario must play out something like this:

- Every 2D insulating crystal is characterized by an integer C characterizing the excess number Δn of up-crossings on any vacuum-terminated edge.
- Those belonging to the class with $C = 0$, as in Fig. 1.10(a), will be referred to as "normal" or "trivial," with the others denoted as "nontrivial." As a special case, the vacuum is classified as a trivial insulator, as is any insulating crystal with TR symmetry.
- Any vacuum-exposed surface of a sample with integer C has C more down-going than up-going edge channels as measured for k in the positive sense circulation.
- Any interface between insulating regions A and B has an excess of $C_A - C_B$ down-going versus up-going edge channels as measured in the positive sense of circulation about region A.
- Every vacuum surface of a nontrivial insulator, and every interface between inequivalent insulators (as classified by their C), is necessarily metallic.
- The circulating edge current of a nontrivial insulator at fixed Fermi energy varies with electric potential[21] as $\partial I / \partial \phi = G = Ce^2/h$, or equivalently,

$$\frac{\partial M_{\text{orb}}}{\partial \phi} = \frac{1}{c} G = C \frac{e^2}{hc}. \tag{1.28}$$

- An adiabatic variation of the bulk crystal Hamiltonian that does not close the gap cannot result in a change of C.

---

[21] We argued at the end of Section 1.1.1 that the distinction between "bound" and "free" charges is not always sharp. In the present context we see that the same is true of "bound" and "free" currents, since it is equally valid to regard the edge current here as resulting from the addition of free carriers, or from a change of the bulk magnetization.

We have here the seeds of a notion of a *topological classification* of insulators. To attach physical significance to our integer index C, however, we need to introduce a physical observable, the anomalous Hall conductivity, that will turn out to be closely related to it.

1.3.2 Anomalous Hall Conductivity

We have argued that a 2D ferromagnetic insulator is characterized by a bulk integer index C that controls the number of edge channels that must occur on every edge through Eq. (1.26). Is this index C related to some measurable physical quantity? We shall show that it is directly related to the anomalous Hall conductivity (AHC); indeed, the fact that edge channels are intrinsically quantized implies that the AHC of an insulator is also quantized.

The Hall conductivity, and its close cousin the AHC, will be introduced more systematically in Section 5.1. For now, we simply define the conductivity tensor of a 2D system as the 2×2 matrix $\sigma_{ij} = dK_i/d\mathcal{E}_j$, where **K** is the sheet current induced to first order by an in-plane electric field $\mathcal{E}$.[22]

But aren't such sheet currents impossible, by definition, in an insulator? Not quite. Our definition of an "insulator" is a crystal whose filled and empty states are separated by a finite energy gap. We are working here in the low-temperature limit; this implies that no electron or hole carriers are excited, so these cannot be responsible for a current. However, this does not rule out the possible flow of dissipationless currents not related to free carriers. For example, consider a 2D insulating sample with some gradient of chemical composition or other variable along x such that $M_{\text{orb}}(x)$ varies with x; then there is a dissipationless sheet current $K_y(x) = -dM_{\text{orb}}/dx$ running in the y direction. Here we are looking for a similar kind of dissipationless current flow, but one that appears in a spatially uniform sample to the first order in an applied electric field.

In our case we can immediately conclude that the symmetric part of the 2×2 σ_{ij} matrix must vanish. A nonzero σ_{xx}, for example, would imply the existence of a current K_x parallel to $\mathcal{E}_x$, depositing energy into the system. However, this would entail a violation of energy conservation, since there are no low-energy excitations capable of dissipating this energy. Then if the symmetric part of σ vanishes, the only freedom left is the antisymmetric AHC part[23]

$$\sigma_{\text{AHC}} = \sigma_{yx} = -\sigma_{xy}, \tag{1.29}$$

[22] Throughout this book the symbol $\mathcal{E}$ (magnitude or component) or $\boldsymbol{\mathcal{E}}$ (vector) is used to denote an electric field so as to avoid confusion with the symbol E for energy. While $\mathcal{E}$ is often used elsewhere to denote electromotive force, this is *not* our usage here.

[23] The convention adopted here is such that the current response is measured in the positive sense of circulation relative to the driving $\mathcal{E}$-field. Be aware that the opposite convention is also commonly used in the literature.

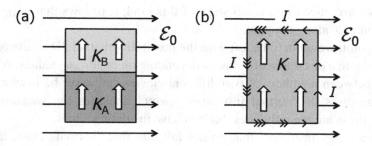

Figure 1.11 Behavior of sheet currents $\mathbf{K}$ and edge currents I for samples subjected to a uniform electric field $\mathcal{E}_0$ in the x direction. (a) A sample consisting of two adjoined materials A and B having two different positive values of σ_{AHC}. (b) A detailed view of the edge currents I around a single sample with positive σ_{AHC}.

or equivalently $\sigma_{ij} = -\epsilon_{ij}\sigma_{\text{AHC}}$, where ϵ_{ij} is the 2×2 antisymmetric tensor. We have no reason a priori to rule out such a response.

Thus we should take the possibility seriously and ask: what conditions must be satisfied if an AHC is present in an insulator? A severe constraint arises when we consider possible charge accumulation at surfaces and interfaces. Consider a situation like that shown in Fig. 1.11(a), which depicts a finite rectangular sample of material A adjoined to material B (still in 2D). A small electric field $\mathcal{E} = \mathcal{E}_0\hat{x}$ is applied, such that a current $K_A = \sigma_{\text{AHC}}^A \mathcal{E}_0$ flows upward in region A and $K_B = \sigma_{\text{AHC}}^B \mathcal{E}_0$ flows upward in region B. We also ensure that $\mathcal{E}_0$ is small enough that the induced potential drop from left to right of the sample is much less than the band gap, which is always possible since we are discussing just the linear response to a first-order-small $\mathcal{E}$. Thus, the system has a well-defined ground state with a Fermi energy in a global gap, and we can consider the stationary ground state of the system. In a stationary state, all observable expectation values must be independent of time, and in particular the charge density at the top edge of the sample must be independent of time. This is inconsistent with the arrival of charge from the sheet current K_B if this edge is insulating, and we conclude that a nonzero σ_{AHC} requires a metallic edge. Conversely, the presence of an insulating surface implies that σ_{AHC} vanishes in the bulk. A similar argument allows us to show that, if a boundary such as the horizontal AB boundary in Fig. 1.11(a) is insulating, then we must have $\sigma_{\text{AHC}}^A = \sigma_{\text{AHC}}^B$, since otherwise charge would be accumulating at this boundary.

Similar reasoning can be used to demonstrate that a continuous change in the crystal Hamiltonian of an insulator cannot change σ_{AHC} at all. Let this change be parametrized by λ, and let materials A and B in Fig. 1.11(a) correspond to parameter values λ_0 and $\lambda_0 + \Delta\lambda$, respectively. For small enough $\Delta\lambda$, the AB boundary will certainly be insulating, so σ_{AHC}^B is exactly equal to σ_{AHC}^A. Since any insulating path in λ can be discretized into small steps of this kind, it follows that σ_{AHC} must be independent of λ along the path.

At this point, we seem to be left with the possibility that all 2D insulators can be categorized into a discrete set of classes depending on their σ_{AHC} values. A physical boundary between members of two different classes can never be insulating. One class is the class of "normal insulators" (which includes the vacuum) having $\sigma_{AHC} = 0$; these are the only ones that can have insulating edges.

If there are any insulators that do not fall into this "normal" class, then their metallic edge states must be able to carry off just the right amount of current to cancel the effects of the arriving sheet current and prevent any charge accumulation at the edge. This suggests a connection to edge channels, so it is time to recall the discussion in Section 1.3.1, where we found that a given 2D insulator is characterized by an integer C that expresses the net number of clockwise versus counterclockwise edge channels through Eq. (1.26). Clearly the "normal" insulators are those with $C=0$ and $\sigma_{AHC}=0$. We now show that everything falls into place if we assume that σ_{AHC} is required to be an integer multiple of e^2/h, where the integer is just C. That is, we can sort all 2D insulators into classes labeled by an integer C, which we call "chirality" for now, with $\sigma_{AHC} = Ce^2/h$; those having $C \neq 0$ are known as "quantum anomalous Hall" (QAH) insulators.[24]

Let's see how this works out. Consider a rectangular sample of a QAH material with $C=1$ immersed in a small uniform electric field $\mathcal{E}$ as shown in Fig. 1.11(b), with one chiral edge state circulating around its boundary in the clockwise direction consistent with Eq. (1.26). So, for example, there is one right-going edge state on the top edge.[25] Since we are considering a system in its stationary ground state, the charge density must everywhere be independent of time, so that the current must be divergence-free. The $\mathcal{E}$-induced current arriving at a segment of length dx on the top edge of the sample is $K_y\, dx$, so this will be drained off only if the positively circulating edge current I on the top edge of the sample is x-dependent with $dI/dx = -K_y$ as sketched in Fig. 1.11(b). But using Eqs. (1.25) and (1.27) we have

$$K_y = -\frac{dI}{dx} = -\frac{dI}{d\phi}\frac{d\phi}{dx} = C\frac{e^2}{h}\mathcal{E}_x \qquad (1.30)$$

or, in short,

$$\sigma_{AHC} = C\frac{e^2}{h}. \qquad (1.31)$$

This important result demonstrates the quantization of the AHC in a 2D insulator!

---

[24] As we shall see in Chapter 5, C is actually a Chern number, and QAH insulators are also known as "Chern insulators." Note that sign conventions are not agreed upon in the literature. Our conventions are that $\sigma_{AHC} = \sigma_{yx} = Ce^2/h$.

[25] It is important to understand that this statement does not determine the sign of the *current* flowing on the edge, although it does determine the sign of the *change* of current with ϕ or E_F.

The existence of a nonzero AHC in a 2D system has two interesting consequences. The first has already been mentioned on p. 29, namely that the orbital magnetization varies linearly with electrostatic potential ϕ (at fixed E_F). Combining Eqs. (1.28) and (1.31), this becomes

$$\frac{\partial M_{orb}}{\partial \phi} = \frac{1}{c} \sigma_{AHC}. \tag{1.32}$$

This result is connected with the change in population of the chiral edge channels as a function of ϕ.

A second consequence is that the density of electrons in the interior of the 2D sample varies linearly with a perpendicularly applied magnetic field $B_\perp$, a fact that is most easily understood in terms of Faraday induction. Imagine a disk of material as the magnetic field is slowly turned on from zero. This generates an azimuthal $\mathcal{E}$-field according to $\nabla \times \mathcal{E} = (-1/c)\, d\mathbf{B}/dt$ (Gaussian units have been adopted), which in turn drives a current $\mathbf{K} = \sigma_{AHC}\, \hat{\mathbf{n}} \times \mathcal{E}$ in the radial direction. Conservation of charge in the form $-e\dot{n} = -\nabla \cdot \mathbf{K}$, where n is density of electrons per unit area, then gives

$$\dot{n} = \frac{\sigma_{AHC}}{e} \nabla \cdot (\hat{\mathbf{n}} \times \mathcal{E}) = \frac{-\sigma_{AHC}}{e} \hat{\mathbf{n}} \cdot (\nabla \times \mathcal{E}) = \frac{\sigma_{AHC}}{ec} \dot{B}_\perp \tag{1.33}$$

or, after integrating with respect to time, an overall change of density

$$\Delta n = \frac{\sigma_{AHC}}{ec} B_\perp \tag{1.34}$$

relative to the case when $B_\perp$ is absent.

This is essentially the *Středa formula*, derived by Středa (1982) for the more general case of a metal at fixed chemical potential. For a QAH insulator it implies $\Delta n = (e/hc)\, C B_\perp$. At first sight this formula looks problematic for an insulating disk, since the extra electron density in the interior has to come from somewhere, but it is precisely the presence of the chiral edge channel that saves the day: electrons are sucked out of the edge states and pumped into the interior as the magnetic field strength is turned up.[26]

Actually, the two properties just mentioned – the variation of M_{orb} with ϕ and the variation of n with $B_\perp$ – are intimately related by a kind of Maxwell relation as follows. If we consider the energy per unit area $E(\phi, B_\perp)$ of the system at fixed E_F, we can make the interpretations $\partial E/\partial B_\perp = -M_{orb}$ and $\partial E/\partial \phi = \sigma_{charge} = -en$. Then starting with Eq. (1.32) and equating mixed partials, we obtain

$$\frac{1}{c}\sigma_{AHC} = \frac{\partial M_{orb}}{\partial \phi} = -\frac{\partial}{\partial \phi}\frac{\partial E}{\partial B_\perp} = -\frac{\partial}{\partial B_\perp}\frac{\partial E}{\partial \phi} = e\frac{\partial n}{\partial B_\perp}. \tag{1.35}$$

---

[26] As mentioned on p. 14, the classical division into "bound" and "free" charges is not always apt. The excess charge density in Eq. (1.34) is closer in spirit to a bound charge, but is not associated with the divergence of a polarization.

This makes the connection and provides an alternative derivation of the Středa formula in Eq. (1.34).

1.3.3 Discussion

Of course, we still have not shown that insulators with nonzero C, or equivalently, with unbalanced right- and left-going edge channels, actually exist in nature – only that they are a logical possibility, consistent with charge conservation in a stationary state. The possibility of the existence of such states began to be taken seriously after 1988, when Haldane (1988) wrote down an explicit model Hamiltonian having $C = \pm 1$ in some regions of its parameter space. In 2013, a first experimental synthesis of a QAH system was demonstrated by the group of Qi-Kun Xue at Tsinghua University (Chang et al., 2013). Furthermore, the physics here is very similar to that of the quantum Hall state discovered in 1980; in that case the insulating state and the broken TR symmetry are both delivered by an external magnetic field, but the quantization of σ_{AHC} in units of e^2/h and the presence of chiral edge channels (see, for example, Beenakker, 1990) are just like what has been described in this section. Thus, there is no doubt that such states *can* exist.

Once again, notice that the arguments given here are quite general, and should be robust against the presence of interactions and/or disorder. For example, if we let regions A and B in Fig. 1.11(a) correspond to a certain system with and without interactions [$\mu = 1$ and $\mu = 0$, respectively, in Eq. (1.21)], then a first-order $\mathcal{E}$-field applied parallel to the boundary is only consistent with charge conservation if the two regions have identical σ_{AHC} values. So, with the possible exception of highly correlated states that cannot be adiabatically connected to a noninteracting insulating ground state, such as the 1/3 fractional quantum Hall state mentioned on p. 22, we conclude that σ_{AHC} is quantized to integer multiples of e^2/h even in the presence of interactions.

Exercises

1.4 (a) Argue that the result in Eq. (1.24) remains exact in the presence of weak interactions and/or disorder.

(b) Argue that Eq. (1.26) remains true in the presence of weak interactions and/or disorder. You may construct arguments in which the strength of interactions or disorder is turned on or off in various regions of the sample.

1.5 At the end of Section 1.3.1 it is stated that "an adiabatic variation of the bulk crystal Hamiltonian that does not close the gap cannot result in a change of C." Justify this statement.

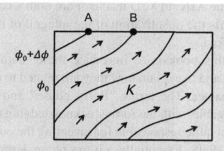

Figure 1.12 Rectangular sample of a QAH insulator having chirality $C > 0$. Lines show contours of constant electrostatic potential ϕ; arrows indicate sheet current **K**.

1.6 Figure 1.12 shows a contour plot of the electrostatic potential in a certain 2D sample of a QAH insulator.

    (a) Show that charge is locally conserved in the interior of the sample by showing that $\nabla \cdot \mathbf{K} = \partial_i K_i$ vanishes. [*Hint*: recall $K_i = \sigma_{ij}\mathcal{E}_j$, use Eq. (1.29), and express $\mathcal{E}$ in terms of ϕ.]

    (b) Show that the currents induced by σ_{AHC} always run parallel to the contour lines, and that the total current carried by the ribbon between contour ϕ_0 and contour $\phi_0 + \Delta\phi$ is $C(e^2/h)\Delta\phi$. (*Hint*: compute the bulk current arriving at edge segment AB.)

    (c) Referring to Eqs. (1.24–1.27), show that the edge current I flowing at point A exceeds that at point B by the same amount $C(e^2/h)\Delta\phi$, thus conserving charge at the edge.

1.4 Recap and Preview

In this chapter we have conducted a brief survey of some physical phenomena, mostly connected with the physics of crystalline insulators, that have a topological character and sometimes lead to quantization of certain observables. This includes electric polarization and its relation to surface charges and adiabatic currents, the quantization of charge pumping over an adiabatic loop, orbital magnetization and its relation to surface currents, and the quantized AHC and associated chiral edge channels.

It turns out that these physical properties are closely related to a cluster of mathematical concepts related to geometric phases, especially Berry phases, Berry curvatures, and Chern numbers. At least at the single-particle level, the phenomena described in this chapter are all represented most naturally in this language. For example, we shall see that the quantization of the adiabatic charge transport in a

1D insulator and of the AHC in a 2D insulator are both a consequence of the Chern theorem, which asserts the quantization of the integral of the Berry curvature over any closed 2D manifold.

The purpose of this book is to introduce these mathematical concepts in a pedagogical fashion, and to explain how they are related to the physical phenomena discussed in this chapter. The terms "Berry phase" and "geometric phase" are essentially interchangable, with the former acknowledging the influence of Michael Berry, who, while not quite responsible for inventing the concept, systematized and popularized it in the 1980s. Essentially, a Berry phase describes the phase evolution of a vector in a complex vector space as a function of some parameter or set of parameters. The connection to the physics discussed in this book arises when the vectors are associated with occupied Bloch functions $|\psi_{n\mathbf{k}}\rangle$ of a crystalline insulator and the parameter space corresponds to the wavevector $\mathbf{k}$ in the Brillouin zone (BZ), possibly augmented by adiabatic parameters describing the slow evolution of the system. The topological nature of the problem emerges because the BZ can be regarded as a closed manifold – that is, a loop, 2-torus, or 3-torus in one, two, or three dimensions, respectively. This fact is at the heart of the theory of topological insulators, which will be discussed in one of the later chapters.

As has been emphasized in this chapter, many of the physical phenomena under discussion, such as the quantization of charge transport or AHC, are robust against the presence of interactions and disorder, at least if these are sufficiently weak that they do not destroy the insulating state or otherwise cause a phase transition. This strongly suggests that some principles must be at work that transcend the single-particle restriction. An alternative viewpoint, and one that more naturally encompasses interactions and disorder, is the field-theory approach. Electromagnetism, being a gauge field theory, automatically conserves charge, so that many of the arguments relying on charge conservation given earlier in this chapter follow automatically. The 2D QAH state can be described by augmenting the usual electromagnetic field theory with a Chern–Simons action term, and the quantization of the AHC and other properties then follow naturally from the resulting Lagrangian. Even in cases where the microscopic Hamiltonian is too complex to solve, or for experimental systems for which the microscopic Hamiltonian is poorly known, we may be able to argue that the system is characterized at the macroscopic level by an effective field theory such as the Chern–Simons one. While this field-theory viewpoint is an enormously powerful one, it is not the focus of the present book. For some introduction to this interesting subject, the reader is referred to a foundational paper by Qi, Hughes, and Zhang (2008), the review by Qi and Zhang (2011), and the book by Bernevig (2013).

Having provided some motivation and physical context in this chapter, we are now ready to begin our journey on the road to understanding how the physical properties introduced in this chapter can be understood in terms of mathematical concepts related to Berry phases and curvatures. We begin by going back to basics, starting with Bloch's theorem, and introduce the tight-binding approach to electronic structure theory in Chapter 2. Next, the definitions and basic development of the theory of Berry phases, and its application to Bloch functions in the BZ, are presented in Chapter 3, where Wannier functions are also introduced. Chapter 4 is then devoted to the Berry-phase theory of polarization, a subject of broad importance and interest in modern computational electronic structure theory. Chapter 5 deals with the theory of topological insulators, beginning with a discussion of the quantum anomalous Hall state in two dimensions and then moving on to a discussion of other kinds of topological states in two and three dimensions. Finally, the theory of orbital magnetization and magnetoelectric coupling is developed in Chapter 6, culminating in a discussion of the Chern–Simons axion coupling and axion electrodynamics, topics that are closely related to the theory of topological insulators.

2

Review of Electronic Structure Theory

The basic concepts of the electronic band structure of crystalline materials, notably Bloch's theorem and its consequences, are developed in any standard solid state physics text. These concepts are briefly reviewed in the first part of this chapter, in part to establish the notation to be used later. For the reader interested in modern computational methods for solving for the electronic structure, the book by Martin (2004) is recommended. The second part of this chapter provides an introduction to one particularly simple approach, that of empirical tight-binding theory, which is often capable of giving a useful qualitative picture in terms of bonding between localized orbitals. The concepts to be introduced in the later chapters of this book will often be illustrated via examples that are based on the tight-binding description. Finally, the last section introduces linear response theory, which will also play an important role in our later considerations.

2.1 Electronic Hamiltonian and Bloch Functions

The Hamiltonian for a system of electrons in a crystalline solid can be regarded as known. All the difficulties are associated with the need to simplify the Hamiltonian to a tractable form.

2.1.1 Reduction to a Single-Particle Hamiltonian

We begin with the Hamiltonian for a finite system such as a molecule or cluster, which can be written as

$$H = T_e + T_n + V_{e\text{-}e} + V_{n\text{-}e} + V_{n\text{-}n} + \cdots \qquad (2.1)$$

where T_e and T_n are electron and nuclear kinetic energies, respectively, and $V_{e\text{-}e}$, $V_{n\text{-}e}$, and $V_{n\text{-}n}$ are the electron–electron, electron–nuclear, and nuclear–nuclear Coulomb interactions, respectively. Explicitly,

$$T_e = \sum_i \frac{p_i^2}{2m}, \tag{2.2}$$

$$T_n = \sum_I \frac{P_I^2}{2M_I}, \tag{2.3}$$

$$V_{e\text{-}e} = \frac{1}{2} \sum_{ij} \frac{e^2}{|\mathbf{r}_i - \mathbf{r}_j|}, \tag{2.4}$$

$$V_{n\text{-}e} = -\sum_{iI} \frac{e^2 Z_I}{|\mathbf{r}_i - \mathbf{R}_I|}, \tag{2.5}$$

$$V_{n\text{-}n} = \frac{1}{2} \sum_{IJ} \frac{e^2 Z_I Z_J}{|\mathbf{R}_I - \mathbf{R}_j|}, \tag{2.6}$$

where i runs over the electrons in the system with coordinates $\mathbf{r}_i$, momenta $\mathbf{p}_i$, charge $-e$, and mass m, while I runs over the nuclei with coordinates $\mathbf{R}_I$, momenta $\mathbf{P}_I$, charge eZ_I, and mass M_I. Gaussian units are used throughout this book. The spin of the electrons has been suppressed for the moment, and the ellipsis '$\cdots$' in Eq. (2.1) indicates other terms that have been neglected, including relativistic corrections, interactions with external electric and magnetic fields, and interactions involving nuclear magnetic moments.

We first apply the *adiabatic approximation*, which states that the nuclear coordinates can be treated as slow variables because the nuclei are so much heavier than the electrons. Adopting this framework, we now treat the nuclear coordinates $\mathbf{R}_I$ as classical variables that are fixed by hand, or else are evolving slowly according to some classical equations of motion, and concentrate on solving the electronic Hamiltonian

$$H_{elec} = T_e + V_{n\text{-}e} + V_{e\text{-}e}. \tag{2.7}$$

Defining

$$V_{ext}(\mathbf{r}) = \sum_I \frac{e^2 Z_I}{|\mathbf{r} - \mathbf{R}_I|} \tag{2.8}$$

to be the bare potential felt by the electrons in the Coulomb field of the nuclei, which are regarded as "external" to the electron system, this can be written as

$$H_{elec} = \sum_i \left[\frac{p_i^2}{2m} + V_{ext}(\mathbf{r}_i) \right] + \frac{1}{2} \sum_{ij} \frac{e^2}{|\mathbf{r}_i - \mathbf{r}_j|}, \tag{2.9}$$

which appeared earlier as Eq. (1.19), albeit with a few small changes of notation.

Equation (2.9) is famously impossible to solve when the number of electrons N becomes large. The problem does not reside in the kinetic-energy or external-potential terms, which are single-particle operators, but rather in the Coulomb interaction term, which is a two-particle operator. As a consequence, the many-body electron wave function $\Psi(\mathbf{r}_1, \ldots, \mathbf{r}_N)$ cannot be written as a single Slater determinant or otherwise factored into any elementary form, so that the complexity of the wave function, and the time to an approximate solution, grow exponentially with N. In the quantum chemistry community, this problem is typically tackled by expanding Ψ as a linear combination of some subset of Slater determinants that are deemed most relevant, with an artful truncation of small terms in the expansion. While successful for atoms and molecules of modest size, these approaches still tend to become intractable for larger systems such as biomolecules and solids.

For those systems, a very widely used approach is that of *density-functional theory* (DFT), which reduces Eq. (2.9) to an approximate one-particle mean-field Hamiltonian. DFT is based upon a proof by Hohenberg and Kohn (1964), which states that the many-body ground-state wave function can be regarded as a functional of the ground-state electron density in the context of the electronic structure problem. The ground-state energy is then also a functional of the density, and takes its minimum at the true ground state. Remarkably, this means that it is possible in principle to solve the many-body problem by minimizing this energy functional without reference to any many-body wave function. The subsequent paper of Kohn and Sham (1965) provided an iterative prescription for such a solution, based on the introduction of a noninteracting Hamiltonian constructed so as to have the same ground-state density as the true interacting system. Unfortunately, this Kohn–Sham procedure is only as good as the energy functional it is based on. While the Hohenberg–Kohn theorem provides an existence proof for it, the exact functional remains far out of reach to this day.

Nevertheless, useful approximations to this energy functional have been developed over the years. The critical piece that is left over after accounting for the known direct Coulomb and kinetic-energy terms is the "exchange-correlation" functional. Thankfully, we now have in hand numerous local and semilocal approximations to this functional that are of sufficient quality to allow accurate calculations of many properties of materials. An excellent review of this subject appears in Part II of the book by Martin (2004), while Parr and Weitao (1989) provide a good treatment of some of the more formal aspects of the theory. The review article by Jones (2015) gives an excellent overview from a modern perspective.

In the DFT approach, one constructs an effective single-particle Kohn–Sham Hamiltonian

$$H_{\text{KS}} = \frac{p^2}{2m} + V_{\text{KS}}^{[n]}(\mathbf{r}) \tag{2.10}$$

where the superscript $[n]$ on the Kohn–Sham potential V_{KS} indicates that it is a functional of the electron density $n(\mathbf{r})$. Specifically,

$$V_{KS}^{[n]}(\mathbf{r}) = V_{ext}(\mathbf{r}) + V_H^{[n]}(\mathbf{r}) + V_{XC}^{[n]}(\mathbf{r}) \tag{2.11}$$

where

$$V_H^{[n]}(\mathbf{r}) = e^2 \int \frac{n(\mathbf{r'})}{|\mathbf{r} - \mathbf{r'}|} d^3 r' \tag{2.12}$$

is the Hartree potential, reflecting the Coulomb repulsion from the electron cloud, and

$$V_{XC}^{[n]}(\mathbf{r}) = \frac{\delta E_{XC}[n]}{\delta n(\mathbf{r})} \tag{2.13}$$

is the exchange-correlation potential, which encodes nonclassical many-body effects. Here V_{XC} is written formally as a functional derivative of the exchange-correlation energy $E_{XC}[n]$, which plays a central role in the DFT formalism. The eigensolution of the Kohn–Sham Hamiltonian of Eq. (2.10) is then

$$H_{KS} |\psi_i\rangle = E_i |\psi_i\rangle \tag{2.14}$$

where E_i and $|\psi_i\rangle$ are the single-particle Kohn–Sham eigenvalues and eigenvectors.[1] The ground-state solution is given by identifying the lowest N eigenvalues and constructing the charge density

$$n(\mathbf{r}) = \sum_{i=1}^{N} |\psi_i(\mathbf{r})|^2, \tag{2.15}$$

which is then plugged back into Eqs. (2.12) and (2.13) to obtain a new Kohn–Sham potential H_{KS}. The process is iterated until a self-consistent solution for $n(\mathbf{r})$ and $V_{KS}(\mathbf{r})$ is obtained. The most time-consuming step is typically the eigensolution of Eq. (2.14), but since this is now only a single-particle problem, it can be carried out efficiently even for large molecules and solids.

This gain in efficiency comes at the cost of introducing an approximation for the exchange-correlation functional $E_{XC}[n]$ appearing in Eq. (2.13). The simplest is the local-density approximation (LDA) in which the exchange-correlation energy is assumed to have the form

$$E_{XC}^{[n]}(\mathbf{r}) = \int n(\mathbf{r}) \, \epsilon_{XC}(n(\mathbf{r})) \, d^3 r \tag{2.16}$$

where $\epsilon_{XC}(n)$ is the exchange-correlation energy per particle of a uniform electron gas of density n, and this function is evaluated at the local density $n(\mathbf{r})$. In this case,

---

[1] Note that i now runs over one-particle states, not over electrons as in Eq. (2.9).

V_{XC} takes a local form $V_{XC}(\mathbf{r}) = \mu_{XC}(n(\mathbf{r}))$, where $\mu_{XC}(n) = \epsilon_{XC}(n) + n\epsilon'_{XC}(n)$. The functions $\epsilon_{XC}(n)$ and $\mu_{XC}(n)$ are extracted once and for all from high-level many-body calculations on the uniform electron gas.

Many more sophisticated functionals have since been invented, among which variants of the generalized-gradient approximation (GGA) are especially popular. Again, the reader is referred to Part II of the excellent book by Martin (2004) for an overview.

For our purposes, all we really have to know is that the many-electron problem can be reduced to an approximate one-particle form, as given by Eq. (2.10), with reasonable accuracy for materials that are not too strongly correlated. We shall work under this approximation henceforth. We shall also assume that the expectation value of any one-particle observable $\mathcal{O}$ (e.g., dipole operator, momentum or current operator) is given by

$$\langle \mathcal{O} \rangle = \sum_{i=1}^{N} \langle \psi_i | \mathcal{O} | \psi_i \rangle \tag{2.17}$$

where the $|\psi_i\rangle$ are the occupied Kohn–Sham orbitals.

One other approximation that is commonly used in practice is the *pseudopotential approximation*, in which the attraction of an electron to nucleus I in Eq. (2.8) is replaced by an effective attraction $V_I^{PS}(|\mathbf{r}-\mathbf{r}_I|)$. The pseudopotential V^{PS} is constructed, once and for all for each atomic species needed for the calculation, in such a way as to reproduce the properties of the valence electrons as closely as possible while avoiding the need to include core electrons in the calculation. Again, the text by Martin (2004) is recommended for an overview of this topic.

2.1.2 Spin, Spin-Orbit Coupling, and External Fields

The development in Section 2.1.1 was written for the case of spinless electrons. In reality, of course, electrons have to be treated as spinors. For materials that do not contain heavy atoms from the bottom of the Periodic Table, spin-orbit coupling (SOC) can usually be safely neglected; in this case the spin-up and spin-down systems of electrons can be treated independently.[2] The Kohn–Sham equations are then easily generalized by attaching spin labels to the wave functions, potentials, and densities. That is, spin-up electrons obey

$$\left(\frac{p^2}{2m} + V_{ext} + V_H + V_{XC,\uparrow} \right) |\psi_{i\uparrow}\rangle = E_{i\uparrow} |\psi_{i\uparrow}\rangle \tag{2.18}$$

---

[2] As an exception, the presence of spin frustration can sometimes lead to noncollinear structures even when the SOC is negligible.

and similarly for down electrons. The Hartree potential remains as written in Eq. (2.12), where $n(\mathbf{r}) = n_\uparrow(\mathbf{r}) + n_\downarrow(\mathbf{r})$ is now the total electron density. After the exchange-correlation energy of Eq. (2.16) has been generalized to account for two spin species, a pair of spin-dependent potentials $\mu_\uparrow^{\text{XC}}(n_\uparrow, n_\downarrow)$ and $\mu_\downarrow^{\text{XC}}(n_\uparrow, n_\downarrow)$ are again extracted from the electron-gas results, but this time on a uniform spin-polarized electron gas. This theory is known as the *local spin-density approximation* (LSDA). For nonmagnetic systems, the spin-up and spin-down systems are identical; one can then treat the electrons as though they were spinless, just inserting a factor of 2 in the right-hand sides of Eqs. (2.15) and (2.17) to account for the spin degeneracy.

When present, the SOC term mixes spin-up and spin-down electron states in such a way that a true spinor treatment is unavoidable. This is a relativistic effect; it becomes important for heavy elements, roughly those with atomic numbers $Z > 50$, because the electrons travel at relativistic speeds in the core region of the atom. When it is included, the Kohn–Sham equation takes the form

$$\left[\frac{p^2}{2m} + V_{\text{ext}} + V_{\text{H}} + V_{\text{XC}} + \mathbf{W}_{\text{XC}} \cdot \boldsymbol{\sigma} + h_{\text{SOC}}\right] \begin{pmatrix} \psi_{i\uparrow} \\ \psi_{i\downarrow} \end{pmatrix} = E_i \begin{pmatrix} \psi_{i\uparrow} \\ \psi_{i\downarrow} \end{pmatrix} \qquad (2.19)$$

where V_{XC} and $\mathbf{W}_{\text{XC}}$ are the scalar and spin-dependent portions of the exchange-correlation potential, respectively, and $\boldsymbol{\sigma} = (\sigma_x, \sigma_y, \sigma_z)$ is the vector of Pauli spin matrices

$$\sigma_x = \begin{pmatrix} 0 & 1 \\ 1 & 0 \end{pmatrix}, \quad \sigma_y = \begin{pmatrix} 0 & -i \\ i & 0 \end{pmatrix}, \quad \sigma_z = \begin{pmatrix} 1 & 0 \\ 0 & -1 \end{pmatrix}. \qquad (2.20)$$

The spin-orbit Hamiltonian h_{SOC} is derived from the nonrelativistic limit of the Dirac equation and can be written as

$$h_{\text{SOC}} = \frac{\hbar}{4m^2c^2} \boldsymbol{\sigma} \cdot \nabla V(\mathbf{r}) \times \mathbf{p} \qquad (2.21)$$

where $V(\mathbf{r}) = -e\phi(\mathbf{r})$ is the electrostatic potential felt by the electron. (Recall that Gaussian units are being used.) Since $\nabla\phi$ is strongest in the core regions, h_{SOC} is often approximated by a sum of on-site "atomic" contributions of the form $\xi(r)\mathbf{L}\cdot\mathbf{S}$, where $\mathbf{L}$ and $\mathbf{S}$ are orbital and spin angular momenta on the site in question and $\xi(r)$ is a radial function related to $d\phi/dr$ in the core via

$$\xi(r) = \frac{-e}{2m^2c^2} \frac{1}{r} \frac{d\phi}{dr} \qquad (2.22)$$

(see Ex. 2.1). The SOC will play an important role in the discussion of topological insulators and semimetals in Chapter 5.

In magnetic materials, the spontaneously broken time-reversal (TR) symmetry almost invariably arises in the spin sector. In other words, the system finds that it can lower its energy relative to a fully spin-paired nonmagnetic state by developing unequal populations of spin-up and spin-down electrons on some atomic sites. Typically these magnetic moments develop a well-defined magnetic order (e.g., ferromagnetic or antiferromagnetic) so as to minimize the energy of exchange interactions between them. This process can often be well described at the LSDA level. If SOC is not included in the description, then the spin-up and spin-down sectors "don't know" that TR has been broken. It is only when SOC is included that the information about broken TR symmetry is transmitted to the orbital sector, and orbital currents develop.[3] Orbital currents are central to many of the phenomena to be described in Chapter 5, such as the anomalous Hall conductivity discussed in Section 5.1. As we shall see, then, the SOC will have an important role to play in the discussion of those phenomena.

For the case of a crystal under the influence of external electric and magnetic fields, expressed as $\mathcal{E} = -\nabla\phi - c^{-1}d\mathbf{A}/dt$ and $\mathbf{B} = \nabla \times \mathbf{A}$ in terms of the scalar and vector potentials $\phi(\mathbf{r}, t)$ and $\mathbf{A}(\mathbf{r}, t)$, Eq. (2.19) needs to be extended in three ways: (1) the addition of a term $-e\phi(\mathbf{r})$ to describe the electrostatic potential; (2) the canonical replacement $\mathbf{p} \rightarrow \mathbf{p} + e\mathbf{A}(\mathbf{r})/c$; and (3) the addition of a Zeeman interaction $-\gamma\mathbf{B}(\mathbf{r})\cdot\mathbf{S}$, where γ is the gyromagnetic ratio ($\gamma = -e/mc$ in Gaussian units when taking the g-factor of the electron to be exactly $g_e=2$, or $\gamma = -g_e e/2mc$ more generally). In the simplest case of a free electron in the presence of external fields, the electronic Hamiltonian then takes the form

$$H = \frac{1}{2m}\left(\mathbf{p} + \frac{e}{c}\mathbf{A}(\mathbf{r})\right)^2 - e\phi(\mathbf{r}) + \frac{\hbar e}{2mc}\mathbf{B}(\mathbf{r}) \cdot \boldsymbol{\sigma}\,. \tag{2.23}$$

In the following discussion, we will typically suppress the spin indices and neglect external fields and SOC for simplicity.

2.1.3 Crystal Potential, Bloch's Theorem, and Reciprocal Space

Taking the crystal Hamiltonian to be given by Eq. (2.10), we simplify the notation further to

$$H = \frac{p^2}{2m} + V(\mathbf{r})\,. \tag{2.24}$$

For a 3D crystal, the potential obeys the periodicity conditions

$$V(\mathbf{r} + \mathbf{R}_\ell) = V(\mathbf{r}) \tag{2.25}$$

---

[3] In principle, one could imagine that the magnetic instability could occur directly in the orbital sector, and then get transmitted to the spin sector by the SOC. Various proposals for this kind of "flux state" have been put forward over the years, but have typically been controversial. If such flux states do exist, it is safe to say that they are rare.

for any lattice vector

$$\mathbf{R}_\ell = \ell_1 \mathbf{a}_1 + \ell_2 \mathbf{a}_2 + \ell_3 \mathbf{a}_3 , \tag{2.26}$$

where $\ell = (\ell_1, \ell_2, \ell_3)$ is a triplet of integers and $\mathbf{a}_1$, $\mathbf{a}_2$, and $\mathbf{a}_3$ are three primitive lattice vectors. Then the system is invariant under any of the three translation operators $T_{\mathbf{a}_j}$, where $T_\mathbf{u}$ is defined as the translation operator that shifts the entire system by displacement $\mathbf{u}$ – that is, $T_\mathbf{u} f(\mathbf{r}) = f(\mathbf{r} - \mathbf{u})$. Thus, the commutators $[H, T_{\mathbf{a}_j}]$ all vanish. Moreover, the three translation operators commute with one another, so it follows that the eigenfunctions of H can be chosen to be simultaneous eigenfunctions of all three of the $T_{\mathbf{a}_j}$. Because $T_{\mathbf{a}_j}$ is unitary, its eigenvalues must have unit norm and can be written as $e^{-i\kappa_j}$ for some phase angle κ_j.

This leads directly to Bloch's theorem. The preceding conditions imply that the eigenvectors of H can be labeled as $\psi_{n\kappa}$, where $\kappa = (\kappa_1, \kappa_2, \kappa_3)$ is a triplet of phase angles and n is an additional label (the "band index") that counts states of the same κ in order of increasing energy, such that

$$H|\psi_{n\kappa}\rangle = E_{n\kappa} |\psi_{n\kappa}\rangle . \tag{2.27}$$

The label κ indicates that the state transforms under translations as $T_{\mathbf{a}_j}|\psi_{n\kappa}\rangle = e^{-i\kappa_j} |\psi_{n\kappa}\rangle$, or equivalently,

$$\psi_{n\kappa} (\mathbf{r} + \mathbf{R}_\ell) = e^{i\kappa \cdot \ell} \psi_{n\kappa} (\mathbf{r}) . \tag{2.28}$$

Note that κ_j and $\kappa_j + 2\pi$ correspond to the same eigenvalue $e^{-i\kappa_j}$ of $T_{\mathbf{a}_j}$ and, therefore, label the same state, so that κ_j is best regarded as living on the circumference of a unit circle. By the same token, the set of labels (κ_1, κ_2) in 2D has the topology of a torus, and the triplet of labels $(\kappa_1, \kappa_2, \kappa_3)$ in 3D has the topology of a 3-torus. The fact that these are closed spaces has strong implications for the existence of topological properties, as will be discussed in later chapters.

To cast Bloch's theorem into a more familiar form, we introduce the reciprocal lattice. Its primitive reciprocal lattice vectors $\mathbf{b}_j$ are defined so as to be dual to the $\mathbf{a}_j$ in the sense that $\mathbf{a}_i \cdot \mathbf{b}_j = 2\pi \delta_{ij}$. Explicitly, $\mathbf{b}_1 = 2\pi (\mathbf{a}_2 \times \mathbf{a}_3)/V_{\text{cell}}$ (and similarly for $\mathbf{b}_2$ and $\mathbf{b}_3$ by permutation of indices) since the cell volume is $V_{\text{cell}} = \mathbf{a}_1 \cdot \mathbf{a}_2 \times \mathbf{a}_3$.[4] An arbitrary reciprocal lattice vector is of the form

$$\mathbf{G}_\mathbf{m} = m_1 \mathbf{b}_1 + m_2 \mathbf{b}_2 + m_3 \mathbf{b}_3 , \tag{2.29}$$

and the real and reciprocal lattices are easily seen to obey the important condition

$$e^{i\mathbf{G}_\mathbf{m} \cdot \mathbf{R}_\ell} = 1 \tag{2.30}$$

---

[4] In 2D, the corresponding construction is $\mathbf{b}_1 = 2\pi (\mathbf{a}_2 \times \hat{\mathbf{n}})/A_{\text{cell}}$ and $\mathbf{b}_2 = 2\pi (\hat{\mathbf{n}} \times \mathbf{a}_1)/A_{\text{cell}}$, where $\hat{\mathbf{n}}$ is the unit normal in the direction of $\mathbf{a}_1 \times \mathbf{a}_2$.

for any $\mathbf{G_m}$ and any $\mathbf{R}_\ell$. Finally, we define the wavevector $\mathbf{k}$ to be

$$\mathbf{k} = \frac{\kappa_1}{2\pi}\mathbf{b}_1 + \frac{\kappa_2}{2\pi}\mathbf{b}_2 + \frac{\kappa_3}{2\pi}\mathbf{b}_3, \tag{2.31}$$

and adopt $\mathbf{k}$ in place of κ as the label of the eigenvector. Bloch's theorem then takes the familiar form

$$H|\psi_{n\mathbf{k}}\rangle = E_{n\mathbf{k}}|\psi_{n\mathbf{k}}\rangle \tag{2.32}$$

where $\psi_{n\mathbf{k}}$ transforms under translations as

$$\psi_{n\mathbf{k}}(\mathbf{r} + \mathbf{R}) = e^{i\mathbf{k}\cdot\mathbf{R}}\,\psi_{n\mathbf{k}}(\mathbf{r}) \tag{2.33}$$

for any lattice vector $\mathbf{R}$. The groups of states with the same n index are referred to as "bands," and bands that are nowhere degenerate with the next lower or higher band are denoted as "isolated bands."

Note that $\mathbf{k}$, which is just a rescaled version of κ, also labels states redundantly. Indeed, the wavevectors $\mathbf{k}$ and $\mathbf{k} + \mathbf{G}$ are duplicate labels for the same state for any reciprocal lattice vector $\mathbf{G}$, so a single unit cell in reciprocal space is enough to label each state once and only once. The space of unique wavevector labels $\mathbf{k}$ is known as the *Brillouin zone* (BZ). Historically this term was usually understood to refer to the Wigner–Seitz reciprocal-space cell, which by definition is the locus of reciprocal-space points closer to the origin than to any other $\mathbf{G}$-vector. In this book, by contrast, I adopt a broader definition that includes any suitable unit cell in reciprocal space. For example, it could be a parallelopiped formed from the three reciprocal lattice vectors $\mathbf{b}_1$, $\mathbf{b}_2$, and $\mathbf{b}_3$, either centered on the origin or with a corner at the origin. If we regard the BZ in this way – that is, as an open region with boundaries – then the states on the boundaries must be related by

$$|\psi_{n,\mathbf{k}+\mathbf{G}}\rangle = e^{i\eta(\mathbf{k})}|\psi_{n\mathbf{k}}\rangle \tag{2.34}$$

where $\mathbf{G}$ is a reciprocal lattice vector relating two states on the boundary. The phase factor $e^{i\eta(\mathbf{k})}$ is allowed since it does not affect the physical interpretation of the state. If we insist that this phase factor should be unity, this defines the *periodic gauge condition* under which

$$|\psi_{n,\mathbf{k}+\mathbf{G}}\rangle = |\psi_{n\mathbf{k}}\rangle. \tag{2.35}$$

Again, we emphasize that $\mathbf{k}$, like κ, can most naturally be regarded as living in a closed manifold with the topology of a 3-torus. From this viewpoint, the BZ is a closed manifold without boundary. If the phase of the wave function $|\psi_{n\mathbf{k}}\rangle$ is smoothly defined everywhere on this manifold, then Eq. (2.35) is automatically satisfied, which is to say that a periodic gauge is implicit.

Equations (2.27–2.28) and Eqs. (2.32–2.33) are equivalent statements of Bloch's theorem, with the first pair being written in "internal coordinates" while the second pair is in the usual Cartesian coordinates. The former representation is often more convenient in practical calculations, but in any case we shall go back and forth freely between the two as needed in the subsequent discussions.[5]

The presentation to this point has mainly been framed in terms of ordinary 3D crystals, but the generalization to 2D or 1D is straightforward. It is worth adding a word here, however, about the terminology for dimensionality to be used in this book. Systems are classified here by the dimension of the lattice (or, equivalently, of the reciprocal lattice). Thus, "1D systems" are taken to include periodic polymers or nanotubes embedded in 3D space, or nanoribbons embedded in 2D space, in addition to "true" 1D systems described by Hamiltonians of the form $p^2/2m + V(x)$. Similarly, graphene and MoS_2 sheets are regarded as "2D systems" even though they have some extent in 3D, and molecules and clusters are classified as 0D. In this terminology, therefore, the dimensionality of the system is the same as the dimension of the BZ.

2.1.4 Electron Counting

To count electron states, it is convenient to impose periodic boundary conditions on a supercell built from $N = N_1 \times N_2 \times N_3$ primitive cells, with the understanding that the limit $N_j \rightarrow \infty$ will be taken later. Then k-space becomes discretized in the sense that only $\psi_{n\mathbf{k}}$ with $\mathbf{k} = (n_1/N_1)\mathbf{b}_1 + (n_2/N_2)\mathbf{b}_2 + (n_3/N_3)\mathbf{b}_3$ for integers $n_j = 1, \ldots, N_j$ are consistent with these boundary conditions. For any given band, there are N states in the supercell of volume NV_{cell}, or one per V_{cell}. This establishes the principle that a filled band contributes a density of exactly one electron per unit cell.[6]

It also implies that the density of states per unit volume of k-space is $V_{\text{cell}}/(2\pi)^3$. A useful heuristic relation embodying this principle is

$$\sum_{\mathbf{k}} \longleftrightarrow \frac{V}{(2\pi)^3} \int d^3k \qquad (2.36)$$

where $V = NV_{\text{cell}}$ is the supercell volume. Then the average of some quantity $g(\mathbf{k})$ over the BZ can be written as

$$\bar{g} = \frac{1}{N} \sum_{\mathbf{k}}^{\text{BZ}} g(\mathbf{k}) = \frac{V_{\text{cell}}}{(2\pi)^3} \int_{\text{BZ}} g(\mathbf{k}) \, d^3k. \qquad (2.37)$$

---

[5] The term 'BZ' is also commonly used later to designate the unit cell in κ space.
[6] When spin degeneracy is present, a "band" is often taken to comprise both spins; in this case, it contributes two electrons per cell.

For example, the density of states $\rho(E)$, defined as the number of states per unit energy interval per unit volume, is given by inserting a factor of $\delta(E - E_{n\mathbf{k}})/V_{\text{cell}}$ into Eq. (2.37) (δ is the Dirac delta function) to get

$$\rho(E) = \frac{1}{(2\pi)^3} \sum_n \int_{\text{BZ}} \delta(E - E_{n\mathbf{k}}) \, d^3k \,. \tag{2.38}$$

Similarly, the electron density $n(\mathbf{r})$ can be written in terms of the Bloch functions $\psi_{n\mathbf{k}}$ (normalized to a unit cell) as

$$n(\mathbf{r}) = \frac{1}{N} \sum_{n\mathbf{k}} f_{n\mathbf{k}} \, |\psi_{n\mathbf{k}}(\mathbf{r})|^2 = \frac{V_{\text{cell}}}{(2\pi)^3} \sum_n \int_{\text{BZ}} f_{n\mathbf{k}} \, |\psi_{n\mathbf{k}}(\mathbf{r})|^2 \, d^3k \,, \tag{2.39}$$

where $f_{n\mathbf{k}}$ is the Fermi occupation function, and the electronic contribution to the charge density is just $\rho_{\text{elec}}(\mathbf{r}) = -en(\mathbf{r})$.

2.1.5 Cell-Periodic Bloch Functions

The Bloch functions $\psi_{n\mathbf{k}}(\mathbf{r})$ obey twisted periodic boundary conditions on the primitive cell as specified in Eq. (2.33). This means that Bloch functions with different $\mathbf{k}$ obey different boundary conditions – a feature that would be awkward for many purposes. Thus, it is often convenient to work instead with the *cell-periodic Bloch functions*

$$u_{n\mathbf{k}}(\mathbf{r}) = e^{-i\mathbf{k}\cdot\mathbf{r}} \psi_{n\mathbf{k}}(\mathbf{r}) \tag{2.40}$$

which obey ordinary periodic boundary conditions

$$u_{n\mathbf{k}}(\mathbf{r} + \mathbf{R}) = u_{n\mathbf{k}}(\mathbf{r}) \,. \tag{2.41}$$

We can then think of the Bloch function $\psi_{n\mathbf{k}}(\mathbf{r}) = e^{i\mathbf{k}\cdot\mathbf{r}} u_{n\mathbf{k}}(\mathbf{r})$ as a product of a plane-wave envelope function $e^{i\mathbf{k}\cdot\mathbf{r}}$ times an underlying cell-periodic function $u_{n\mathbf{k}}(\mathbf{r})$, as illustrated in Fig. 2.1. From Eq. (2.35), it follows that

$$u_{n,\mathbf{k}+\mathbf{G}}(\mathbf{r}) = e^{-i\mathbf{G}\cdot\mathbf{r}} u_{n\mathbf{k}}(\mathbf{r}) \tag{2.42}$$

in a periodic gauge.

Unlike the original Bloch functions, the $u_{n\mathbf{k}}$ all belong to the same Hilbert space of periodic functions defined on the primitive unit cell. This has many desirable properties. For example, derivatives such as $\nabla_{\mathbf{k}} u_{n\mathbf{k}}$ are also well-defined periodic functions belonging to the same Hilbert space, whereas $\nabla_{\mathbf{k}} \psi_{n\mathbf{k}}(\mathbf{r}) = e^{i\mathbf{k}\cdot\mathbf{r}}[\nabla_{\mathbf{k}} u_{n\mathbf{k}}(\mathbf{r}) + i\mathbf{r}\, u_{n\mathbf{k}}(\mathbf{r})]$ has a piece that blows up linearly with distance from the origin. For these reasons it will turn out to be essential to use the $u_{n\mathbf{k}}$ and not the $\psi_{n\mathbf{k}}$ when defining Berry phases and related quantities for the Bloch functions in Section 3.4.

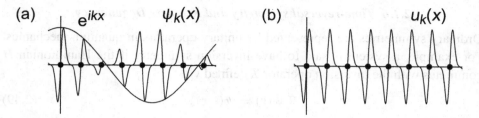

Figure 2.1 (a) Bloch wave function $\psi_k(x)$ showing how it is modulated by an envelope function e^{ikx}. (b) Definition of cell-periodic Bloch function $u_k(x)$, defined in such a way that the envelope function has been factored out.

It is also conventional to define the **k**-dependent effective Hamiltonian

$$H_{\mathbf{k}} = e^{-i\mathbf{k}\cdot\mathbf{r}} H e^{i\mathbf{k}\cdot\mathbf{r}} \tag{2.43}$$

such that

$$H_{\mathbf{k}}\,|u_{n\mathbf{k}}\rangle = E_{n\mathbf{k}}\,|u_{n\mathbf{k}}\rangle . \tag{2.44}$$

For the special case of the simplest crystal Hamiltonian $H = p^2/2m + V(\mathbf{r})$, this takes the form

$$H_{\mathbf{k}} = \frac{(\mathbf{p} + \hbar\mathbf{k})^2}{2m} + V(\mathbf{r}), \tag{2.45}$$

but only Eq. (2.43) is generally correct. In the same spirit, any one-particle operator $\mathcal{O}$ can be converted into a **k**-dependent operator in this representation via $\mathcal{O}_{\mathbf{k}} = e^{-i\mathbf{k}\cdot\mathbf{r}}\mathcal{O}e^{i\mathbf{k}\cdot\mathbf{r}}$, such that expectation values (per unit cell) take the form[7]

$$\langle\mathcal{O}\rangle = \frac{1}{N}\sum_{n\mathbf{k}}\langle\psi_{n\mathbf{k}}|\mathcal{O}|\psi_{n\mathbf{k}}\rangle_{\text{cell}} = \frac{1}{N}\sum_{n\mathbf{k}}\langle u_{n\mathbf{k}}|\mathcal{O}_{\mathbf{k}}|u_{n\mathbf{k}}\rangle \tag{2.46}$$

where N is the number of **k**-points in the BZ mesh. Of particular interest is the velocity operator

$$\mathbf{v} = \frac{-i}{\hbar}[\mathbf{r}, H], \tag{2.47}$$

which is the Heisenberg-picture time derivative of the position operator $\mathbf{r}$; in the transformed representation, $\mathbf{v}_{\mathbf{k}}$ obeys the useful formula

$$\mathbf{v}_{\mathbf{k}} = \frac{1}{\hbar}\nabla_{\mathbf{k}}H_{\mathbf{k}} \tag{2.48}$$

(see Ex. 2.2).

---

[7] The "cell" subscript on $\langle\psi|\mathcal{O}|\psi\rangle_{\text{cell}}$ indicates that the spatial integral is to be carried out over one unit cell only. See the discussion of normalization conventions at the end of Section 3.5.1.

2.1.6 Time-reversal symmetry and Kramers Degeneracy

Ordinary symmetries are represented by unitary operators in quantum mechanics. For example, a system is said to have inversion symmetry if the Hamiltonian H commutes with the inversion operator $\mathcal{I}$ defined via

$$\mathcal{I}\psi(\mathbf{r}) = \psi(-\mathbf{r})\,. \tag{2.49}$$

Then, by a standard theorem of quantum mechanics, the eigenvectors of H can be chosen so that they are also eigenvectors of $\mathcal{I}$, and states can be labeled as "even" or "odd" under inversion (i.e., of even or odd parity) depending on whether the inversion eigenvalue is ± 1.

The situation is more complicated for the case of TR symmetry, which is instead described in terms of an *antiunitary* operator $\mathcal{T}$. Like a unitary operator, an antiunitary one preserves the norm of any vector $|\psi\rangle$ that it acts on, but is also *antilinear*, which implies that $\mathcal{T}c|\psi\rangle = c^*\mathcal{T}|\psi\rangle$ for any scalar c. A discussion of the special properties of antiunitary operators, and the TR operator in particular, can be found in most standard quantum mechanics texts, although a particularly careful treatment is given in Section 4.4 of the well-known book of Sakurai (1994). When acting on a system of scalar particles, $\mathcal{T}$ is just the complex-conjugation operator $\mathcal{K}$,

$$\mathcal{T}\psi(\mathbf{r}) = \mathcal{K}\psi(\mathbf{r}) = \psi^*(\mathbf{r})\,, \tag{2.50}$$

with $\mathcal{T}^2 = 1$. For spinors, the situation is more complex: Now $\mathcal{T}^2 = -1$ is required,[8] and it is conventional to write the TR operator as

$$\mathcal{T} = i\sigma_2\mathcal{K} \tag{2.51}$$

where σ_2 is the second of the three Pauli matrices. We still say that a system conserves TR symmetry if the Hamiltonian commutes with $\mathcal{T}$,

$$H\mathcal{T} = \mathcal{T}H\,, \tag{2.52}$$

but the consequences are somewhat different. For example, we cannot simply classify states into those that are even or odd under the TR symmetry operator as we did for inversion.

In this book we are concerned with electrons, which really are spinors, but there are times when we will treat them as though they were spinless scalar particles. For example, in the LSDA treatment described on p. 43, the spin-up and spin-down electrons are treated as though they are two species of spinless particles, and the Haldane model, which will be introduced in Section 5.1.1, is a well-known model

---

[8] The TR operator squares to $+1$ for bosons, which necessarily have integer spin, and to -1 for fermions, which must have half-integer spin.

of spinless electrons that exemplifies the quantum anomalous Hall state. We shall speak of *spinful* electrons when we want to emphasize that we are treating them correctly as spinors, in contrast to *spinless* electrons when we treat them as scalars.

A crucial property of TR symmetry when applied to spinful electrons is expressed by the *Kramers theorem*, which states that all energy eigenvalues are at least two-fold degenerate. The standard demonstration is via proof by contradiction, as follows. Suppose that $|u\rangle$ is a nondegenerate eigenvector of H with eigenvalue E. Let $|v\rangle = \mathcal{T}|u\rangle$; it is also an eigenvector of H with the same eigenvalue, since $H(\mathcal{T}|u\rangle) = \mathcal{T}H|u\rangle = \mathcal{T}\epsilon|u\rangle = \epsilon(\mathcal{T}|u\rangle)$, using the fact that ϵ is real. Since we assumed nondegeneracy, this must be the same state up to a phase: $|v\rangle = e^{i\varphi}|u\rangle$. But then $\mathcal{T}^2|u\rangle = \mathcal{T}e^{i\varphi}|u\rangle = e^{-i\varphi}\mathcal{T}|u\rangle = |u\rangle$, which is inconsistent with $\mathcal{T}^2 = -1$. Thus $|u\rangle$ and $|v\rangle$ must be two different but degenerate states, related by $\mathcal{T}|u\rangle = |v\rangle$ and $\mathcal{T}|v\rangle = -|u\rangle$. When SOC is weak, we can think of these as "spin-up" and "spin-down" partners, but for strong SOC we should just regard them as time-reversed partners.

For crystalline systems, TR symmetry implies that a Bloch eigenstate $|\psi_{n\mathbf{k}}\rangle$ is degenerated with a time-reversed partner at $-\mathbf{k}$ related by

$$\mathcal{T}|\psi_{n\mathbf{k}}\rangle = e^{i\varphi}|\psi_{n,-\mathbf{k}}\rangle \tag{2.53}$$

for some ($n\mathbf{k}$-dependent) phase angle φ. The same holds for the cell-periodic Bloch function, $\mathcal{T}|u_{n\mathbf{k}}\rangle = e^{i\varphi}|u_{n,-\mathbf{k}}\rangle$, and the corresponding relation for the $\mathbf{k}$-dependent effective Hamiltonian of Eq. (2.43) is

$$\mathcal{T}H_{\mathbf{k}}\mathcal{T}^{-1} = H_{-\mathbf{k}}. \tag{2.54}$$

In the special case that $-\mathbf{k}$ and $\mathbf{k}$ are duplicate labels, which occurs if $-\mathbf{k} = \mathbf{k}$ modulo a reciprocal lattice vector, we have that $\mathcal{T}H_{\mathbf{k}}\mathcal{T}^{-1} = H_{\mathbf{k}}$, so that the Kramers theorem applies, and all states at these special $\mathbf{k}$-points must be doubly degenerate. These special wavevectors have come to be known as *time-reversal invariant momenta* (TRIM). In terms of the reduced wavevectors κ introduced in Eq. (2.27), these occur when each κ_j is either 0 or π, so that there are two TRIM in 1D, four in 2D, and eight in 3D. These special points will play an important role in the discussion of topological states in Chapter 5.

The TRIM also play a role for inversion symmetry, since the inversion operator also relates states at $\mathbf{k}$ to those at $-\mathbf{k}$:

$$\mathcal{I}|\psi_{n\mathbf{k}}\rangle = e^{i\varphi}|\psi_{n,-\mathbf{k}}\rangle. \tag{2.55}$$

Here, however, they induce no degeneracy; it is just that states at the TRIM carry definite parity labels, while states at arbitrary $\mathbf{k}$ do not. If TR and inversion are *both* present,[9] the combined operator $\mathcal{I} * \mathcal{T}$ is an antiunitary operator that maps $\mathbf{k}$ to itself at all $\mathbf{k}$, and we can repeat the Kramers-theorem argument to show that all bands are doubly degenerate everywhere in the BZ. This is what we would expect from spin degeneracy in a nonmagnetic system in the absence of SOC; the Kramers theorem tells us that the bands remain doubly degenerate even in the presence of SOC if inversion is also present.

Exercises

2.1 Starting from Eq. (2.21) and assuming a radial potential $V(r) = -e\phi(r)$, derive that $h_{\text{SOC}} = \xi(r)\mathbf{L} \cdot \mathbf{S}$ with $\xi(r)$ given by Eq. (2.22).

2.2 (a) Using only the very general definition of the velocity operator given in Eq. (2.47), show that the transformed velocity operator $\mathbf{v_k} = e^{-i\mathbf{k}\cdot\mathbf{r}}\mathbf{v}e^{i\mathbf{k}\cdot\mathbf{r}}$ is given by Eq. (2.48).

   (b) For an electron (mass m, charge $-e$) in the presence of an external scalar potential $\phi(\mathbf{r})$ and vector potential $\mathbf{A}(\mathbf{r})$, the Hamiltonian is

$$H = \frac{1}{2m}\left(\mathbf{p} + \frac{e}{c}\mathbf{A}(\mathbf{r})\right)^2 - e\phi(\mathbf{r})$$

   Find $\mathbf{v}$, convert it to $\mathbf{v_k}$, and check for consistency with the expression in part (a).

2.3 In Chapter 3 we will encounter the concept of a *gauge freedom* in the choice of the Bloch functions. That is, given a set of Bloch functions $\psi_{n\mathbf{k}}(\mathbf{r})$, define a new set via

$$|\tilde{\psi}_{n\mathbf{k}}\rangle = e^{-i\beta_n(\mathbf{k})}|\psi_{n\mathbf{k}}\rangle,$$

where β_n is a real function of $\mathbf{k}$. The transformation from the old set to the new set is known as a "gauge transformation."

   (a) A "periodic gauge" is one for which $|\psi_{n,\mathbf{k}+\mathbf{G}}\rangle = |\psi_{n\mathbf{k}}\rangle$ (i.e., equal with the same phase). What has to be true about $\beta_n(\mathbf{k})$ if it is to preserve the periodicity of the gauge?

   (b) Show that the $|u_{n\mathbf{k}}\rangle$ transform in the same way as the $|\psi_{n\mathbf{k}}\rangle$ under a gauge transformation.

   (c) In Chapter 3 we will introduce the "Berry connection" $\mathbf{A}_n(\mathbf{k}) = i\langle u_{n\mathbf{k}}|\nabla_{\mathbf{k}}u_{n\mathbf{k}}\rangle$. Derive an expression that describes how this transforms under a gauge transformation. That is, express $\tilde{\mathbf{A}}_n(\mathbf{k}) = i\langle\tilde{u}_{n\mathbf{k}}|\nabla_{\mathbf{k}}\tilde{u}_{n\mathbf{k}}\rangle$ in terms of $\mathbf{A}_n(\mathbf{k})$ plus a correction.

   (d) Show that $\nabla_{\mathbf{k}} \times \mathbf{A}_n$ is gauge-invariant.

---

[9] Actually, all we require here is that the composed operator $\mathcal{I} * \mathcal{T}$ is a symmetry, which may occur even when $\mathcal{I}$ and $\mathcal{T}$ are not symmetries individually.

2.4 An operator $\mathcal{O}$ is said to be "even under TR" if it commutes with $\mathcal{T}$ (i.e., $\mathcal{O}\mathcal{T} = \mathcal{T}\mathcal{O}$), and "odd" if it anticommutes (i.e., $\mathcal{O}\mathcal{T} = -\mathcal{T}\mathcal{O}$). A similar classification applies to inversion symmetry.

(a) Which of the operators $\mathbf{r}$, $\mathbf{p}$, $p^2/2m$, $\mathbf{v}$, $\mathbf{S} = \hbar\boldsymbol{\sigma}/2$, and $\mathbf{L}$ are even, and which are odd, under TR?

(b) Repeat part (a), but answer the question for inversion.

(c) Show that h_{SOC} as given by Eq. (2.21), and also as given by the expression in Ex. 2.1, conserves both TR and inversion symmetry.

2.2 Tight-Binding Model Hamiltonians

Using modern methods of computational electronic-structure theory, the Kohn–Sham equations introduced in Section 2.1.1 can be solved to very good precision; the book by Martin (2004) gives a good overview of the technical methods employed to do so. However, one is often interested not so much in the full results of an accurate calculation, but in some simplified representation that is more likely to provide physical and chemical understanding. In such a situation, localized-orbital and tight-binding methods are frequently adopted.

2.2.1 Finite Systems

We consider first a localized system such as a molecule or cluster consisting of atoms indexed by μ located at positions $\boldsymbol{\tau}_\mu$. We choose some atomic-like orbitals $\varphi_{\mu\alpha}$ on site μ, where α runs over the orbitals on a given atom. These will typically be s-, p-, or d-like orbitals taking the form of a radial function times the appropriate $Y_{lm}(\theta, \phi)$ spherical harmonics. Introducing a compound index $j = \{\mu\alpha\}$ that runs over all M localized orbitals in the basis, we can write the trial Hamiltonian eigenstates as

$$\psi_n(\mathbf{r}) = \sum_j C_{nj}\, \varphi_j(\mathbf{r} - \boldsymbol{\tau}_j) \tag{2.56}$$

where the C_{nj}, the expansion coefficients of the nth eigenstate on the jth basis orbital, are to be determined. Plugging this into the Schrödinger equation $H|\psi_n\rangle = E_n|\psi_n\rangle$), the solutions are given by solving the matrix equation

$$(H - E_n S)\, C_n = 0 \tag{2.57}$$

where H and S are the $M \times M$ matrices

$$H_{ij} = \langle \varphi_i | H | \varphi_j \rangle \,,$$
$$S_{ij} = \langle \varphi_i | \varphi_j \rangle \,, \tag{2.58}$$

and C_n is the M-component column vector of coefficients C_{nj}. The diagonal elements of the Hamiltonian matrix H are referred to as "site energies" and the off-diagonal ones are "hopping matrix elements" (or just "hoppings"), while the S matrix is called the "overlap matrix." Equation (2.57) is a generalized eigenvalue problem that can be solved using standard computational methods, and the Hamiltonian eigenstates can be constructed using Eq. (2.56).

Different philosophies may be applied during the implementation of this approach, depending on the desired accuracy. At the accurate end of the spectrum are the methods of quantum chemistry, where one chooses a fairly extensive set of atomic orbitals as basis functions, typically including two or more radial functions per angular-momentum channel, and the calculations are carried out in the context of Hartree–Fock or DFT at a minimum, and more often using correlated wave-function approaches. The GAUSSIAN code is a well-known implementation of this kind. At the intermediate level are methods that sometimes go under the name of "linear combination of atomic orbitals" (LCAO), in which one adopts a minimal basis, typically just one radial function per angular-momentum channel, but still computes the matrix elements and overlaps of Eq. (2.58) explicitly, usually at the DFT level, as implemented, for example, in the SIESTA code package. Here we shall be interested in an implementation that lies at the low-accuracy end of this spectrum, known as the *empirical tight-binding* approach, where the idea is to construct a simple model containing the essential ingredients so as to enhance physical understanding.

The philosophy of empirical tight binding is, first, to focus just on those basis orbitals φ_j needed to describe the valence and low-lying conduction states of the system of interest. But second, these orbitals are never explicitly constructed; instead, the Hamiltonian and overlap matrix elements between them, given by Eq. (2.58), are parametrized in a model-building sense. Typically the hopping matrix elements between orbitals are truncated to connect only nearest-neighbor, or perhaps second-neighbor, sites, and the overlap matrix is often taken to be the unit matrix ("orthogonal tight binding"). Equation (2.57) is again solved using standard computational methods to get the energies E_n, but the corresponding eigenstates $\psi_n(\mathbf{r})$ are never actually constructed; the vector of coefficients C_n plays the role of eigenstate instead.

As an example, let's set up and solve an empirical tight-binding model of the water molecule. We take the relevant orbitals to be $|s\rangle$, $|p_x\rangle$, $|p_y\rangle$, $|p_z\rangle$, $|h_1\rangle$, and $|h_2\rangle$, as illustrated in Fig. 2.2(a), and assume they are orthonormal. The first four are located on the oxygen atom and are assigned site energies (E_s, E_p, E_p, E_p), while the last two are s orbitals on the hydrogen atoms and have site energies (E_h, E_h). We also let there be nonzero hoppings $t_s = \langle s|H|h_1\rangle$ between the oxygen s orbital and a hydrogen, and t_p between an oxygen p orbital and a hydrogen toward which it is directed. Then the 6×6 Hamiltonian matrix is

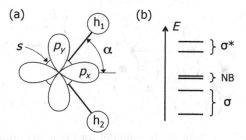

Figure 2.2 (a) Empirical tight-binding model for the water molecule. The oxygen p_z orbital, directed out of the plane, is not shown. (b) Eigenspectrum showing unoccupied antibonding (σ^*) orbitals and occupied nonbonding (NB) and bonding (σ) orbitals.

$$H_{H_2O} = \begin{pmatrix} E_s & 0 & 0 & 0 & t_s & t_s \\ 0 & E_p & 0 & 0 & t_p \cos\alpha & t_p \cos\alpha \\ 0 & 0 & E_p & 0 & t_p \sin\alpha & -t_p \sin\alpha \\ 0 & 0 & 0 & E_p & 0 & 0 \\ t_s & t_p \cos\alpha & t_p \sin\alpha & 0 & E_h & 0 \\ t_s & t_p \cos\alpha & -t_p \sin\alpha & 0 & 0 & E_h \end{pmatrix} \qquad (2.59)$$

where α is the half-bond-angle sketched in Fig. 2.2(a). We thus arrive at simple but instructive six-parameter model. The eigenenergies E_j are the six solutions of the secular equation $\det[H - E\mathbb{1}] = 0$, where $\mathbb{1}$ is the 6×6 unit matrix.

While this looks difficult to solve by hand, the problem can be simplified by noting that the matrix can be block-diagonalized on the basis of symmetry. The orbital $|p_z\rangle$ is the only one that is odd in z, so it forms an eigenvector by itself. The remaining five orbitals that are even in z can be grouped into three that are even and two that are odd in y, after reorganizing the H orbitals into linear combinations $[|h_1\rangle \pm |h_2\rangle]/\sqrt{2}$. Even so, a closed-form solution requires the roots of a cubic equation.

Note, however, that a direct solution on the computer is quite straightforward. Let us take a set of parameter values $E_s = -1.5$, $E_p = -1.2$, $E_h = -1.0$, $t_s = -0.4$, $t_p = -0.3$, and $\alpha = 52°$ (energies in Ry) as being roughly representative of H_2O. This parameter set yields the spectrum shown in Fig. 2.2(b) with corresponding eigenvalues and eigenvectors

n	eigval	eigvec					
1	-1.896	[0.802	0.201	0.000	0.000	0.398	0.398]
2	-1.458	[-0.000	0.000	0.800	0.000	0.424	-0.424]
3	-1.242	[-0.342	0.927	0.000	0.000	0.110	0.110]
4	-1.200	[0.000	-0.000	-0.000	1.000	-0.000	0.000]
5	-0.742	[-0.000	0.000	0.600	0.000	-0.566	0.566]
6	-0.562	[0.490	0.317	-0.000	-0.000	-0.574	-0.574]

The four lowest-energy states are doubly occupied (counting spin) and the last two are empty.

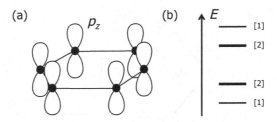

Figure 2.3 (a) Empirical tight-binding model for the π manifold on the benzene molecule (hydrogen atoms not shown). (b) Eigenspectrum of the model; numbers in brackets indicate the degeneracy.

In the spirit of empirical tight binding, we do not expect these results to be highly accurate, but they provide some useful qualitative information and insight. For example, the two lowest eigenstates can be seen to be bonding combinations of O *sp* hybrids and H *s* orbitals, while the unoccupied ones are the corresponding antibonding orbitals, and the middle states have an O nonbonding character. The symmetries are also evident from an inspection of the eigenvectors; for example, the state at E_1 is fully symmetric in both y and z.

As a second example, suppose we are interested in the manifold of π states on the benzene molecule, made from linear combinations of p_z orbitals on the C atoms, as shown in Fig. 2.3(a). These are distinguished from the C–C and C–H σ-bonding and antibonding states by being odd under the mirror M_z and being closer to the Fermi energy, so we discard all other orbitals and just model the behavior of the π manifold. This time we write the Hamiltonian in the notation of raising and lowering operators as

$$H = E_p \sum_j c_j^\dagger c_j + t \left(\sum_j c_{j+1}^\dagger c_j + \text{h.c.} \right) \tag{2.60}$$

where j runs over the six orbitals around the ring ($j + 1$ is interpreted mod 6), and E and t are site energies and nearest-neighbor hoppings, respectively. This model is easily solved; using the six-fold symmetry, the eigenvectors are of the form $C_{nj} = e^{2\pi i n j / 6}$. The spectrum of states is sketched in Fig. 2.3(b); the three lowest states are occupied.

Once the tight-binding Hamiltonian matrix elements H_{ij} are specified, no other information is needed to compute the spectrum of energy eigenstates. For some other purposes, however, additional information may be needed. For example, in general one should specify the position matrix elements

$$X_{ij} = \langle \varphi_i | x | \varphi_j \rangle, \quad Y_{ij} = \langle \varphi_i | y | \varphi_j \rangle, \quad Z_{ij} = \langle \varphi_i | z | \varphi_j \rangle, \tag{2.61}$$

which may be needed, for example, for computing electric dipoles. In the spirit of the empirical tight-binding method, we typically make the simplest possible approximation – namely that these position matrices are diagonal with $\langle \varphi_i | \mathbf{r} | \varphi_j \rangle = \delta_{ij} \boldsymbol{\tau}_j$, where $\boldsymbol{\tau}_j$ is the nuclear coordinate of the atom on which the orbital is centered.[10]

2.2.2 The PYTHTB Package

It turns out that the concepts at the heart of this book – Berry phases, Berry curvature, electric polarization, magnetoelectric couplings, and topological insulators – can all be illustrated insightfully using tight-binding models. Such models play an important role because they are often capable of capturing the essential qualitative features of interest with a minimum of complexity. Many of the seminal papers in the development of the theory of topological insulators, such as those by Haldane (1988), Kane and Mele (2005b), and Fu et al. (2007), illustrated the central concepts of their work by introducing tight-binding models. We shall encounter these models later in Chapter 5.

This was the motivation for the development of the PYTHTB software package, which originated in the Department of Physics and Astronomy at Rutgers University in 2012 and has undergone several revisions since then.[11] PYTHTB is a software package written in the PYTHON programming language, and designed to allow the user to construct and solve tight-binding models of the electronic structure, not only for finite systems such as those introduced earlier, but also for crystals, slabs, ribbons, polymer chains, and other configurations that display periodicity in one or more dimensions. Likewise, it is rich with features for computing Berry phases and related properties.

The PYTHTB package is described in some detail in Appendix D, but we introduce it here by using it to solve the H_2O and benzene model systems discussed above. The code needed to solve the water-molecule model can be found in Appendix D.1, but is short enough to be reproduced in its entirety here:

```
#!/usr/bin/env python
from __future__ import print_function # python3 style print

# ---------------------------------------------------------------
# tight-binding model for H2O molecule
# ---------------------------------------------------------------
```

---

[10] The diagonal approximation misses some important physics in the case of orbitals of mixed parity on the same site. For example, a matrix element of the form $\langle s|x|p_x \rangle$ between s and p_x orbitals on the same site is needed to describe the fact that the center of charge of an sp^3 hybrid is displaced from the atomic center.

[11] The PYTHTB package is available at www.physics.rutgers.edu/pythtb.

```
# import the pythtb module
from pythtb import *
import numpy as np

# geometry: bond length and half bond-angle
b=1.0; angle=54.0*np.pi/180

# site energies [O(s), O(p), H(s)]
eos=-1.5; eop=-1.2; eh=-1.0

# hoppings [O(s)-H(s), O(p)-H(s)]
ts=-0.4; tp=-0.3

# define frame for defining vectors: 3D Cartesian
lat=[[1.0,0.0,0.0],[0.0,1.0,0.0],[0.0,0.0,1.0]]

# define coordinates of orbitals: O(s,px,py,pz) ; H(s) ; H(s)
orb=[ [0.,0.,0.], [0.,0.,0.], [0.,0.,0.], [0.,0.,0.],
      [b*np.cos(angle), b*np.sin(angle),0.],
      [b*np.cos(angle),-b*np.sin(angle),0.] ]

# define model
my_model=tbmodel(0,3,lat,orb)
my_model.set_onsite([eos,eop,eop,eop,eh,eh])
my_model.set_hop(ts,0,4)
my_model.set_hop(ts,0,5)
my_model.set_hop(tp*np.cos(angle),1,4)
my_model.set_hop(tp*np.cos(angle),1,5)
my_model.set_hop(tp*np.sin(angle),2,4)
my_model.set_hop(-tp*np.sin(angle),2,5)

# print model
my_model.display()

# solve model
(eval,evec)=my_model.solve_all(eig_vectors=True)

# the model is real, so OK to discard imaginary parts of eigenvectors
evec=evec.real

# optional: choose overall sign of evec according to some specified rule
# (here, we make the average oxygen p component positive)
for i in range(len(eval)):
  if sum(evec[i,1:4]) < 0:
    evec[i,:]=-evec[i,:]

# print results, setting numpy to format floats as xx.xxx
np.set_printoptions(formatter={'float': '{: 6.3f}'.format})
# print eigenvalues and real parts of eigenvectors, one to a line.
print(" n    eigval    eigvec")
for n in range(6):
    print(" %2i  %7.3f  " % (n,eval[n]), evec[n,:])
```

While there is no expectation that you will understand this code in any detail at this point, especially if you are unfamiliar with PYTHON, I hope this example will serve to illustrate the simplicity and power of PYTHTB programming. After the PYTHTB module and the NUMPY numerical library are imported, the next few lines define the various model parameters. The `lat=` line specifies the three unit vectors (here, Cartesian ones) in terms of which other coordinates will be given, while the orb line specifies the locations of the atoms in this coordinate system. The my_model= line calls the tbmodel function to create an instance of a tight-binding model named my_model, based on four input variables. The first two specify the number of periodic dimensions (none, since this is a finite model) and the dimension of the atomic coordinate space (three in this example); the last two are just the lat and orb variables defined earlier. The model is then augmented further by defining the on-site energies and hoppings using the set_onsite and set_hop functions applied to my_model. (Note that PYTHON indexes arrays starting from 0, not 1, so the six orbitals are labeled as {0,1,2,3,4,5}.) Finally, the solve_all function is applied to my_model and the resulting eigenvalues and eigenvectors are returned in arrays eval and evec. The tbmodel, set_onsite, set_hop, and solve_all functions (and many more not used here) are provided by the PYTHTB module that was imported in the first line. The last part of the program converts the eigenvectors from complex to real, fixes their signs according to a somewhat arbitrary rule, and prints the results as follows:

```
n    eigval    eigvec
0    -1.896    [ 0.802   0.201   0.000  -0.000   0.398   0.398]
1    -1.458    [-0.000   0.000   0.800  -0.000   0.424  -0.424]
2    -1.242    [-0.342   0.927  -0.000   0.000   0.110   0.110]
3    -1.200    [ 0.000  -0.000   0.000   1.000   0.000  -0.000]
4    -0.742    [ 0.000   0.000   0.600   0.000  -0.566   0.566]
5    -0.562    [ 0.490   0.317  -0.000  -0.000  -0.574  -0.574]
```

This was the basis of the results presented earlier on p. 55.[12]

A similar PYTHTB code for solving the benzene model of Eq. (2.60) can be found in Appendix D.2. These two examples should serve to illustrate the mode in which the PYTHTB package is typically used. A tight-binding model is specified, often in just a few lines of user-written code, and the solution is then created automatically by the package subroutines. Numerous examples and exercises that rely on the PYTHTB package will appear later in this book, so the interested reader is encouraged to download and install the package as described in Appendix D. A good way to familiarization yourself with the package is to download and run the example programs in Appendix D and experiment with making modifications to them.

2.2.3 Extended Systems

We shall encounter the PYTHTB package again in Section 2.2.4, but for the moment we return to the formal development of tight-binding theory and apply it to the case

---

[12] In the raw output shown here, the band index n starts from zero in PYTHON style; on p. 55 it was shifted by 1 to make the table look more conventional.

of extended periodic systems. The presentation is done here for a 3D crystal, but the formulation for 2D layers such as graphene, or 1D systems such as polymers, is straightforward. Readers who are impatient to learn about Berry phases could skip forward and read the first two sections of Chapter 3 at this point, but material appearing later in this chapter will be needed by the time we arrive at Sections 3.3 and 3.4.

In a 3D-periodic crystal, the tight-binding orbitals

$$\phi_{\mathbf{R}j}(\mathbf{r}) = \varphi_j(\mathbf{r} - \mathbf{R} - \tau_j) \tag{2.62}$$

are indexed by the lattice vector $\mathbf{R}$, which specifies the unit cell, and j, which runs over the M orbitals in the cell. We again assume orthonormality,

$$\langle \phi_{\mathbf{R}i} | \phi_{\mathbf{R}'j} \rangle = \delta_{\mathbf{R}\mathbf{R}'}\, \delta_{ij}, \tag{2.63}$$

and that the position matrix has the simplest possible form,

$$\langle \phi_{\mathbf{R}i} | \mathbf{r} | \phi_{\mathbf{R}'j} \rangle = (\mathbf{R} + \tau_j)\, \delta_{\mathbf{R}\mathbf{R}'}\, \delta_{ij}. \tag{2.64}$$

The Hamiltonian matrix elements are defined via

$$H_{ij}(\mathbf{R}) = \langle \phi_{\mathbf{R}'i} | H | \phi_{\mathbf{R}'+\mathbf{R},j} \rangle = \langle \phi_{\mathbf{0}i} | H | \phi_{\mathbf{R}j} \rangle \tag{2.65}$$

where $\mathbf{R}$ is now a relative lattice-vector index. That is, the matrix element appearing in this expression corresponds to a hopping from orbital j in cell $\mathbf{R}' + \mathbf{R}$ to orbital i in cell $\mathbf{R}'$, or equivalently using translational symmetry, from orbital j in cell $\mathbf{R}$ to orbital i in the home unit cell $\mathbf{R} = \mathbf{0}$.[13]

To make the transition to Bloch states, we first construct Bloch-like basis functions and then compute the Hamiltonian matrix elements in this basis. There are, however, two distinct ways of doing this that differ in detail, which we shall refer to as being associated with Convention I or II defined as follows.

In Convention I, Bloch-like basis functions are constructed as

$$|\chi_j^{\mathbf{k}}\rangle = \sum_{\mathbf{R}} e^{i\mathbf{k}\cdot(\mathbf{R}+\tau_j)} |\phi_{\mathbf{R}j}\rangle \tag{2.66}$$

with the understanding that the normalization is to a single unit cell:

$$\langle \chi | \chi' \rangle \equiv \int_{\text{cell}} \chi^*(\mathbf{r})\, \chi'(\mathbf{r})\, d^3r. \tag{2.67}$$

It follows from Eq. (2.63) that

$$\langle \chi_i^{\mathbf{k}} | \chi_j^{\mathbf{k}} \rangle = \delta_{ij}. \tag{2.68}$$

---

[13] Matrix-element expressions such as those in Eq. (2.65) are most naturally read from right to left. In second quantized notation, this hopping would have the form $c_{\mathbf{R}'i}^{\dagger} c_{\mathbf{R}'+\mathbf{R},j}$.

The Bloch eigenstates are then expanded as

$$|\psi_{nk}\rangle = \sum_j C_j^{nk} |\chi_j^{\mathbf{k}}\rangle \tag{2.69}$$

and the Hamiltonian matrix $\langle \chi_i^{\mathbf{k}}|H|\chi_j^{\mathbf{k}}\rangle$ expressed in the basis of the Bloch-like functions $|\chi_j^{\mathbf{k}}\rangle$ is, after a few lines of algebra,

$$H_{ij}^{\mathbf{k}} = \langle \chi_i^{\mathbf{k}}|H|\chi_j^{\mathbf{k}}\rangle = \sum_{\mathbf{R}} e^{i\mathbf{k}\cdot(\mathbf{R}+\tau_j-\tau_i)} H_{ij}(\mathbf{R}) . \tag{2.70}$$

The eigenvalue equation to be solved is

$$H_{\mathbf{k}} \cdot C_{n\mathbf{k}} = E_{n\mathbf{k}} C_{n\mathbf{k}} \tag{2.71}$$

where $H_{\mathbf{k}}$ is the $M \times M$ matrix of elements $H_{ij}^{\mathbf{k}}$ and $C_{n\mathbf{k}}$ is the column vector of elements C_j^{nk}.

Standard numerical packages can easily be used to solve this eigenvalue problem, which amounts to diagonalizing $H_{\mathbf{k}}$, thereby giving the tight-binding solution for the energy eigenvalues and eigenvectors. If only the eigenvalues are needed, these can be obtained from the secular equation

$$\det(H_{\mathbf{k}} - E_{n\mathbf{k}}) = 0 . \tag{2.72}$$

Of course, this tight-binding solution produces only M bands, where M is the number of tight-binding basis orbitals per cell, representing an approximation to the M bands of the crystal that are built from these tight-binding orbitals (usually these are the M lowest valence and conduction bands).

In Convention II, the phase factor $e^{i\mathbf{k}\cdot\tau_j}$ is not included in the definition of the Bloch-like basis functions. Using tildes to denote objects defined in Convention II, we get

$$|\widetilde{\chi}_j^{\mathbf{k}}\rangle = \sum_{\mathbf{R}} e^{i\mathbf{k}\cdot\mathbf{R}} |\phi_{\mathbf{R}j}\rangle , \tag{2.73}$$

$$|\psi_{nk}\rangle = \sum_j \widetilde{C}_j^{nk} |\widetilde{\chi}_j^{\mathbf{k}}\rangle , \tag{2.74}$$

$$\widetilde{H}_{ij}^{\mathbf{k}} = \langle \widetilde{\chi}_i^{\mathbf{k}}|H|\widetilde{\chi}_j^{\mathbf{k}}\rangle = \sum_{\mathbf{R}} e^{i\mathbf{k}\cdot\mathbf{R}} H_{ij}(\mathbf{R}) , \tag{2.75}$$

and the secular equation is

$$\widetilde{H}_{\mathbf{k}} \cdot \widetilde{C}_{n\mathbf{k}} = E_{n\mathbf{k}} \widetilde{C}_{n\mathbf{k}} . \tag{2.76}$$

The quantities in the two conventions are related via

$$\widetilde{H}_{ij}^{\mathbf{k}} = e^{i\mathbf{k}\cdot(\tau_i-\tau_j)} H_{ij}^{\mathbf{k}} \tag{2.77}$$

and

$$\tilde{C}_j^{nk} = e^{ik\cdot\tau_j}\, C_j^{nk}\,. \tag{2.78}$$

The two conventions are essentially just related by a unitary rotation in the M-dimensional space.

When reading the literature, it is worth taking the time to determine which convention the authors are using. Convention II is probably the more commonly used option, because the extra factors of $e^{ik\cdot\tau_j}$ can be ignored, and the relation between the Hamiltonian matrix elements expressed in Eq. (2.75) is that of a simple discrete Fourier transform. However, Convention I is in many ways more natural for our purposes, as we shall see later, and it is the convention adopted in the PYTHTB code.

One way of understanding the distinction between the two conventions is to draw an analogy between the Bloch function $\psi_{nk}(r)$ and $\tilde{C}_j^{nk}$, and between the cell-periodic Bloch function $u_{nk}(r)$ and C_j^{nk}. Recalling that

$$\psi_{nk}(\mathbf{r}) = e^{ik\cdot r} u_{nk}(\mathbf{r})\,, \tag{2.79}$$

and temporarily adopting the change of notation $C_j^{nk} \to C_{nk}(j)$, and similarly for $\tilde{C}$, we can write

$$
\begin{aligned}
|\psi_{nk}\rangle &= \sum_R \int_{\text{cell}} d^3r\; u_{nk}(\mathbf{r})\; e^{ik\cdot(\mathbf{R}+r)}\,|\mathbf{R}+r\rangle\,, \\
&= \sum_R \sum_j \quad C_{nk}(j)\; e^{ik\cdot(\mathbf{R}+\tau_j)}\,|\phi_{\mathbf{R}j}\rangle
\end{aligned}
\tag{2.80}
$$

while

$$
\begin{aligned}
|\psi_{nk}\rangle &= \sum_R \int_{\text{cell}} d^3r\; \psi_{nk}(\mathbf{r})\; e^{ik\cdot R}\,|\mathbf{R}+r\rangle\,, \\
&= \sum_R \sum_j \quad \tilde{C}_{nk}(j)\; e^{ik\cdot R}\,|\phi_{\mathbf{R}j}\rangle\,.
\end{aligned}
\tag{2.81}
$$

The vertical alignment in Eqs. (2.80–2.81) is intentional and is designed to highlight the correspondence between the terms in the continuum framework (the first line in each case) and the analogous ones in the tight-binding framework (the second line). Thus we can see that Eq. (2.80) identifies C_j^{nk} (Convention I) as the tight-binding analogue of the cell-periodic function $u_{nk}(r)$, while Eq. (2.81) identifies the $\tilde{C}_j^{nk}$ (Convention II) with the ordinary Bloch function $\psi_{nk}(r)$.

As hinted earlier on p. 48, the cell-periodic functions u_{nk} will play a more central role than the Bloch functions ψ_{nk} in the formulation of Berry-phase quantities such as electric polarization in the following chapters. It is largely for this reason that we have adopted Convention I for use here and for the PYTHTB implementation.

2.2.4 Examples

The tight-binding formalism for extended systems is best understood via illustrative examples. We first consider the dispersion of the s-like states of an alkali metal in the body-centered cubic (bcc) crystal structure, taking Li (lattice constant $a = 3.5$ Å) for definiteness. We construct the simplest possible tight-binding model with one s orbital with site energy E_s on each atom, and a hopping of amplitude t between each atom and its eight nearest neighbors. The lattice vectors can be chosen as $\mathbf{a}_1 = (a/2)(-1, 1, 1)$, $\mathbf{a}_2 = (a/2)(1, -1, 1)$, and $\mathbf{a}_3 = (a/2)(1, 1, -1)$, in which case the nearest neighbors are located at $\pm\mathbf{a}_1$, $\pm\mathbf{a}_2$, $\pm\mathbf{a}_3$, and $\pm(\mathbf{a}_1 + \mathbf{a}_2 + \mathbf{a}_3)$, with $V_{cell} = a^3/2$. Since we have only one orbital per unit cell, we discard indices ij in the formalism of Section 2.2.3. The real-space Hamiltonian of Eq. (2.65) is just $H(\mathbf{0}) = E_s$, $H(\mathbf{R}) = t$ for the eight nearest-neighbor vectors, and zero otherwise. Then $H_\mathbf{k}$ of Eq. (2.70) becomes

$$H_\mathbf{k} = E_0 + 2t \cos[(-k_x + k_y + k_z)a/2] + 3 \text{ more terms.} \tag{2.82}$$

Since $H_\mathbf{k}$ is a 1×1 matrix, Eq. (2.72) implies that $E_\mathbf{k} = H_\mathbf{k}$, and we have obtained a closed-form solution for the energy bands.

Since it is not easy to visualize a function of three variables (k_x, k_y, k_z), the convention is to plot the energy bands along some high-symmetry lines in reciprocal space. Let's do this first along the line $\mathbf{k} = (k, 0, 0)$, where we find $E(k) = E_0 + 8t \cos(ka/2)$. This is plotted for $k \in [0, 2\pi/a]$ in the right panel of Fig. 2.4 for parameters $E_s = 4.5$ eV, $t = -1.4$ eV. Then again, we can look along $\mathbf{k} = (k, k, k)$, and obtain $E(k) = E_0 + 6t \cos(ka/2) + 2t \cos(3ka/2)$, which is plotted from right to left in the left panel of Fig. 2.4 for $k \in [0, \pi/a]$. Each Li atom contributes one valence electron per unit cell. Since each spin-degenerate band can accommodate two electrons per cell according to Section 2.1.4, the band is half-occupied. The horizontal line at $E = 0$ in Fig. 2.4 indicates the approximate position of the Fermi level. A PYTHTB program that implements this model is provided in Appendix D.3.

This model gives a qualitatively correct description of the Fermi surface, which is roughly spherical, filling half the volume of the BZ. However, it overestimates the departures from sphericity and fails to give the correct energy and degeneracy for states at the BZ boundary. These features can be improved by including more atomic orbitals in the basis; for example, the addition of p and d orbitals on each

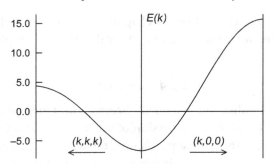

Figure 2.4 Band structure obtained from the nearest-neighbor tight-binding model of s orbitals in bcc Li. The line at $E=0$ indicates the approximate position of the Fermi level.

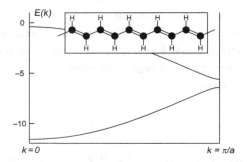

Figure 2.5 Dispersion of bands obtained from the tight-binding model of p_z orbitals on C atoms in $(CH)_x$ (polyacetylene) for parameter values of $E_p = -6.0\,\text{eV}$, $t = -2.8\,\text{eV}$, $\Delta = 0$, and $\delta = -0.2\,\text{eV}$ (see text). The inset shows a plan view of the molecule, which lies in the x-y plane; the C–C bonds alternate between double and single bonds as indicated in the sketch.

site would give a nine-band model whose lowest band would resemble that in Fig. 2.4, but with a dispersion in better agreement with experiment.[14]

As a second example, we consider a tight-binding model of the highest occupied and lowest unoccupied bands of polyacetylene. This planar 1D polymer is composed of CH units, as shown in the inset of Fig. 2.5, which we take to be lying in the x-y plane. Because carbon prefers to be fourfold coordinated, the chain adopts a dimerized structure of alternating "double" and "single" bonds as sketched in the figure. In physical terms, this translates to "short" and "long" C–C bonds. Since the entire molecule has M_z mirror symmetry, the Bloch bands cleanly separate into those arising from states of even M_z symmetry (principally s orbitals on H and

---

[14] Actually, the nearly free electron model, described in any solid state physics text, gives a more accurate and concise description of the energy bands of alkali metals such as Li. We have focused on the tight-binding approach here for pedagogical purposes.

s, p_x, and p_y orbitals on C) and those of odd M_z symmetry (mainly p_z on C). The states of even M_z symmetry form Bloch bands corresponding to covalent C–C and C–H bonds, with a large band gap between the lower-energy σ-bonding and higher-energy σ-antibonding bands. In contrast, the π manifold made out of the p_z orbitals provides states that are nearly metallic, as we shall see.

We therefore concentrate on constructing a tight-binding model only for these p_z states. We assign nearest-neighbor hopping strengths $t_s = t + \delta$ and $t_l = t - \delta$ to the short and long bonds, respectively, where δ reflects the difference in hopping strength arising from the dimerization. We also allow the assignment of different site energies $E_1 = E_p + \Delta$ and $E_2 = E_p - \Delta$ to the two C p_z orbitals, where Δ is a measure of the asymmetry between the two C sites. This is not chemically motivated – we would have $\Delta=0$ for ordinary polyacetylene – but a model that includes the asymmetry Δ will prove useful to us later. (You can think of this asymmetry as caused by the application of an electric field along the y-direction if you like.)

We take $\mathbf{a}_1 = a\hat{x}$ and assume the C atoms in the home unit cell are located at positions $\boldsymbol{\tau}_1 = (-a/4, b/2, 0)$ and $\boldsymbol{\tau}_2 = (a/4, -b/2, 0)$. (Note that we have neglected the dimerization in setting the atomic positions, assuming for the sake of simplicity that it is already adequately represented by the δ term in the Hamiltonian.) There are two orbitals per cell so that $H(\mathbf{R})$ is a 2×2 matrix, and we find

$$H(0) = \begin{pmatrix} E_p + \Delta & t + \delta \\ t + \delta & E_p - \Delta \end{pmatrix},$$

$$H(\mathbf{a}_1) = \begin{pmatrix} 0 & 0 \\ t - \delta & 0 \end{pmatrix},$$

$$H(-\mathbf{a}_1) = \begin{pmatrix} 0 & t - \delta \\ 0 & 0 \end{pmatrix}. \tag{2.83}$$

We then evaluate Eq. (2.70) to get the 2×2 $H_{\mathbf{k}}$ matrix at each $\mathbf{k}$. Here the reciprocal space is one-dimensional, since the chain is periodic in only one direction. Changing the notation from $H_{\mathbf{k}}$ to $H(k)$, Eq. (2.70) becomes

$$H_{ij}(k) = \sum_{\ell} e^{ik(\ell a + \tau_{j,x} - \tau_{i,x})} H_{ij}(\ell \mathbf{a}_1) \tag{2.84}$$

where ℓ is an integer cell index. For H_{12}, only the $\ell=0$ and $\ell=-1$ terms survive, and the corresponding exponential factors in Eq. (2.84) are $e^{ika/2}$ and $e^{-ika/2}$, respectively, so that $H_{12}(k) = 2t\cos(ka/2) + 2i\delta\sin(ka/2)$. A similar analysis shows that $H_{21}(k)$ is the complex conjugate of this, so that we get

$$H(k) = \begin{pmatrix} E_p + \Delta & 2t\cos(ka/2) + 2i\delta\sin(ka/2) \\ \text{c.c.} & E_p - \Delta \end{pmatrix}. \tag{2.85}$$

Before plugging into the secular equation of Eq. (2.72), it is instructive to rewrite this in the canonical form

$$H = f_0 \, \mathbb{1} + f_x \, \sigma_x + f_y \, \sigma_y + f_z \, \sigma_z = f_0 \, \mathbb{1} + \mathbf{f} \cdot \boldsymbol{\sigma} \tag{2.86}$$

where $\mathbb{1}$ is the 2×2 identity and the σ_j are the Pauli matrices of Eq. (2.20). An elementary calculation shows that $\det(H - E) = 0$ yields eigenvalues $E = f_0 \pm \sqrt{f_x^2 + f_y^2 + f_z^2}$. In our case we have $f_0 = E_p$, $f_x(k) = 2t \cos(ka/2)$, $f_y(k) = -2\delta \sin(ka/2)$, and $f_z = \Delta$, so that

$$E(k) = E_p \pm \sqrt{4t^2 \cos^2(ka/2) + \Delta^2 + 4\delta^2 \sin(ka/2)} \,. \tag{2.87}$$

The first term in the square root is the dominant one, since we have in mind that Δ (the site asymmetry) is small or zero, and δ (the bond asymmetry) is weaker than the bond itself. At the zone boundary $k = \pi/a$, however, that term vanishes, and $E(\pi/a) = E_p \pm \sqrt{\Delta^2 + 4\delta^2}$. An electron counting argument shows that this band subspace is half-filled in $(CH)_x$, so it follows that the system is an insulator with a global gap only if Δ and/or δ is nonzero. A numerical solution for the case of $E_p = -6.0\,\mathrm{eV}$, $t = -2.8\,\mathrm{eV}$, $\Delta = 0$, and $\delta = -0.2\,\mathrm{eV}$ is shown in Fig. 2.5.

Once again, this simple tight-binding model has its limitations. A more careful treatment including nonorthogonality or farther-neighbor overlaps would modify the shape of these bands slightly. More importantly, the inclusion of σ-bonding and -antibonding orbitals of even mirror symmetry would give additional occupied and unoccupied bands that overlap to some degree with the two π bands. Nevertheless, the model provides a good qualitative and semiquantitative description of the low-energy excitations controlling the gap and the valence and conduction band edges. It is the foundation for the well-known model developed by Su et al. (1979), which also includes electron-phonon interactions to describe the spontaneous dimerization and the formation of defects in this interesting system.

Working in the opposite direction, we can simplify this model to its bare essentials to obtain what we shall refer to as the "alternating site model," also known as the Rice-Mele model (Rice and Mele, 1982), sketched in Fig. 2.6. We have neglected the existence of the H atoms and the buckled nature of the chain to arrive at a completely linear model. There are two tight-binding sites per unit cell having alternating site energies Δ and $-\Delta$, and these are connected by hoppings

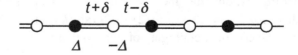

Figure 2.6 Simplified alternating site model. There are two sites per unit cell in a linear geometry with alternating site energies $\pm\Delta$ and alternating hoppings $t \pm \delta$. Single and double lines represent weak and strong bonds, respectively.

$t + \delta$ and $t - \delta$ on alternating bonds. The model is mathematically identical to that of Eq. (2.83) except that E_p has been set to zero. We typically take $t < 0$ as we did earlier, since bonding in solids is characterized by negative hoppings; the figure is drawn with $\delta < 0$ as well, so that the strong and weak bonds occur as shown. A PYTHTB implementation of this model is given in Appendix D.4, and the resulting band structure plot appears as Fig. D.2. This simplified model will be used repeatedly to illustrate concepts in later chapters.

Another model that we shall make frequent use of in various extended forms is the honeycomb model shown in Fig. 2.7(a). This is essentially a model of a single graphene layer in which only the p_z orbitals (i.e., normal to the plane) on the carbon atoms are included. Just as for the models of Figs. 2.3 and 2.5, this focus on the π manifold is justified on the basis that these states are the closest to the fundamental gap and determine the "low-energy" electronic behavior of the material. In the simplest version of this model, all sites have a common site energy E_p and there are only nearest-neighbor hoppings t. The corresponding

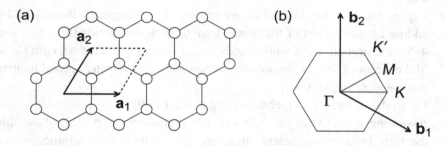

Figure 2.7 Simple tight-binding model of the π orbitals making up a graphene sheet. (a) Real space view; lattice vectors are $\mathbf{a}_1$ and $\mathbf{a}_2$, with dashed lines completing one unit cell. (b) Reciprocal space view.

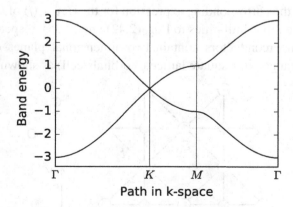

Figure 2.8 Band structure of π manifold of graphene as generated by the graphene.py program of Appendix D.5. The band structure is plotted along the path from Γ to K to M and then back to Γ (see Fig. 2.7).

reciprocal-space BZ is shown in Fig. 2.7(b), with some of the high-symmetry points
identified by their conventional labels.

A PYTHTB implementation of this model is provided in Appendix D.5, and the
resulting band structure is plotted in Fig. 2.8 along a path chosen to connect several
high-symmetry points in the 2D BZ. The results confirm the well-known result that
graphene has a semimetallic band structure as a result of a gap closure at certain
locations (the K and K' points) in the 2D BZ.

Exercises

2.5 Take the program `benzene.py` in Appendix D.2 and modify it by
considering the presence of a uniform electric field along $\hat{x}$ or $\hat{y}$ (your
choice), and by letting the site energies be raised or lowered in a way that
is linear in their spatial coordinates. Make a plot of the six eigenvalues
versus the field strength. Does the high-field behavior make sense? Explain
why.

2.6 Repeat Ex. 2.5, but instead of an external field, augment the model by
adding an alternation of the bond strength such that bonds 0–1, 2–3, and
4–5 have strength $t + \delta$ while bonds 1–2, 3–4, and 5–0 have strength $t - \delta$.
Make a plot of the six eigenvalues versus δ. Does anything special happen
when $\delta = t$? Explain why.

2.7 Fill in the steps in the algebra leading to Eq. (2.70).

2.8 Assuming a periodic gauge (recall the definition on p. 46), show that
the tight-binding coefficients in Convention I have the reciprocal-space
translational property

$$C_{n,\mathbf{k}+\mathbf{G}}(j) = e^{-i\mathbf{G}\cdot\boldsymbol{\tau}_j}\, C_{n\mathbf{k}}(j)$$

and derive the corresponding expression for the $\tilde{C}_{n,\mathbf{k}+\mathbf{G}}(j)$ of Convention II.
Comment on the similarities to Eqs. (2.42) and (2.35), respectively.

2.9 High-T_c superconductors contain two-dimensional planes of Cu and O
orbitals arranged in a square lattice (3 orbitals/cell) as shown:

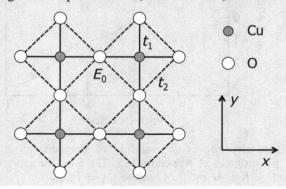

(The orbitals are really Cu $3d_{x^2-y^2}$ and oxygen $2p_z$ orbitals, but please ignore these details, and just treat them as illustrated in the figure.) Let $E_{Cu} = 0$ by convention, and let the parameters of the model be the oxygen site energy E_O, the nearest-neighbor Cu–O hopping t_1, and the farther-neighbor O–O hopping t_2. Write a PYTHTB program to calculate the bands for this model. Show your program and a sample output for some parameter values chosen by you (realistically, t_1 should be quite a bit larger than the other parameters).

2.10 Invent a tight-binding model according to your own imagination. It can be in 3D, 2D, or 1D. Provide a sketch defining the parameters of the model, a printout of your Python program, and a sample output. You can model your program after one of the example programs in the PYTHTB package, but if you take this route, be sure to modify it significantly.

2.3 Linear Response Theory

Before concluding this chapter, we briefly introduce linear response theory. This is essentially just a slight extension of ordinary first-order perturbation theory, but is usually omitted in elementary quantum mechanics texts. Here again, the formalism we develop now will prove useful for what follows in the later chapters.

We start with ordinary perturbation theory as it applies to a state $|n(\lambda)\rangle$ satisfying $H(\lambda)|n(\lambda)\rangle = E_n(\lambda)|n(\lambda)\rangle$ for a Hamiltonian depending smoothly on a parameter λ. Hereafter, we drop the explicit λ dependence and write $(E_n - H)|n\rangle = 0$. Taking the first derivative $\partial_\lambda = d/d\lambda$ of this expression and keeping terms at first order in the perturbation yields

$$(E_n - H)|\partial_\lambda n\rangle = \partial_\lambda(H - E_n)\,|n\rangle \tag{2.88}$$

where $\partial_\lambda H$ represents the perturbation (often written as V or H_1 in quantum mechanics texts). Applying $\langle n|$ from the left on Eq. (2.88), the left-hand side vanishes and we get

$$\partial_\lambda E_n = \langle n|(\partial_\lambda H)|n\rangle\,. \tag{2.89}$$

Then the right-hand side of Eq. (2.88) becomes $(1 - |n\rangle\langle n|)(\partial_\lambda H)|n\rangle$, and we can write Eq. (2.88) as

$$(E_n - H)|\partial_\lambda n\rangle = \mathcal{Q}_n\,(\partial_\lambda H)\,|n\rangle \tag{2.90}$$

where

$$\mathcal{Q}_n = \sum_{m \neq n} |m\rangle\langle m| = 1 - \mathcal{P}_n \tag{2.91}$$

is the projection operator onto all states other than state n – that is, the complement of the projector $\mathcal{P}_n = |n\rangle\langle n|$.

Equation (2.90), sometimes known as a *Sternheimer equation*, is an inhomogeneous linear equation for the perturbed wave function $|\partial_\lambda n\rangle$. Its solution is

$$|\partial_\lambda n\rangle = -iA_n|n\rangle + \sum_{m \neq n} \frac{|m\rangle\langle m|}{E_n - E_m}(\partial_\lambda H)|n\rangle \qquad (2.92)$$

for any real A_n, as can be checked by plugging it into Eq. (2.90) and multiplying on the left by an arbitrary eigenstate $\langle m|$ (see Ex. 2.11). In the case that $\langle m| = \langle n|$, this readily yields $\langle n|\partial_\lambda n\rangle = -iA_n$, since $\langle n|m\rangle = 0$ for $m \neq n$. Since $\partial_\lambda \langle n|n\rangle = 2\text{Re}\,\langle n|\partial_\lambda n\rangle$ must vanish, the coefficient

$$A_n(\lambda) = i\langle n|\partial_\lambda n\rangle \qquad (2.93)$$

is real. This object, known as the *Berry connection*, is one that we shall come to know well in Chapter 3.[15]

One way to think about the first term in Eq. (2.92) is to observe what happens if H actually has no dependence on λ at all. In that case the last term in Eq. (2.92) vanishes, so that its solution is

$$|n(\lambda)\rangle = e^{i\int A_n(\lambda)\,d\lambda}|n_0\rangle . \qquad (2.94)$$

Thus, the *physical state* does not change with λ, but its *representation as a vector* changes because of a phase rotation in the definition of $|n\rangle$ as a function of λ. Because it does not represent a physical change of state, this term is dropped in most elementary quantum texts, and we shall do so here as well. However, to clarify that we have done so, we include a $\mathcal{Q}_n$ factor explicitly in rewriting Eq. (2.92) as

$$\mathcal{Q}_n|\partial_\lambda n\rangle = \sum_{m \neq n} \frac{|m\rangle\langle m|}{E_n - E_m}(\partial_\lambda H)|n\rangle. \qquad (2.95)$$

We can simplify the notation further by introducing the definition[16]

$$T_n = \sum_{m \neq n} \frac{|m\rangle\langle m|}{E_n - E_m}, \qquad (2.96)$$

in which case we arrive at a very concise expression for the perturbed wave function:

$$\mathcal{Q}_n|\partial_\lambda n\rangle = T_n\,(\partial_\lambda H)\,|n\rangle . \qquad (2.97)$$

---

[15] We also encountered the Berry connection in **k**-space in Ex. 2.3.
[16] Note that $\mathcal{Q}_n T_n = T_n \mathcal{Q}_n = T_n$ since T_n has no projection on $|n\rangle$.

In the linear response theory, we also consider how the expectation value of some Hermitian observable $\mathcal{O}$ changes as a result of the perturbation. We have

$$
\begin{aligned}
\partial_\lambda \langle \mathcal{O} \rangle_n &= \partial_\lambda \langle n|\mathcal{O}|n\rangle \\
&= \langle \partial_\lambda n|\mathcal{O}|n\rangle + \langle n|\mathcal{O}|\partial_\lambda n\rangle \\
&= 2\mathrm{Re}\, \langle n|\mathcal{O}|\partial_\lambda n\rangle \\
&= 2\mathrm{Re}\, \langle n|\mathcal{O}\mathcal{Q}_n|\partial_\lambda n\rangle
\end{aligned} \tag{2.98}
$$

where the last equality follows because the contribution from the first term in Eq. (2.92) is purely imaginary. If we wish, we can combine this with Eq. (2.97) to get

$$
\partial_\lambda \langle \mathcal{O} \rangle_n = 2\mathrm{Re}\, \langle n|\mathcal{O}\, T_n\, (\partial_\lambda H)|n\rangle \tag{2.99}
$$

at the expense of introducing the sum over states encoded in T_n. Both Eqs. (2.98) and (2.99) are valid and equivalent expressions.

Two comments are in order. First, in the context of an independent-particle treatment of an electron system such as a molecule, the expression

$$
\partial_\lambda \langle \mathcal{O} \rangle = 2\mathrm{Re}\, \sum_n^{\mathrm{occ}} \langle n|\mathcal{O}\mathcal{Q}_n|\partial_\lambda n\rangle \tag{2.100}
$$

for the total change to $\langle \mathcal{O} \rangle$ coming from all occupied states can be replaced by the more convenient expression

$$
\partial_\lambda \langle \mathcal{O} \rangle = 2\mathrm{Re}\, \sum_n^{\mathrm{occ}} \langle n|\mathcal{O}\mathcal{Q}|\partial_\lambda n\rangle , \tag{2.101}
$$

where $\mathcal{Q}_n$ has been replaced by

$$
\mathcal{Q} = \sum_m^{\mathrm{unocc}} |m\rangle\langle m| = 1 - \sum_n^{\mathrm{occ}} |n\rangle\langle n| , \tag{2.102}
$$

the projection onto the unoccupied-state manifold. This follows because

$$
\begin{aligned}
2\mathrm{Re}\, \sum_n^{\mathrm{occ}} \langle n|\, \mathcal{O}(\mathcal{Q}_n - \mathcal{Q})|\partial_\lambda n\rangle &= 2\mathrm{Re}\, \sum_{n\neq n'}^{\mathrm{occ}} \langle n|\mathcal{O}|n'\rangle\langle n'|\partial_\lambda n\rangle \\
&= \sum_{n\neq n'} \Big(\langle n|\mathcal{O}|n'\rangle\langle n'|\partial_\lambda n\rangle + \langle n'|\mathcal{O}|n\rangle\langle \partial_\lambda n|n'\rangle \Big) \\
&= \sum_{n\neq n'} \Big(\langle n|\mathcal{O}|n'\rangle\langle n'|\partial_\lambda n\rangle - \langle n'|\mathcal{O}|n\rangle\langle n|\partial_\lambda n'\rangle \Big) \\
&= 0
\end{aligned} \tag{2.103}
$$

where the second line is obtained from $2\text{Re}(z) = z + z^*$, the third line comes from $\partial_\lambda \langle n'|n\rangle = \partial_\lambda \delta_{nn'} = 0$, and the last line follows from the interchange of dummy labels n and n' in the last term. Intuitively, insofar as the λ perturbation induces only a unitary rotation between occupied states $|n\rangle$ and $|n'\rangle$, this has no influence on the overall $\langle \mathcal{O}\rangle$; it is only when the perturbation induces an occupied state $|n\rangle$ to acquire character on an unoccupied state $|m\rangle$ that $\langle \mathcal{O}\rangle$ is perturbed. As a result, we do not really care about $\mathcal{Q}_n|\partial_\lambda n\rangle$; it is enough to calculate $\mathcal{Q}|\partial_\lambda n\rangle$, which is the component of the perturbed wave function $|n\rangle$ that projects into the unoccupied space; this is the only piece that enters the linear response theory. It is also convenient to replace Eq. (2.96) by a revised definition

$$T_n = \sum_m^{\text{unocc}} \frac{|m\rangle\langle m|}{E_n - E_m},\qquad(2.104)$$

which we will take as the definition of T_n going forward. Then

$$\mathcal{Q}|\partial_\lambda n\rangle = T_n\,(\partial_\lambda H)\,|n\rangle\qquad(2.105)$$

are the first-order wave-function perturbations that are needed for insertion into Eq. (2.101).[17]

Second, while $\mathcal{Q}$ is easily expressed as a finite sum via $\mathcal{Q} = 1 - \sum_m^{\text{occ}}|m\rangle\langle m|$, the same trick cannot be applied to T_n, which still has to be computed as a sum over all unoccupied states. This is often problematic; in practice, one has to truncate the infinite sum to a finite number of terms, and the resulting calculations are tedious and require careful convergence tests. Fortunately, there is an alternative approach, in which the Sternheimer equation of Eq. (2.90), now rewritten as $(E_n - H)|\partial_\lambda n\rangle = \mathcal{Q}(\partial_\lambda H)|n\rangle$, is solved directly by an iterative algorithm for each occupied state n. The resulting vectors $|\partial_\lambda n\rangle$ are then inserted into Eq. (2.101) to find the change in the expectation of operator $\mathcal{O}$.

Frequently one considers two perturbations corresponding to two operators $A = \partial H/\partial\lambda_A$ and $B = \partial H/\partial\lambda_B$, or equivalently, $H = H_0 + \lambda_A A + \lambda_B B$. Then Eq. (2.99) yields

$$\frac{\partial^2 E}{\partial\lambda_A\,\partial\lambda_B} = \frac{\partial\langle A\rangle}{\partial\lambda_B} = \frac{\partial\langle B\rangle}{\partial\lambda_A} = \sum_n^{\text{occ}} 2\text{Re}\,\langle n|AT_n B|n\rangle.\qquad(2.106)$$

For example, λ_A and λ_B could refer to two different atomic displacements in a molecule, in which case Eq. (2.106) yields the force-constant matrix connecting

---

[17] In alternative formulations, the perturbation of $\langle \mathcal{O}\rangle$ can equivalently be expressed in terms of Greens functions, which essentially embody the action of Eq. (2.104), or, following Niu and Thouless (1984), in terms of the one-particle density matrix (see Ex. 2.12).

these two displacements, as would be needed for a calculation of the vibrational frequencies. Written explicitly, Eq. (2.106) takes the probably more familiar form

$$\frac{\partial^2 E}{\partial \lambda_A \, \partial \lambda_B} = \sum_n^{\text{occ}} \sum_m^{\text{unocc}} \frac{\langle n|A|m\rangle \langle m|B|n\rangle}{E_n - E_m} + \text{c.c.} \tag{2.107}$$

where 'c.c.' indicates the complex conjugate. For $A = B$, the numerator becomes $|\langle n|A|m\rangle|^2$ and Eq. (2.107) reduces to the well-known result of standard second-order perturbation theory. In practice, the infinite sum over unoccupied states is usually avoided by replacing $T_n B|n\rangle$ by $\mathcal{Q}|\partial n/\partial \lambda_B\rangle$, or else replacing $\langle n|AT_n$ by $\langle \partial n/\partial \lambda_A|\mathcal{Q}$, where the iterative Sternheimer approach is used to obtain the needed wave function derivative.

The preceding equations form the basis of linear response theory as it is implemented in many common code packages, such as ABINIT or QUANTUM ESPRESSO in the solid-state community. These equations were formulated for the case of a single-particle Hamiltonian; when implemented in the context of DFT, additional terms corresponding to the self-consistently induced variation of the Hartree and exchange-correlation potentials also need to be included. For further details on the formalism and computational implementation of this approach, known as "density functional perturbation theory," the reader is referred to the excellent review by Baroni et al. (2001), Chapter 19 of the book by Martin (2004), and the papers of Gonze (1997) and Gonze and Lee (1997). The formalism can also be generalized to describe dynamical response functions – that is, the ratio of a response $A(\omega)$ to a perturbation $B(\omega)$ at frequency ω, with the energy denominators modified to become $1/(E_n - E_m \pm \hbar\omega)$.

Exercises

2.11 (a) Fill in the steps to show that Eq. (2.92) satisfies Eq. (2.90). Also use the fact that $\langle n|n\rangle$ is independent of λ to show that A_n is real.

(b) Fill in the steps needed to demonstrate Eq. (2.103).

2.12 An alternative approach to the calculation of the linear response properties of a quantum system defines the one-particle density matrix

$$\rho = \sum_n^{\text{occ}} |n\rangle \langle n|$$

in terms of which the expectation of any one-particle operator $\mathcal{O}$ is

$$\langle \mathcal{O} \rangle = \text{Tr}\,[\rho \mathcal{O}].$$

(a) Recalling from Eq. (2.92) that

$$|\partial_\lambda n\rangle = -iA_n|n\rangle + T_n(\partial_\lambda H)|n\rangle \,,$$

and using the notation of Eq. (2.96), show that the linear response of the density matrix to the perturbation is given by

$$\partial_\lambda \rho = \sum_n^{\text{occ}} \left(T_n(\partial_\lambda H)|n\rangle\langle n| + |n\rangle\langle n|(\partial_\lambda H)T_n \right) \,,$$

where the Berry connection $A_n(\lambda)$ has dropped out. The density matrix is thus free of the "gauge ambiguity" that appears as the first term on the right of Eq. (2.92).

(b) Using the result of part (a), show that the perturbation of an operator expectation value, $\partial_\lambda \langle \mathcal{O} \rangle = \text{Tr}\,[(\partial_\lambda \rho)\mathcal{O}]$, is consistent with Eq. (2.99).

3

Berry Phases and Curvatures

A *Berry phase* is a phase angle (i.e., running between 0 and 2π) that describes the global phase evolution of a complex vector as it is carried around a path in its vector space. It can also be referred to as a "geometric phase" or a "Pancharatnam phase," with the latter term being derived from early work by Pancharatnam (1956). The concept was systematized and popularized in the 1980s by Sir Michael Berry, notably in a seminal paper (Berry, 1984), and by Berry and others in a series of subsequent publications that are well represented in the edited volume of Wilczek and Shapere (1989). A formal discussion of Berry phases and related concepts of algebraic geometry (e.g., fiber bundles, connections, Berry curvatures) can be found in modern texts on topological physics such as those by Frankel (1997), Nakahara (2003), and Eschrig (2011). Berry phases have found broad application in diverse areas including atomic and molecular physics, nuclear physics, classical optics, photonics, and condensed-matter physics.

Our goal here is to introduce the concept of the Berry phase and explain how it enters into the quantum-mechanical band theory of electrons in crystals. We will begin by introducing the Berry phase in its abstract mathematical form, and then discuss its application to the adiabatic dynamics of finite quantum systems. After these preliminaries, we will turn to the main theme of this book, where the complex vector in question is a Bloch wavevector, and the path lies in the space of wavevectors **k** within the Brillouin zone.

3.1 Berry Phase, Gauge Freedom, and Parallel Transport

As indicated previouisly, a geometric or Berry phase is a quantity that describes how a global phase accumulates as some complex vector is carried around a closed loop in a complex vector space. Since we are interested only in phases, we can take these complex vectors to be unit vectors, and we will typically identify them with the ground-state wave function of some quantum system. For example, the vector

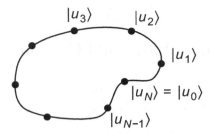

Figure 3.1 Illustration of the evolution of some complex unit vector $|u\rangle$ around a path in parameter space. The first and last points $|u_0\rangle$ and $|u_N\rangle$ are identical.

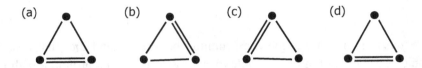

Figure 3.2 Triangular molecule going though a sequence of distortions in which first the bottom bond, then the upper-right bond, and then the upper-left bond is the shortest and strongest of the three. The configurations in (a) and (d), representing the beginning and end of the loop, are identical.

could represent the ground state of the electrons in a molecule with fixed nuclear coordinates, or that of a spinor in an external magnetic field. We then consider a gradual variation of the nuclear coordinates or field such that the system returns to its starting point at the end of the loop. This situation is sketched in Fig. 3.1, where states $|u_0\rangle$, ..., $|u_7\rangle$ correspond to the $N = 8$ points around the loop (with $|u_8\rangle = |u_0\rangle$).

For a concrete example, consider the triatomic molecule shown in Fig. 3.2. It is almost equilateral, but a distortion has been introduced so as to shorten one of the three bonds slightly, as indicated by the "double bond" in the figure. We have in mind a continuous deformation path, with each of the three bonds being gradually shortened and lengthened in such a way that the illustrations in Fig. 3.2 represent snapshots along the way.[1] We then wish to consider the phase evolution of the ground state of this molecule as it is carried around the loop. For another example, consider the evolution of the ground state of a spinor (e.g., an electron or proton) in an external magnetic field as the direction of this field varies around some closed loop on the unit sphere. In each case, the Berry phase will encode some information about the phase evolution of the ground state along the path in question.

---

[1] Note that the molecule itself is not rotating, but only the pattern of shortened bonds. This is often referred to as a "pseudorotation" and plays an important role in the physics of Jahn–Teller effects in molecules.

3.1.1 Discrete Formulation

Let's start with a discrete formulation, in which N representative vectors $|u_0\rangle$ to $|u_{N-1}\rangle$ are chosen around this loop, as, for example, with $N = 3$ in Fig. 3.2. Note that $|u_N\rangle$ and $|u_0\rangle$ are identical. The Berry phase ϕ is then defined to be

$$\phi = -\text{Im} \ln \left[\langle u_0|u_1\rangle \langle u_1|u_2\rangle \ldots \langle u_{N-1}|u_0\rangle \right]. \tag{3.1}$$

Recall that for a complex number $z = |z|\, e^{i\varphi}$, the expression $\text{Im} \ln z = \varphi$ just takes the complex phase and discards the magnitude. Thus, the Berry phase ϕ is minus the complex phase of the product of inner products of the state vectors at neighboring points around the loop. [Note that the sign convention is not universal; some authors define the Berry phase without the minus sign in Eq. (3.1).]

First consider a simple example based on the triatomic molecule of Fig. 3.2. Suppose that there are two degenerate states $|1\rangle$ and $|2\rangle$ when the triangle is undistorted, and that the distortion breaks this degeneracy everywhere along the loop. Suppose furthermore that the lower-energy of the two states for the snapshots shown in Fig. 3.2 are

$$|u_a\rangle = |u_d\rangle = \tfrac{1}{\sqrt{2}} \begin{pmatrix} 1 \\ 1 \end{pmatrix}, \quad |u_b\rangle = \tfrac{1}{\sqrt{2}} \begin{pmatrix} 1 \\ e^{2\pi i/3} \end{pmatrix}, \quad |u_c\rangle = \tfrac{1}{\sqrt{2}} \begin{pmatrix} 1 \\ e^{4\pi i/3} \end{pmatrix}. \tag{3.2}$$

where the top and bottom elements in the column vector are the amplitudes on basis states $|1\rangle$ and $|2\rangle$, respectively. A trivial computation then shows that the corresponding Berry phase is

$$\phi = -\text{Im} \ln \left[\langle u_a|u_b\rangle \langle u_b|u_c\rangle \langle u_c|u_a\rangle \right] = -\text{Im} \ln \left[\left(\frac{e^{\pi i/3}}{2} \right)^3 \right] = -\pi \tag{3.3}$$

(or equivalently $\phi = \pi$, since a phase is only well defined modulo 2π). Incidentally, you can see that the setting in a *complex* vector space is important; in the case of *real* vectors, the global product in Eq. (3.1) is always real, so the Berry phase is always 0 or π depending on the sign of that product.[2]

It is probably not yet obvious why the Berry phase defined in this way is a useful quantity, but at least it is mathematically well defined in the sense that it is *independent of the choices made for the phases of the individual* $|u_j\rangle$. That is, suppose we introduce a new set of N states

$$|\tilde{u}_j\rangle = e^{-i\beta_j} |u_j\rangle \tag{3.4}$$

---

[2] It happens that $\phi = \pi$ for the complex vectors in Eq. (3.2), but this is an artifact of a symmetry, specifically time reversal (TR) symmetry, implicit in the model.

(β_j is real) related to the old ones by a j-dependent phase rotation β_j, an operation that is known as a *gauge transformation* in the Berry phase context.[3] Then the Berry phase ϕ is unaffected, since any given vector, such as $|u_2\rangle$, appears in Eq. (3.1) once in a ket and once in a bra, so that the phases $e^{\pm i\beta_j}$ cancel out. For example, we can replace $|u_c\rangle$ in Eq. (3.2) by the physically equivalent vector

$$|u_c\rangle = \tfrac{1}{\sqrt{2}}\begin{pmatrix} e^{2\pi i/3} \\ 1 \end{pmatrix} \tag{3.5}$$

and confirm that the final result of the computation in Eq. (3.3) is the same. The gauge invariance of the Berry phase strongly hints that it may be connected with some physically observable phenomena.

We have passed over a subtlety in the preceding discussion, namely the need to impose a branch choice on the definition of Im ln z, as by restricting it to the interval $-\pi < \phi \le \pi$. In this case, Eq. (3.1) always results in a Berry phase lying in this interval, while the nominally equivalent expression

$$\phi = -\sum_{j=0}^{N-1} \text{Im} \ln \langle u_j | u_{j+1} \rangle \tag{3.6}$$

can yield a result that differs by an integer multiple of 2π. If we take the viewpoint that ϕ is just a shorthand for a phase angle, so that only $\cos \phi$ and $\sin \phi$ matter, then this distinction can be safely ignored. However, in any practical implementation the phase angles are normally mapped onto some interval on the real axis, and we can claim only that the Berry phase should be gauge-invariant modulo 2π in the context of an expression like that of Eq. (3.6).

You may be wondering about the magnitude information that has been discarded in Eqs. (3.1) and (3.6). Each inner product has a magnitude somewhat smaller than unity, so a partner function $-\text{Re} \ln \prod_j \langle u_j | u_{j+1} \rangle$ would measure the extent to which the *character* of the states varies from point to point along the loop, whereas the Berry phase ϕ is instead related to the relative *phases* along the loop.

The concept of a Berry phase is also naturally described in terms of *parallel transport*, defined in the present context as follows. Suppose we have a chain of states $|u_0\rangle, |u_1\rangle, \ldots, |u_N\rangle$ with no special phase relations between them. We define a new set of "parallel transported" states $|\bar{u}_0\rangle, |\bar{u}_1\rangle, \ldots$ to be the same as the previous set, except with their phases adjusted as follows. Set $|\bar{u}_0\rangle = |u_0\rangle$. Then choose $|\bar{u}_1\rangle$ to be $|u_1\rangle$ times a phase chosen such that $\langle \bar{u}_0 | \bar{u}_1 \rangle$ is real and positive. Similarly, choose $|\bar{u}_2\rangle$ such that $\langle \bar{u}_1 | \bar{u}_2 \rangle$ is also real and positive, and continue in this way around the loop, imposing the constraint

---

[3] The name is chosen in close analogy to the use of the same term in the theory of electromagnetism. A particular choice of gauge may influence the intermediate results of a calculation, but should not affect any physically meaningful prediction.

$$\text{Im ln } \langle \bar{u}_j | \bar{u}_{j+1} \rangle = 0 \tag{3.7}$$

on each link connecting neighboring points. Conclude by choosing $|\bar{u}_N\rangle$ such that its product with $\langle \bar{u}_{N-1} |$ is real and positive. This generates what is known as a *parallel-transport gauge*.[4]

Assuming that the states form a closed loop as in Fig. 3.1, the two vectors $|u_N\rangle$ and $|u_0\rangle$ are identical. By contrast, while the two vectors $|\bar{u}_N\rangle$ and $|\bar{u}_0\rangle$ describe the same physical state, *they generally differ by a phase*. In fact, the phase mismatch between $|\bar{u}_0\rangle$ and $|\bar{u}_N\rangle$ is nothing other than the Berry phase! To see this, recall that Eq. (3.1) is gauge invariant, so we can evaluate it using the parallel-transport gauge for the states $0, \ldots, N-1$, i.e., $\phi = -\text{Im ln} [\langle \bar{u}_0 | \bar{u}_1 \rangle \ldots \langle \bar{u}_{N-1} | \bar{u}_0 \rangle]$. Since $|\bar{u}_0\rangle$ and $|\bar{u}_N\rangle$ differ only by a phase, we can replace $|\bar{u}_0\rangle$ at the end of the product by $|\bar{u}_N\rangle \langle \bar{u}_N | \bar{u}_0 \rangle]$ to get $\phi = -\text{Im ln} [\langle \bar{u}_0 | \bar{u}_1 \rangle \ldots \langle \bar{u}_{N-1} | \bar{u}_N \rangle \langle \bar{u}_N | \bar{u}_0 \rangle]$. Then all inner products are real and positive except the last, so that

$$\phi = -\text{Im ln } \langle \bar{u}_N | \bar{u}_0 \rangle . \tag{3.8}$$

For the case of Eq. (3.2), for example, we get

$$|\bar{u}_a\rangle = \begin{pmatrix} 1 \\ 1 \end{pmatrix}, \quad |\bar{u}_b\rangle = \begin{pmatrix} e^{-\pi i/3} \\ e^{\pi i/3} \end{pmatrix}, \quad |\bar{u}_c\rangle = \begin{pmatrix} e^{-2\pi i/3} \\ e^{2\pi i/3} \end{pmatrix}, \quad |\bar{u}_d\rangle = \begin{pmatrix} -1 \\ -1 \end{pmatrix} \tag{3.9}$$

(where we have now dropped the irrelevant normalization prefactors), so that $\phi = -\text{Im ln } \langle \bar{u}_d | \bar{u}_0 \rangle = \pi$ as before.

Note that the parallel-transport gauge is not quite unique, since there is still the freedom to choose the phase of the initial vector $|\bar{u}_0\rangle$. Since this choice of initial phase also propagates into $|\bar{u}_N\rangle$, however, it does not affect the value of ϕ coming from Eq. (3.8).

For a closed loop of the kind that we are considering here, the parallel-transport gauge is somewhat unsatisfying, in that it has a discontinuity where the end of the loop rejoins the starting point. We can smooth out this discontinuity by constructing a *twisted parallel-transport gauge* by starting from the parallel-transport gauge and applying phase twists

$$|\tilde{u}_j\rangle = e^{-ij\phi/N} |\bar{u}_j\rangle . \tag{3.10}$$

The new gauge no longer has the discontinuity at the end of the loop. It has the property that $\text{Im ln } \langle \tilde{u}_j | \tilde{u}_{j+1} \rangle$ has the uniform value $-\phi/N$ at every point on the loop, which is manifestly consistent with Eq. (3.6). In other words, we have distributed

---

[4] The term "parallel transport" comes from differential geometry, where the basic idea is that one chooses a local orthonormal basis of vectors at each point along a path on a curved manifold in such a way that the basis is "as aligned as possible" with its neighbors everywhere along the path. Here, the phrase "as aligned as possible" should be reinterpreted in terms of phase alignment.

the phase evolution uniformly along the loop in such a way as to iron out the gauge discontinuity that would otherwise occur at the end of the loop.

While the freedom in the choice of the twisted parallel-transport gauge is still strongly restricted, it is less restricted than for a true parallel-transport gauge for the following important reason. In addition to rotating the phase of the starting state $|\tilde{u}_0\rangle$ (which amounts to a global rotation of all phases), we now have the possibility of replacing ϕ by $\phi + 2\pi m$ (for some integer m) in Eq. (3.10). Taking $m = 1$, for example, this changes all the $\text{Im}\ln\langle\tilde{u}_j|\tilde{u}_{j+1}\rangle$ by $-2\pi/N$, which is still much less than 2π for large N. In other words, we are free to choose different ways of unwinding the phase discontinuity such that $\text{Im}\ln\langle\tilde{u}_j|\tilde{u}_{j+1}\rangle$ is identical for each pair of neighbors, and each of these is a different but equally valid twisted parallel-transport gauge. We will usually choose the one such that $|\text{Im}\ln\langle\tilde{u}_j|\tilde{u}_{j+1}\rangle|$ is at its minimum, but this is not a fundamental restriction. The gauge choice of Eq. (3.2) is an example of a twisted parallel-transport gauge.

3.1.2 Continuous Formulation and Berry Connection

Another hint that the Berry phase formula may be physically meaningful arises from the fact that it has a well-defined continuum limit, shown in Fig. 3.3(c), obtained by increasing the density of points along the path as sketched in Fig. 3.3(a–b). In the continuum formulation, we can take the path to be parametrized by a real variable λ such that $|u_\lambda\rangle$ traverses the path as λ evolves from 0 to 1, with $|u_{\lambda=0}\rangle \equiv |u_{\lambda=1}\rangle$. (Such a convention should be familiar from Chapter 1.) We assume here that $|u_\lambda\rangle$ is a smooth and differentiable function of λ. To derive the continuum expression for the Berry phase that corresponds to Eq. (3.6), we note that

$$\ln\langle u_\lambda|u_{\lambda+d\lambda}\rangle = \ln\langle u_\lambda|\left(|u_\lambda\rangle + d\lambda\,\frac{d|u_\lambda\rangle}{d\lambda} + \cdots\right)$$
$$= \ln(1 + d\lambda\,\langle u_\lambda|\partial_\lambda u_\lambda\rangle + \cdots)$$
$$= d\lambda\,\langle u_\lambda|\partial_\lambda u_\lambda\rangle + \cdots$$

Figure 3.3 (a) Evolution of a state vector $|u\rangle$ in N discrete steps around a closed loop, as in Fig. 3.1. (b) Approach to the continuum limit by increasing the density of points around the loop. (c) Continuum limit, in which the parameter runs over $\lambda \in [0, 1]$ with $|u_{\lambda=0}\rangle = |u_{\lambda=1}\rangle$.

where ∂_λ is a shorthand for $d/d\lambda$ and '...' indicates terms of second order and higher in $d\lambda$. The latter can be discarded in taking the continuum limit of Eq. (3.6), such that we obtain

$$\phi = -\text{Im} \oint \langle u_\lambda | \partial_\lambda u_\lambda \rangle \, d\lambda. \tag{3.11}$$

In fact, $\langle u_\lambda | \partial_\lambda u_\lambda \rangle$ is purely imaginary since

$$2\text{Re} \, \langle u_\lambda | \partial_\lambda u_\lambda \rangle = \langle u_\lambda | \partial_\lambda u_\lambda \rangle + \langle \partial_\lambda u_\lambda | u_\lambda \rangle = \partial_\lambda \, \langle u_\lambda | u_\lambda \rangle = 0,$$

so Eq. (3.11) can also be written as

$$\phi = \oint \langle u_\lambda | i \partial_\lambda u_\lambda \rangle \, d\lambda. \tag{3.12}$$

This is the famous expression for a geometric Berry phase in the continuous formulation (Berry, 1984; Wilczek and Shapere, 1989).

The integrand on the right-hand side of Eq. (3.12) is known as the *Berry connection* or *Berry potential*,[5]

$$A(\lambda) = \langle u_\lambda | i \partial_\lambda u_\lambda \rangle = -\text{Im} \, \langle u_\lambda | \partial_\lambda u_\lambda \rangle, \tag{3.13}$$

in terms of which the Berry phase is

$$\phi = \oint A(\lambda) \, d\lambda. \tag{3.14}$$

Let us understand how these quantities vary under a gauge transformation, which now takes the form

$$|\tilde{u}_\lambda\rangle = e^{-i\beta(\lambda)} |u_\lambda\rangle \tag{3.15}$$

where $\beta(\lambda)$ is some continuous real function of λ. We find

$$\tilde{A}(\lambda) = \langle \tilde{u}_\lambda | i \partial_\lambda | \tilde{u}_\lambda \rangle = \langle u_\lambda | e^{i\beta(\lambda)} i \partial_\lambda \, e^{-i\beta(\lambda)} | u_\lambda \rangle = \langle u_\lambda | i \partial_\lambda | u_\lambda \rangle + \beta'(\lambda)$$

where $\beta'(\lambda) = d\beta/d\lambda$. Thus the Berry connection is *not gauge-invariant*; it is transformed under a gauge change according to

$$\tilde{A}(\lambda) = A(\lambda) + \beta'(\lambda). \tag{3.16}$$

But what about the Berry phase? Recall that since $\lambda = 0$ and $\lambda = 1$ label the same state, we must insist that $|\tilde{u}_{\lambda=1}\rangle = |\tilde{u}_{\lambda=0}\rangle$, just as was the case for $|u_\lambda\rangle$. But this implies that

$$\beta(1) = \beta(0) + 2\pi m \tag{3.17}$$

---

[5] These terms are used interchangeably. "Connection" is a term taken from differential geometry, while "potential" invokes an analogy with the vector potential of electromagnetism (see p. 89) and other gauge field theories.

for some integer m. Then

$$\int_0^1 \beta'(\lambda)\, d\lambda = \beta_{\lambda=1} - \beta_{\lambda=0} = 2\pi m \qquad (3.18)$$

so that replacing A by $\widetilde{A}$ in Eq. (3.14) and using Eq. (3.16) yields

$$\widetilde{\phi} = \phi + 2\pi m. \qquad (3.19)$$

That is, the Berry phase ϕ is gauge-invariant modulo 2π, or in other words, gauge-invariant when regarded as a phase angle!

Once again, we can think of the Berry phase as the phase that is "left over" after parallel transport around the loop. In the continuous case, a parallel-transport gauge is one in which the Berry connection $A(\lambda)$ vanishes:

$$\bar{A}(\lambda) = \langle \bar{u}_\lambda | i \partial_\lambda \bar{u}_\lambda \rangle = 0. \qquad (3.20)$$

If we impose such a gauge, then the Berry phase is just the phase mismatch at the end of the loop,

$$\phi = -\mathrm{Im}\ln \langle \bar{u}_{\lambda=1} | \bar{u}_{\lambda=0} \rangle, \qquad (3.21)$$

exactly as in Eq. (3.8). We can also construct a twisted parallel-transport gauge as $|\tilde{u}_\lambda\rangle = e^{-i\phi\lambda} |\bar{u}_\lambda\rangle$, in analogy with Eq. (3.10), with the result that $\widetilde{A}_\lambda$ is constant around the loop.

The fact that the Berry phase is gauge-invariant modulo 2π should not come as a surprise, reflecting as it does our experience with the discrete case, but its importance is profound. Because quantum probabilities are proportional to the norm squared of an amplitude, there is a tendency to think that "the phase doesn't matter." On the contrary, phases can lead to interference phenomena that are physically important. For example, if duplicate copies of a system are prepared, subjected to parallel transport along different paths in parameter space, and then recombined, the resulting phase difference can lead to physical and measurable interference effects.

We typically discuss Berry phases in the context of evolution along closed paths, but it is useful to establish some terminology for open paths as well. For an open path such as that shown in Fig. 3.4(a), we can define an open-path Berry phase

$$\phi = \int_{\lambda_i}^{\lambda_f} A(\lambda)\, d\lambda \qquad (3.22)$$

where λ is a scalar running from λ_i to λ_f specifying progress along some path in a higher-dimensional space labeled as (λ_x, λ_y) in the sketch. However, an open-path Berry phase is *not* gauge-invariant; a gauge transformation in the form of Eq. (3.15) changes ϕ by $\beta_f - \beta_i$. Only when the path is closed, as in Fig. 3.4(b),

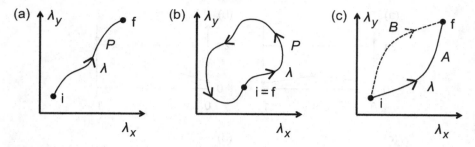

Figure 3.4 (a) Open path P connecting initial point 'i' to final point 'f' in a 2D parameter space. (b) Closed path P in parameter space. (c) Pair of open paths A and B with common initial and final points, such that $A - B$ (that is, A followed by the reverse traversal of B) is a closed path.

is the Berry phase gauge-invariant (modulo 2π). But Fig. 3.4(c) shows another interesting case: If a system is carried from λ_i to λ_f along two *different* paths A and B, the *relative* phase $\Delta\phi = \phi_B - \phi_A$ is again gauge-invariant. This follows trivially from the fact that $\Delta\phi$ is the Berry phase obtained by traversing first path B, then path A in the reverse direction; this is equivalent to circulating around a closed path, as in Fig. 3.4(b).

Returning now to closed paths, note that all possible gauge transformations given by Eq. (3.15) can be classified topologically according to the integer m appearing in Eq. (3.17), which is a *winding number* specifying how many times $e^{-i\beta}$ circulates around the unit circle in the complex plane as λ circulates around the loop. We shall refer to gauge changes characterized by $m = 0$, illustrated in Fig. 3.5(a), as *progressive* gauge transformations. These have the special property that the gauge function $\beta(\lambda)$ can be smoothly deformed to the identity transformation ($\beta = 0$ independent of λ). By contrast, we reserve the term *radical* for gauge changes with a nontrivial winding, as shown in Fig. 3.5(b) and (d). (These are sometimes referred to as "small" and "large" gauge transformations, respectively.[6])

3.1.3 An Example

So far the discussion has been rather general; we have been treating $|u_\lambda\rangle$ as a parametrized path in some complex vector space. For physical applications, we will usually be concerned with the case in which $|u_\lambda\rangle$ is the ground state of some quantum-mechanical Hamiltonian H_λ parametrized by λ. For example, we might

---

[6] Note that the Berry phase itself is not quantized and does not serve as a topological index. Instead, the space of gauge transformations on the loop does admit a topological classification, with the winding number serving as the topological index.

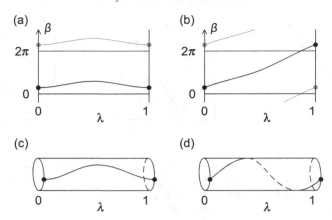

Figure 3.5 Possible behaviors of the function $\beta(\lambda)$ defining a gauge transformation through Eq. (3.15). (a–b) Conventional plots of "progressive" (a) and "radical" (b) gauge transformations, for which β returns to itself or is shifted by a multiple of 2π at the end of the loop, respectively. Gray lines show 2π-shifted periodic images. (c–d) Same as (a–b) but plotted on the surface of a cylinder to emphasize the nontrivial winding of the radical gauge transformation in (b) and (d).

be concerned with the electronic ground state of a molecule as λ, describing the variation of some atomic structural coordinates or the components of some external field, is varied in such a way that both H_λ and $|u_\lambda\rangle$ evolve smoothly in time. In particular, we take this variation to be so slow that the state vector is always well approximated by the static solution $|u_\lambda\rangle$ at the current value of λ. This is known as the *adiabatic approximation*, and we shall often refer to such a process as an *adiabatic evolution*. These terms are defined more precisely in Section 3.3.

A simple and instructive example is the case of a spin-$\frac{1}{2}$ particle, such as an electron or a neutron, at rest in free space and subjected to a uniform magnetic field $\mathbf{B} = B\hat{\mathbf{n}}$ directed along $\hat{\mathbf{n}}$. Its Hamiltonian is just

$$H = -\gamma \mathbf{B} \cdot \mathbf{S} = -\left(\frac{\gamma \hbar B}{2}\right) \hat{\mathbf{n}} \cdot \boldsymbol{\sigma} \qquad (3.23)$$

where γ is the gyromagnetic ratio (assumed to be positive), $\mathbf{S} = \hbar\boldsymbol{\sigma}/2$ is the spin, and σ_j are the Pauli matrices. The ground state $|u_B\rangle$ is an eigenstate of $\hat{\mathbf{n}} \cdot \boldsymbol{\sigma}$ with its spin directed along $\hat{\mathbf{n}}$, and is therefore completely independent of the *magnitude* of $\mathbf{B}$. Thus it is natural to write it as $|u_{\hat{\mathbf{n}}}\rangle$, emphasizing that it depends only on the field direction $\hat{\mathbf{n}}$. We can then ask: What is the Berry phase of $|u_{\hat{\mathbf{n}}}\rangle$ as $\hat{\mathbf{n}}$ is carried around a loop in magnetic-field orientation space, as illustrated in Fig. 3.6(a)?

We shall see shortly that there is an elegant answer to this question, even for a curved loop such as that shown in Fig. 3.6(a). First, however, let us consider a

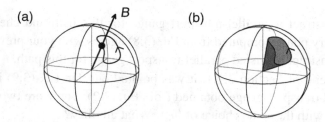

Figure 3.6 (a) Evolution of an applied magnetic field around a closed loop in **B** space. (b) Shaded region shows the solid angle swept out on the unit sphere in **B** space.

simpler "triangular" loop in the discretized approximation. We let $\hat{\mathbf{n}}$ start along $\hat{\mathbf{z}}$, then rotate it to $\hat{\mathbf{x}}$, then to $\hat{\mathbf{y}}$, and then back to $\hat{\mathbf{z}}$, thereby tracing out one octant of the unit sphere in spin space, also known as the "Bloch sphere." From Eq. (3.1), the Berry phase is

$$\phi = -\text{Im} \ln \left[\langle \uparrow_z | \uparrow_x \rangle \langle \uparrow_x | \uparrow_y \rangle \langle \uparrow_y | \uparrow_z \rangle \right]$$

where $|\uparrow_n\rangle$ is the spinor that is "spin up in direction $\hat{\mathbf{n}}$." As shown in standard quantum mechanics texts, such a spinor can be represented as

$$|\uparrow_{\hat{\mathbf{n}}}\rangle = \begin{pmatrix} \cos(\theta/2) \\ \sin(\theta/2)e^{i\varphi} \end{pmatrix}, \tag{3.24}$$

where (θ, φ) are the polar and azimuthal angles of $\hat{\mathbf{n}}$. Thus $|\uparrow_x\rangle = \frac{1}{\sqrt{2}}\begin{pmatrix} 1 \\ 1 \end{pmatrix}$, $|\uparrow_y\rangle = \frac{1}{\sqrt{2}}\begin{pmatrix} 1 \\ i \end{pmatrix}$, and $|\uparrow_z\rangle = \begin{pmatrix} 1 \\ 0 \end{pmatrix}$. We can ignore the normalization factors when inserting these values into the expression for ϕ, obtaining $\phi = -\text{Im} \ln \left[(1)(1 + i)(1)\right] = -\pi/4$. As it happens, this result is exact; a more careful treatment using a dense mesh of intermediate points along each great-circle arc does not change this result. This will become clear when we obtain the general solution to the magnetic-field loop problem after introducing the concept of Berry curvature, which we do next.

Exercises

3.1 Consider a path through the four spinor states

$$|u_0\rangle = \begin{pmatrix} 1 \\ 0 \end{pmatrix}, \quad |u_1\rangle = \frac{1}{\sqrt{2}}\begin{pmatrix} 1 \\ 1 \end{pmatrix}, \quad |u_2\rangle = \begin{pmatrix} 0 \\ 1 \end{pmatrix}, \quad |u_3\rangle = \frac{1}{\sqrt{2}}\begin{pmatrix} 1 \\ i \end{pmatrix} \tag{3.25}$$

which closes on itself with $|u_4\rangle = |u_0\rangle$. This corresponds to a path in which the spin points along $\hat{\mathbf{z}}$, $\hat{\mathbf{x}}$, $-\hat{\mathbf{z}}$, $\hat{\mathbf{y}}$, and then back to $\hat{\mathbf{z}}$.

(a) Compute the discrete Berry phase for the path around this loop.

(b) Construct a parallel-transport gauge for this path and check that the Berry phase computed from Eq. (3.8) agrees with your previous result.

(c) Construct a twisted parallel-transport gauge for this path.

3.2 At the end of Section 3.1.1, it was pointed out that Eq. (3.9) is a twisted parallel-transport gauge obtained from Eq. (3.2). There are two other such gauges with the same choice of $|\bar{u}_a\rangle$. What are they?

3.3 Consider the sequence of N spinor states described by Eq. (3.24), all with the same θ and with φ taking N equally spaced values from 0 to 2π.

(a) Show that the Berry phase is

$$\phi = -N \tan^{-1}\left[\frac{\sin^2(\theta/2)\sin(2\pi/N)}{\cos^2(\theta/2) + \sin^2(\theta/2)\cos(2\pi/N)}\right]. \qquad (3.26)$$

(b) Find $\phi(\theta)$ in the limit that the discrete path becomes continuous (i.e., as $N \to \infty$).

(c) For $\theta = 45°$, compute ϕ numerically for $N = 3, 4, 6, 12$, and 100, and compare with the continuum limit.

3.4 Use PYTHTB to compute the Berry phase of the ground state of the trimer molecule of Fig. 3.2 under the cyclic series of distortions shown there.

(a) Take the site energies to be zero and the hoppings to be

$$t_{01} = t_0 + s\,\cos(\varphi)$$
$$t_{12} = t_0 + s\,\cos(\varphi - 2\pi/3)$$
$$t_{20} = t_0 + s\,\cos(\varphi - 4\pi/3)$$

with $t_0 = -1.0$ and $s = -0.4$. The cycle is parametrized by φ running from 0 to 2π, with $\varphi = 0, 2\pi/3$, and $4\pi/3$ corresponding to panels (a–c) of Fig. 3.2, respectively. You may wish to use the H_2O or benzene example from Appendix D as a rough starting point, but here you need to compute the ground-state eigenvectors on a mesh of φ values and save them in an array. Then explicitly code Eq. (3.6) (i.e., without using PYTHTB function berry_phase) to compute the Berry phase of the ground state around the cycle. Note: Im $\ln\langle\psi_1|\psi_2\rangle$ can be computed using the NUMPY construction

```
np.angle(np.vdot(psi1,psi2))
```

(b) Did you get zero? In fact, the eigenstates were all real. To get something nonzero, repeat this exercise for the case that the Hamiltonian is made complex by multiplying each of the three hoppings in part (a) by the phase factor $e^{i\alpha/3}$. (If you are familiar with the Aharonov–Bohm effect, this heuristically corresponds to the case that a fraction $\alpha/2\pi$

of a magnetic flux quantum is passing through the triangle. Note that the Hamiltonian is still Hermitian; the PYTHTB code automatically applies a reversed phase when hopping in the reverse direction.) For definiteness, consider $\alpha = \pi/4$ ($\frac{1}{8}$ of a flux quantum) and compute the Berry phase again.

(c) Does the Berry phase reverse sign if you reverse the sign of α?

3.2 Berry Curvature and the Chern Theorem

3.2.1 Berry Curvature and Berry Flux

Consider a two-dimensional parameter space such as that illustrated in Fig. 3.7, so that we have vectors $|u_\lambda\rangle$ as a function of $\lambda = (\lambda_x, \lambda_y)$. Then the definition of the Berry connection in Eq. (3.13) naturally generalizes to that of a λ-dependent 2D vector $\mathbf{A} = (A_x, A_y)$ via

$$A_\mu = \langle u_\lambda | i \partial_\mu u_\lambda \rangle \tag{3.27}$$

where $\partial_\mu = \partial/\partial\lambda_\mu$ ($\mu = \{x, y\}$), and the Berry phase expression of Eq. (3.14) can be written as a line integral around the loop P:

$$\phi = \oint_P \mathbf{A} \cdot d\lambda . \tag{3.28}$$

Then the *Berry curvature* $\Omega(\lambda)$ is simply defined as the Berry phase per unit area in (λ_x, λ_y) space. In a discretized context, as with the λ mesh shown in Fig. 3.7, Ω is identified with the Berry phase around one small plaquette[7] divided by the area of that plaquette. In a continuum framework, it becomes just the curl of the Berry connection,

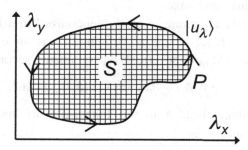

Figure 3.7 Region S of a two-dimensional $\lambda = (\lambda_x, \lambda_y)$ parameter space, bounded by loop P.

---

[7] Because the plaquette is small, the magnitude of the Berry phase around the plaquette can safely be assumed to be $\ll 2\pi$, so there is no ambiguity in its branch choice.

$$\Omega(\lambda) = \partial_x A_y - \partial_y A_x = -2\mathrm{Im}\,\langle\partial_x u|\partial_y u\rangle. \tag{3.29}$$

The last equality follows from a cancellation of terms of the form $\langle u|\partial_x\partial_y u\rangle$ and by noting that $\langle\partial_y u|\partial_x u\rangle^* = \langle\partial_x u|\partial_y u\rangle$.

Once Ω is defined as a curl, we can immediately use Stokes' theorem to relate the "Berry flux" Φ_S through surface patch S to the Berry phase around the boundary P:

$$\begin{aligned}
\Phi_S &= \int_S \Omega(\lambda)\, dS \\
&= \oint_P \mathbf{A} \cdot d\lambda \\
&= \phi_P
\end{aligned} \tag{3.30}$$

where P traces the boundary of region S in the positive sense of circulation. That is, the Berry flux through the surface is equal to the Berry phase around its boundary. In the discrete case, Stokes' theorem is just the statement that if we were to sum up the circulation of the Berry connection $\mathbf{A}$ around all the little plaquettes making up the region shown in Fig. 3.7, we would obtain the Berry phase computed around its boundary, since the contribution from any internal link vanishes as a result of a cancellation between the contributions from the two neighboring plaquettes.

A crucially important property of the Berry curvature is its gauge invariance. That is, under a 2D gauge change $|\tilde{u}_\lambda\rangle = e^{-i\beta(\lambda)}|u_\lambda\rangle$, Eq. (3.16) is generalized to $\tilde{\mathbf{A}} = \mathbf{A} + \nabla\beta$, and since the curl of a gradient is zero, the Berry curvature Ω in Eq. (3.29) is unchanged by the gauge transformation.

In the preceding discussion we implicitly took the Berry curvature Ω to be $\Omega_{xy} = \partial_x A_y - \partial_y A_x$ as in Eq. (3.29), with $\Phi = \Phi^{(xy)}$ being its corresponding Berry flux. However, we can just as well define $\Omega_{yx} = \partial_y A_x - \partial_x A_y$ with its integral being $\Phi^{(yx)}$. Obviously $\Omega_{yx} = -\Omega_{xy}$, so this change of convention has the effect of reversing the signs of Berry curvatures and fluxes.

The concept of Berry curvature is easily generalized to a higher-dimensional parameter space $\lambda = (\lambda_1,\ldots,\lambda_n)$ by defining A_μ as an n-component vector following Eq. (3.27) and $\Omega_{\mu\nu}$ to be a (real) antisymmetric second-rank tensor

$$\Omega_{\mu\nu} = \partial_\mu A_\nu - \partial_\nu A_\mu = -2\,\mathrm{Im}\,\langle\partial_\mu u|\partial_\nu u\rangle. \tag{3.31}$$

Then, for any 2D submanifold S in this higher-dimensional parameter space, Stokes' theorem becomes

$$\phi = \oint_P \mathbf{A} \cdot d\lambda = \int_S \Omega_{\mu\nu}\, ds_\mu \wedge ds_\nu \tag{3.32}$$

where $ds_\mu \wedge ds_\nu$ is an area element on the surface S and P is its boundary. This framework becomes most familiar for a 3D parameter space, where it is natural to use the pseudovector notation $\Omega_z \equiv \Omega_{xy} = -2\text{Im}\,\langle\partial_x u|\partial_y u\rangle$ and so on, which is sometimes written as $\boldsymbol{\Omega} = -\text{Im}\,\langle\nabla_\lambda u| \times |\nabla_\lambda u\rangle$.[8] In this pseudovector notation, Stokes' theorem takes the familiar form

$$\phi = \oint_P \mathbf{A} \cdot d\lambda = \int_S \boldsymbol{\Omega} \cdot \hat{\mathbf{n}}\, dS = \int_S \boldsymbol{\Omega} \cdot d\mathbf{S} \qquad (3.33)$$

where $\hat{\mathbf{n}}$ is a unit vector normal to the surface element of area dS.

There is a close analogy connecting the real-space electromagnetic vector potential $\mathbf{A}(\mathbf{r})$ and its curl, the magnetic field $\mathbf{B}(\mathbf{r})$, with the parameter-space Berry connection $\mathbf{A}(\lambda)$ and its curl $\boldsymbol{\Omega}(\lambda)$. As mentioned earlier, an alternative name for the Berry connection is "Berry potential," and here we see why. In both cases, the "potential" $\mathbf{A}$ is gauge-dependent, while the "field" $\mathbf{B}$ or $\boldsymbol{\Omega}$ is not, essentially because the gauge freedom is removed when the curl is taken.

Let's apply the Berry curvature concept to the case of a spinor subjected to a magnetic field along $\hat{\mathbf{n}}$, as discussed earlier in Section 3.1.3. We again use the representation of Eq. (3.24), which we repeat here for convenience:

$$|\!\uparrow_{\hat{\mathbf{n}}}\rangle = \begin{pmatrix} \cos(\theta/2) \\ \sin(\theta/2)e^{i\varphi} \end{pmatrix}. \qquad (3.34)$$

A gauge choice is implicit in this representation, which makes $|\!\uparrow_{\hat{\mathbf{n}}}\rangle$ smooth and continuous in the vicinity of the "north pole" ($\theta = 0$) of the Bloch sphere.[9] In this vicinity we can regard $|\!\uparrow_{\hat{\mathbf{n}}}\rangle$ as depending on $\lambda = (n_x, n_y)$ through $\hat{\mathbf{n}} = (n_x, n_y, \sqrt{1 - n_x^2 - n_y^2})$. Then the calculation of Ω_{xy} at the north pole is straightforward, since only first derivatives are needed in Eq. (3.31), and a first-order expansion of $|\!\uparrow_{\hat{\mathbf{n}}}\rangle$ gives

$$|\!\uparrow_{\hat{\mathbf{n}}}\rangle \simeq \begin{pmatrix} 1 \\ (n_x + in_y)/2 \end{pmatrix}, \quad |\partial_{n_x}\!\uparrow_{\hat{\mathbf{n}}}\rangle = \tfrac{1}{2}\begin{pmatrix} 0 \\ 1 \end{pmatrix}, \quad |\partial_{n_y}\!\uparrow_{\hat{\mathbf{n}}}\rangle = \tfrac{1}{2}\begin{pmatrix} 0 \\ i \end{pmatrix}. \qquad (3.35)$$

Plugging into Eq. (3.31) yields $\Omega_{xy} = -\tfrac{1}{2}$ at the north pole, or using the pseudo-vector notation, $\Omega_z = -\tfrac{1}{2}$ at $\hat{\mathbf{n}} = \hat{\mathbf{z}}$.

More generally, we can define $\Omega_{\hat{\mathbf{n}}}$ to be the Berry phase per unit solid angle at the point $\hat{\mathbf{n}}$ on the Bloch sphere. Having found that $\Omega_{\hat{\mathbf{n}}} = -\tfrac{1}{2}$ at $\theta = 0$, however, we can argue that it must take the same value everywhere else on the unit sphere. After all, the physics of a spinor in free space is intrinsically isotropic, and we are free to evaluate $\Omega_{\hat{\mathbf{n}}}$ using a coordinate system (x', y', z') with z' aligned with $\hat{\mathbf{n}}$. From this we conclude that $\Omega_{\hat{\mathbf{n}}} = -\tfrac{1}{2}$ everywhere on the unit sphere.

We can easily check the consistency of this result with the Berry phase that we computed on p. 85 for a path tracing one octant of the unit sphere. From Stokes'

---

[8] This notation implies $\Omega_{\mu\nu} = \varepsilon_{\mu\nu\sigma}\Omega_\sigma$ but $\Omega_\mu = \tfrac{1}{2}\varepsilon_{\mu\nu\sigma}\Omega_{\nu\sigma}$, where $\varepsilon_{\mu\nu\sigma}$ is the antisymmetric tensor.
[9] This comes at the expense of introducing a singularity at $\theta = \pi$, where the ket vector has a singular dependence on φ. See the discussion of Eq. (3.39) at the end of Section 3.2.2.

theorem, Eq. (3.30), we would expect that ϕ should be just $-\frac{1}{2}$ times the solid angle $4\pi/8$, or $\phi = -\pi/4$, which is precisely what we found.

We now have the general answer to the problem posed in relation to Fig. 3.6. The Berry phase that results from the evolution of a spinor around the loop in magnetic-field orientation space shown in Fig. 3.6(a) is simply $-\frac{1}{2}$ times the solid angle subtended by the loop, as sketched in Fig. 3.6(b). This is a beautiful and simple result of the mathematical physics of spinors. As a special case, the Berry phase obtained by rotating a spinor around a full great circle, as from $\hat{\mathbf{z}}$ to $\hat{\mathbf{x}}$ to $-\hat{\mathbf{z}}$ to $-\hat{\mathbf{x}}$ and back to $\hat{\mathbf{z}}$, is just $-\pi$, since the solid angle of a hemisphere is 2π. This gives $e^{i\phi} = -1$, reflecting the well-known fact that the parallel transport of a spinor through a full 2π rotation results in a flip of the sign of the spinor wave function.

3.2.2 Chern Theorem

In the previous example, the total Berry flux through the unit sphere is just $\Omega_{\hat{\mathbf{n}}} = -\frac{1}{2}$ times the solid angle of 4π, or $\Phi = -2\pi$. At first sight it may seem puzzling that this could be nonzero. Imagine discretizing the surface of the unit sphere into small triangles or other polygonal plaquettes and calculating the circulation of the Berry connection $\mathbf{A}$ (i.e., the Berry phase ϕ) around each plaquette. Shouldn't the sum of these vanish? One might think so, reasoning that each link between a pair of vertices is traversed once in each direction, leading to a cancellation. But we know that the integrated Berry curvature, which we identify with the total circulation of $\mathbf{A}$, should be -2π. Where have we gone wrong?

Let's look more closely, using a dodecahedral discretization of the sphere as shown in Fig. 3.8. The circulation of $\mathbf{A}$ on path P in Fig. 3.8(a), around a single pentagon, is $1/12$ of -2π or $-\pi/6$. Similarly, the total circulation around path P comprising six pentagons in Fig. 3.8(b) is $-\pi$, and in Fig. 3.8(c) it is $-11\pi/6$. In contrast, path P in Fig. 3.8(c) traces the outline of the bottom outward-directed

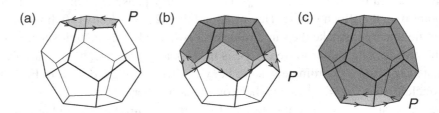

Figure 3.8 Application of Stokes' theorem to portions of the unit sphere in $\mathbf{B}$ space in a discretized approximation. The Berry phase, or circulation of $\mathbf{A}$, around the loop P indicated by arrows must be equal to the sum of circulations of all the enclosed pentagons (shaded), as shown for (a) the top pentagon only; (b) the top six pentagons; and (c) all but the bottom pentagon.

pentagon backward, so the circulation on this path should have been $+\pi/6$, not $-11\pi/6$. Is this a contradiction? No, for we are saved by the fact that a Berry phase is only well defined modulo 2π, according to which $-11\pi/6$ and $+\pi/6$ are identical! You can easily see that this argument generalizes: For any closed surface that is discretized into plaquettes, the total circulation, or equivalently the Berry flux, must be 2π times an integer.

In the continuum limit, this means that the Berry flux Φ_S computed on any closed 2D manifold S is quantized to be 2π times an integer,

$$\oint_S \mathbf{\Omega} \cdot d\mathbf{S} = 2\pi C \qquad (3.36)$$

for some integer C. This is the famous Chern theorem. The integer C is known as the *Chern number* or *Chern index* of the surface,[10] and can be regarded as a "topological index" or "topological invariant" attached to the manifold of states $|u_\lambda\rangle$ defined over the surface S.[11]

Before generalizing this derivation to the continuum case, it may first help to go back and clarify a potentially puzzling aspect of gauge invariance in the context of Stokes' theorem, Eq. (3.33). The right-hand side of that equation represents the flux of Berry curvature passing through surface patch S (i.e., the area integral of the surface-normal component of $\mathbf{\Omega}$ over S); since $\mathbf{\Omega}$ is fully gauge-invariant, the right-hand side is fully determined without any ambiguity. In contrast, the left-hand side is the Berry phase of the curve P that bounds S, and we know that a Berry phase is only well defined modulo 2π. So which is it? Is there a 2π ambiguity, or not?

The answer is that if ϕ is to be determined using a knowledge of $|u_\lambda\rangle$ *only on curve P*, then it is really only well defined modulo 2π. In this case, we can rewrite Eq. (3.33) as

$$\int_S \mathbf{\Omega} \cdot d\mathbf{S} := \oint_P \mathbf{A} \cdot d\lambda \qquad (3.37)$$

using a specialized notation that was introduced in Eq. (1.6). Recall that $:=$ means that the unambiguously defined object on the left-hand side is equal to *one of the values* of the object on the right-hand side, which is ambiguous modulo a quantum. The meaning of Eq. (3.37), then, is that it is possible to make a choice of gauge for the phases of $|u_\lambda\rangle$ around the loop P in such a way that this equation becomes an equality, while other choices will leave a mismatch of some integer multiple of 2π.

---

[10] In some contexts this may also be known as the "TKNN index" after the influential paper of D. J. Thouless et al. (1982).

[11] Another famous topological invariant is the Euler number χ, which arises in geometry as the integral of the Gaussian curvature over a closed surface S and is related to the genus g via $\chi = 2 - 2g$ (Gauss–Bonnet theorem). By contrast, the Chern index depends not on the geometry of the surface, but rather on the manifold of states $|u_\lambda\rangle$ defined over it.

What kind of gauge gives the "correct" answer? Well, if we choose a gauge that is smooth and continuous *everywhere in S*, including on its boundary P, and use this gauge to evaluate the loop Berry phase, then Eq. (3.37) becomes an equality. For in this case, the logic leading to the derivation of Stokes' theorem – summing up circulations around small plaquettes to get the boundary circulation – is sound. While it is possible to make a radical gauge transformation that shifts ϕ_C by 2π when regarding $|u_\lambda\rangle$ as a function defined only in the neighborhood of P, such a gauge change cannot be smoothly continued into the interior S without creating a vortex-like singularity.

Now we are ready to provide a continuum proof of the Chern theorem, Eq. (3.36), which states that the integral of the Berry curvature over any closed 2D manifold is $2\pi C$ for some integer C. We prove this first for a surface having the topology of a simple sphere, as in Fig. 3.9, and divide the sphere into two regions A and B. The loop P forming the boundary between them traverses A in the forward direction and B in the reverse direction, so that by applying Stokes' theorem to each of them, we obtain $\int_A \mathbf{\Omega} \cdot d\mathbf{S} := \phi$ and $-\int_B \mathbf{\Omega} \cdot d\mathbf{S} := \phi$, where ϕ is the Berry phase around P. That is, the results of the two applications of Stokes' theorem must be consistent, but they need to be consistent only modulo 2π. Subtracting these two equations, we get

$$\int_A \mathbf{\Omega} \cdot d\mathbf{S} + \int_B \mathbf{\Omega} \cdot d\mathbf{S} = \oint \mathbf{\Omega} \cdot d\mathbf{S} := 0 \qquad (3.38)$$

which is equivalent to Eq. (3.36).

The same strategy applies to any orientable closed 2D surface, such as a torus. The general strategy is to decompose the surface S into an "atlas" composed of a series of "maps" (A and B in Fig. 3.9), such that a smooth and continuous gauge can be defined in each map. Then Stokes' theorem is applied to each map, and the results are summed. One side of the resulting equation is $\oint \mathbf{\Omega} \cdot d\mathbf{S}$ integrated over the

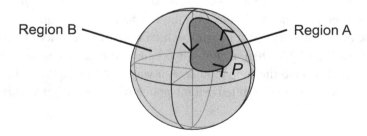

Region B   Region A

Figure 3.9 Proof of the Chern theorem for a manifold S having the topology of a sphere. Closed path P traces the boundaries of subregions A and B in the forward and reverse directions respectively. The uniqueness modulo 2π of the Berry phase around loop P is the key to the proof.

entire surface S, while on the other is the sum of Berry phases along the boundaries, which must cancel modulo 2π. A more careful demonstration of the Chern theorem can be found in a number of topological physics texts such as those by Frankel (1997), Nakahara (2003), and Eschrig (2011). In those texts, the discussion takes place in the language of algebraic topology and differential geometry, specifically in terms of the Chern–Weil theory of topological invariants of vector bundles defined on a smooth manifold.

Looking back, we see that our result $\oint \mathbf{\Omega} \cdot \mathbf{dS} = -2\pi$ over the magnetic Bloch sphere in Section 3.2.1 is consistent with the Chern theorem with $C = -1$. In fact, if we had studied a spin-1 or spin-$\frac{3}{2}$ particle, we would have found $C = -2$ or $C = -3$, respectively. In general, $C = -2s$ for a spin-s particle, and since the Chern theorem requires C to be an integer, it follows that the only allowed spin representations are those with half-integral or integral spin. This is a well-known fact from elementary quantum mechanics, but we have arrived at it here in a novel way.

Note that when the Chern index is nonzero, it is impossible to construct a smooth and continuous gauge over the entire surface S. If such a gauge did exist, then we could apply Stokes' theorem directly to the entire surface and conclude that the Chern number vanishes, in contradiction with the assumption. This is again well illustrated by the case of the spinor on the magnetic unit sphere. If we start from $\hat{\mathbf{n}} = +\hat{\mathbf{z}}$ and construct a gauge that is smooth in the vicinity of $\theta = 0$, and then extend this gauge as smoothly as possible with increasing θ, we get a gauge like that of Eq. (3.24). But while it is smooth in the "northern hemisphere," this gauge has a singularity ("vortex") at $\theta = \pi$, the "south pole." This should remind you of the situation illustrated in Fig. 3.8, where a circulation of order 2π was left at the south pole at the end of the construction. If instead we start at $\hat{\mathbf{n}} = -\hat{\mathbf{z}}$ and work continuously toward the north pole, we can construct an equally valid gauge described by

$$|\uparrow_{\hat{\mathbf{n}}}\rangle = \begin{pmatrix} \cos(\theta/2)e^{-i\varphi} \\ \sin(\theta/2) \end{pmatrix}. \tag{3.39}$$

This gauge, while perfectly well behaved in the southern hemisphere, has a vortex at the north pole. Indeed, there is no possible choice of gauge that is smooth and continuous everywhere on the unit sphere. In such a case, we say that the presence of a nonzero Chern index presents a *topological obstruction* to the construction of a globally smooth gauge.

3.2.3 Berry Flux on a Cylinder or Torus

We will often have reason to consider cases in which the manifold S has the topology of an open cylinder, as in Fig. 3.10(a), or or a closed torus, as in Fig. 3.10(b). Let the parameter space be labeled as (μ, ν), and let ν always be a cyclic parameter

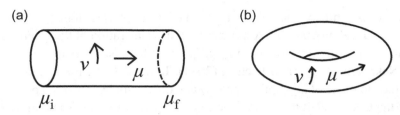

Figure 3.10 (a) Two-dimensional parameter space that is cyclic in ν (with $\nu = 0$ and $\nu = 1$ identified) while extending from $\mu = \mu_i$ to $\mu = \mu_f$. The topology is that of a cylinder. (b) Now μ is also cyclic, such that the topology is that of a torus.

with $\nu = 0$ and $\nu = 1$ identified. We also assume a periodic gauge (see p. 46) in the ν direction, meaning that $|u_{\mu,\nu=0}\rangle$ and $|u_{\mu,\nu=1}\rangle$ are identical, not just equal up to a phase. We adopt the sign convention that $\Omega = \Omega_{\mu\nu} = \partial_\mu A_\nu - \partial_\nu A_\mu$; to emphasize this choice, we write Berry fluxes and Chern numbers as $\Phi^{(\mu\nu)}$ and $C^{(\mu\nu)}$ in what follows. Then the Berry flux $\Phi^{(\mu\nu)} = \oint_S \Omega_{\mu\nu}\, dS$ on the cylinder of Fig. 3.10(a) is

$$\Phi^{(\mu\nu)} = \int_{\mu_i}^{\mu_f} d\mu \int_0^1 d\nu \left(\partial_\mu A_\nu - \partial_\nu A_\mu \right). \tag{3.40}$$

But the periodic gauge condition implies that A_μ is the same at $\nu = 0$ and $\nu = 1$ so that the second term vanishes. Noting that the Berry phase $\phi^{(\nu)}$ in the ν direction is just $\int_0^1 A_\nu\, d\nu$, we find that

$$\Phi^{(\mu\nu)} = \int_{\mu_i}^{\mu_f} (\partial_\mu \phi^{(\nu)})\, d\mu$$
$$= \phi^{(\nu)}(\mu_f) - \phi^{(\nu)}(\mu_i), \tag{3.41}$$

assuming a continuous evolution from μ_i to μ_f. The differential form of this relation is

$$\partial_\mu \Phi^{(\mu\nu)} = \partial_\mu \phi^{(\nu)} \tag{3.42}$$

where it is understood that the derivative is taken with respect to $\mu = \mu_f$. That is, if the region of integration on the cylinder of Fig. 3.10(a) is extended by increasing μ_f, the rate of change of the integrated Berry flux equals the rate of change of the Berry phase at the right end. These equations imply that if we follow the evolution of the Berry phase $\phi^{(\nu)}$ as a function of μ, insisting on a smooth behavior without any 2π jumps, then the Berry flux on the cylinder is just given by the change in $\phi^{(\nu)}$ during the evolution.

This viewpoint provides an important interpretation of the Chern number on a torus, as in Fig. 3.10(b). In this case $\mu = 0$ and $\mu = 1$ are also identified. The Berry phase at these two end points must match modulo 2π, so that at the end of the cycle on μ, $\phi^{(\nu)}$ must have evolved by $2\pi m$ for some integer m. But since the Berry flux on a closed manifold is just 2π times the Chern number, Eq. (3.41) implies that $m = C^{(\mu\nu)}$, the Chern number on the torus. That is, the Chern number is nothing other than the winding number of the Berry phase along ν as we evolve around a cycle in μ. We shall make important use of this principle in the chapters to follow.

Note that we have assumed a smooth and periodic gauge in ν but not necessarily in μ. Alternatively, if we adopted the converse assumption and followed the smooth evolution of $\phi^{(\mu)}$ around a cycle in ν, we would observe a shift of $2\pi C^{(\nu\mu)} = -2\pi C^{(\mu\nu)}$, just the reverse of what was found in the (μ, ν) frame.

3.2.4 A PYTHTB Example

Let us see how this works by returning to the example of the triatomic molecule of Fig. 3.2 and computing the Berry phases and curvatures associated with variations of the parameters. To get a nontrivial behavior, we follow Ex. 3.4(b) in assuming the presence of phase-twisted complex hoppings reflecting broken time-reversal (TR) symmetry, but this time applying all of the twist to a single one of the three hoppings:

$$t_{01} = t_0 + s\,\cos(\varphi)\,,$$

$$t_{12} = t_0 + s\,\cos(\varphi - 2\pi/3)\,,$$

$$t_{23} = [t_0 + s\,\cos(\varphi - 4\pi/3)]\,e^{i\alpha}\,. \tag{3.43}$$

Interpreted in terms of the Aharonov–Bohm effect, this again models the passing of a fraction $\alpha/2\pi$ of a magnetic flux quantum through the triangle, but now in such a way that the Hamiltonian is cyclic in both parameters φ and α. Thus the (α, φ) manifold has the topology of a torus; these parameters play the same role as (μ, ν) in the previous subsection, except that each repeats with period 2π here.

The PYTHTB program `trimer.py` in Appendix D.6 computes Berry phases and curvatures associated with the ground eigenstate of this model. The results are plotted in Fig. 3.11 with parameters $t_0 = -1.0$ and $s = -0.4$. First, the Berry phases $\phi^{(\varphi)}(\alpha) = \oint \langle u_\varphi | i\partial_\varphi u \rangle\, d\varphi$ and $\phi^{(\alpha)}(\varphi) = \oint \langle u_\alpha | i\partial_\alpha u \rangle\, d\alpha$ were calculated individually on a relatively sparse mesh of points, with each phase individually restricted to the interval $[-\pi, \pi]$. These points appear as solid dots in the figure. In Fig. 3.11(a) one can see that $\phi^{(\varphi)}(\alpha)$ starts from zero[12] and decreases to $-\pi$ at $\alpha = \pi$, then jumps discontinuously by $+2\pi$, and subsequently decreases to zero as

---

[12] Recall that the Hamiltonian is real at $\alpha = 0$ and π, so the Berry phase is required to be 0 or π at these points.

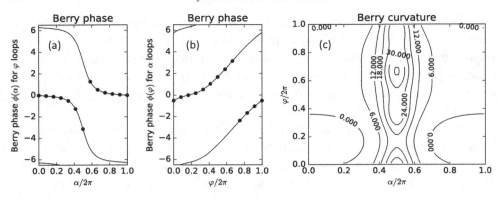

Figure 3.11 Plots produced by PʏᴛʜTB program `trimer.py`, found in Appendix D.6, treating the trimer model of Fig. 3.2 in which φ characterizes a pseudorotation and α represents a TR-breaking magnetic flux according to Eq. (3.43) with $t_0 = -1.0$ and $s = -0.4$. (a) Berry phase around φ loop computed as a function of α. (b) Berry phase around α loop computed as a function of φ. (c) Contour plot of Berry curvature in the (α, φ) plane. In (a–b), solid circles are computed independently and mapped to $[-\pi, \pi]$; solid lines are generated on a finer mesh using continuity to avoid 2π jumps.

α reaches 2π. The solid curve starting at $\phi = 0$ is the result of calculations on a denser mesh using a feature of the PʏᴛʜTB code that enforces continuity from one mesh point to the next; solid curves displaced by $\Delta\phi = \pm 2\pi$ are also plotted for reference. The jump of the plotted points from one branch to the next, displaced by 2π, is now clearly evident. (The value of α at which the jump occurs has no physical significance, but rather is an artifact of the specific branch-cut recipe that was chosen.) The fact that the continuous $\phi(\alpha)$ decreases by 2π implies that the Chern number $C^{(\alpha\varphi)}$ must be -1.

This is confirmed by the corresponding set of plots in Fig. 3.11(b), where the evolution of the α-loop Berry phases are plotted as a function of φ. The Berry phase $\phi(\varphi)$ is now found to *increase* by 2π. This implies that the Chern number $C^{(\varphi\alpha)} = +1$, which of course is consistent with $C^{(\alpha\varphi)} = -1$. The Berry curvature $\Omega^{(\varphi\alpha)}$ is plotted in Fig. 3.11(c); the printed output of the code reports the integrated Berry curvature (i.e., total Berry flux) and confirms that it is indeed $+2\pi$.

The behavior of the Berry phase in this model is largely governed by a degeneracy, often referred to as a "conical intersection" or "diabolical point," between the ground and first excited states of the molecule that occurs at $s = 0$ for a special value of α, as discussed in Ex. 3.7. Because of similar effects that occur in real molecular systems, the study of Berry phases has a long history in the field of molecular chemical spectroscopy.

Exercises

3.5 In Section 3.2.1 we gave a symmetry-based argument that our result $\Omega_{\hat{n}} = -\frac{1}{2}$ for the Berry curvature per unit solid angle for spinors at $\theta = 0$ should also be valid everywhere else on the Bloch sphere. To confirm this, compute $\Omega_{\theta\phi}$ explicitly by applying Eq. (3.29) to Eq. (3.34). (Note that $\Omega_{\hat{n}} = \Omega_{\theta\phi}/\sin\theta$ since the element of solid angle is $\sin\theta \, d\theta d\phi$.)

3.6 Consider the proof of the Chern theorem as in Fig. 3.9 applied to spinors on the Bloch sphere. Let A and B be the upper and lower hemispheres, respectively. Compute the Berry phase around the equator of the Bloch sphere using the gauges of Eqs. (3.34) and (3.39), which are smooth and continuous in A and B, respectively. Do the results differ by a multiple of 2π?

3.7 Consider again the trimer model described in Section 3.2.4. We generalize the definition slightly by $u = s\cos(\varphi)$ and $v = s\sin(\varphi)$ so that

$$t_{01} = t_0 + u,$$
$$t_{12} = t_0 - u/2 + v\sqrt{3}/2,$$
$$t_{23} = \left(t_0 + u/2 - v\sqrt{3}/2\right)e^{i\alpha}. \tag{3.44}$$

Again take $t_0 = -1.0$. We now think of (u, v, α) as a 3D parameter space, whose 2D subspace with $u^2 + v^2 = s^2$ was previously the object of study. Is there a degeneracy point in this 3D space at which the ground and first excited states (bands 0 and 1 in PYTHTB notation) touch? Where is it? Search using analytic arguments, or by exploring using PYTHTB, or both.

3.3 Adiabatic Dynamics

So far, we have been discussing the Berry phase as a property of the slow adiabatic evolution of a quantum system along a certain path in parameter space. It remains, however, to show how this relates to the actual quantum evolution of the system as described by the time-dependent Schrödinger equation.

Consider a Hamiltonian $H(\lambda)$ for some quantum system such as a molecule, with parameter $\lambda(t)$ being a slow function of t. (We shall quantify what is meant by "slow" shortly.) For a given λ, the eigenstates of H are

$$H(\lambda)\,|n(\lambda)\rangle = E_n(\lambda)\,|n(\lambda)\rangle \tag{3.45}$$

where n labels the eigenstates. We start the system in eigenstate n at time $t=0$ and then follow its subsequent time evolution.

If λ did not vary with t at all, the resulting wave function would evolve as $|\psi(t)\rangle = e^{-iE_n t/\hbar}|n\rangle$. In other words, the phase advances by an amount $e^{-iE_n \Delta t/\hbar}$ in a small time interval Δt. Now if λ is slowly changing in time, and if we approximated it as constant in each interval Δt, the the phase evolution would be

$$\prod e^{-iE_n(t)\Delta t/\hbar} = e^{-i\sum E_n(t)\Delta t/\hbar} .$$

In the continuum limit the sum turns into an integral, and we expect the phase evolution to be of the form $|\psi(t)\rangle = e^{-i\gamma(t)}|n(t)\rangle$ with

$$\gamma(t) = \frac{1}{\hbar} \int_0^t E_n(t') \, dt' . \tag{3.46}$$

In the spirit of the adiabatic approximation, this leads to the ansatz

$$|\psi(t)\rangle = c(t) \, e^{-i\gamma(t)} \, |n(t)\rangle \tag{3.47}$$

where $|n(t)\rangle$ on the right-hand side is defined as $|n(\lambda(t))\rangle$; that is, it is just the eigenstate $|n(\lambda)\rangle$ of the *time-independent* problem evaluated at $\lambda = \lambda(t)$. The factor $c(t)$ allows for the possibility that there may be some extra phase evolution beyond the guess based on $\gamma(t)$.

We shall see shortly that the ansatz (3.47) is only the zero-order term in a perturbation expansion in $\dot{\lambda} = d\lambda/dt$. For now we just plug this ansatz into the time-dependent Schrödinger equation

$$\left[i\hbar\partial_t - H(t)\right]|\psi(t)\rangle = 0 \tag{3.48}$$

(where $\partial_t = \partial/\partial t$) to find

$$0 = \dot{c}(t) \, |n(t)\rangle + c(t) \, \partial_t|n(t)\rangle . \tag{3.49}$$

To derive this equation, note that the time derivative ∂_t acts on all three terms in Eq. (3.47), but the term involving $\partial_t e^{-i\gamma(t)}$ cancels against the $H(t)|n(t)\rangle = E_n(t)|n(t)\rangle$ term, leaving the two terms in Eq. (3.49). Acting with $\langle n(t)|$ on the left yields

$$\dot{c}(t) = ic(t)A_n(t) \tag{3.50}$$

where

$$A_n(t) = \langle n(t)|i\partial_t n(t)\rangle . \tag{3.51}$$

Comparing this result with Eq. (3.13), we note that $A_n(t)$ is a "Berry connection in time." The solution of Eq. (3.50) is just $c(t) = e^{i\phi(t)}$ with

$$\phi(t) = \int_0^t A_n(t') \, dt' \tag{3.52}$$

which we immediately recognize as an open-path Berry phase.

Moreover, this Berry phase can be reexpressed in terms of λ instead of t. That is, since $|n(t)\rangle$ is defined as $|n(\lambda(t))\rangle$, application of the chain rule yields $\partial_t |n(t)\rangle = \dot{\lambda}\, \partial_\lambda |n(\lambda)\rangle$. It follows that $A_n(t) = \dot{\lambda}\, A_n(\lambda)$ where

$$A_n(\lambda) = \langle n(\lambda) | i\partial_\lambda n(\lambda)\rangle \qquad (3.53)$$

is the Berry connection in parameter space. Substituting into Eq. (3.52) and using $d\lambda = \dot{\lambda}dt$, we then find

$$\phi(t) = \int_{\lambda(0)}^{\lambda(t)} A_n(\lambda)\, d\lambda . \qquad (3.54)$$

This is a remarkable result; it says that the Berry phase entering into the time-evolving wave function is a function only of the path it has traced in parameter space, and is independent of the rate at which the path is traversed, so long as the parametric evolution is sufficiently slow.[13]

Let's take stock. Our ansatz of Eq. (3.47) is successful only if the extra Berry-phase term is included. The result is that, at leading order in adiabatic perturbation theory, the wave function evolves as

$$|\psi(t)\rangle = e^{i\phi(\lambda(t))}\, e^{-i\gamma(t)}\, |n(t)\rangle \qquad (3.55)$$

where the naively expected dynamical phase $e^{-i\gamma}$ has to be augmented by the Berry phase $e^{i\phi}$ to find the correct phase evolution of the wave function.

As a special case, note that the Berry-phase term is absent in Eq. (3.55) if $|n(\lambda)\rangle$ is chosen to satisfy a parallel-transport gauge as in Eq. (3.20). In other words, our derivation shows that, once the dynamical phase is factored out, the time evolution of the system is such that it follows a parallel-transport gauge.

There may be a tendency to think of the Berry-phase factor in Eq. (3.55) as "only a phase" with little in the way of physical consequences, since probabilities, rather than amplitudes, determine physical observations. However, as mentioned on p. 82, the Berry phase sometimes plays a crucial role by giving rise to interference phenomena. For example, if duplicate copies of a system are prepared and propagated along two paths on which they experience different Berry phases, this difference manifests itself when the systems are recombined. A simple example is discussed in Ex. 3.8.

We hinted earlier that higher-order terms might need to be added to Eq. (3.47) or (3.55) for some purposes. This becomes absolutely crucial for the discussion of *adiabatic charge transport*, which will play an important role for crystalline systems in Chapter 4. For simplicity, however, we consider it here only for a finite

---

[13] This central property helps explain why Berry phases are also called "geometric phases," emphasizing this dependence only on the path in state space and not on such details as the rate of traversal of the path or the specific representation of the states along the path.

system such as an atom or molecule. Recall that the current density for an electron in state $|\psi\rangle$ is $(ie\hbar/2m)[\psi^*(\mathbf{r})\nabla\psi(\mathbf{r}) - \psi(\mathbf{r})\nabla\psi^*(\mathbf{r})]$, which vanishes identically if $\psi(\mathbf{r})$ is real. This will also be true of the wave function in Eq. (3.55) if $|n(\lambda)\rangle$ is real, since the phase factors in front are independent of $\mathbf{r}$. But in a typical case, such as for the ground electronic state of an H_2O molecule as one nucleus is gradually moved, $|n(\lambda)\rangle$ is indeed real. If we assumed Eq. (3.55), then we would conclude that the motion of the nucleus induces no corresponding flow of electron charge. This is clearly nonsense, since the electronic charge density $\rho(\mathbf{r})$ changes with time, which it cannot do if there is no current flow.

The solution to this paradox is to carry the adiabatic perturbation theory to one higher power of $\dot\lambda$. We now expand Eq. (3.55) to become

$$|\psi(t)\rangle = e^{i\phi(\lambda(t))}\, e^{-i\gamma(t)}\left[|n(\lambda(t))\rangle + \dot\lambda\, |\delta n(t)\rangle \right],\tag{3.56}$$

where the extra component $|\delta n(t)\rangle$ is to be determined. We already know that Eq. (3.56) solves the time-dependent Schrödinger equation to order zero in $\dot\lambda$, but we now require that it should also do so at first order. For this purpose we can discard terms that scale like $\ddot\lambda$ or $\dot\lambda^2$, including a $\dot\lambda\partial_t|\delta n\rangle$ term. We find

$$(E_n - H_\lambda)\,|\delta n\rangle = -i\hbar\,(\partial_\lambda + iA_n)\,|n\rangle,\tag{3.57}$$

which can also be written using Eq. (3.53) as

$$(E_n - H_\lambda)\,|\delta n\rangle = -i\hbar Q_n\,|\partial_\lambda n\rangle,\tag{3.58}$$

where $Q_n = 1 - |n\rangle\langle n|$ as defined in Eq. (2.91). Equation (3.58) is very similar to the inhomogeneous Sternheimer equation, Eq. (2.90), for the perturbed wave function that arises in ordinary first-order perturbation theory, except that $(\partial_\lambda H)\,|n\rangle$ has been replaced by $|\partial_\lambda n\rangle$. In fact, these two expressions are just related by Eq. (2.97). Using the methods and notation of Section 2.3, then, the formal solution to Eq. (3.58) can be written either as

$$|\delta n\rangle = -i\hbar T_n|\partial_\lambda n\rangle\tag{3.59}$$

or as

$$|\delta n\rangle = -i\hbar T_n^2(\partial_\lambda H)|n\rangle\,.\tag{3.60}$$

Written explicitly as a sum over Hamiltonian eigenstates $|m(\lambda)\rangle$ at the same λ, this is

$$|\delta n\rangle = -i\hbar \sum_{m\neq n} \frac{\langle m|\partial_\lambda n\rangle}{E_n - E_m}\,|m\rangle\tag{3.61}$$

$$= -i\hbar \sum_{m\neq n} \frac{\langle m|(\partial_\lambda H)|n\rangle}{(E_n - E_m)^2}\,|m\rangle\,.\tag{3.62}$$

Note that time has again disappeared, and we can think in terms of evolution with respect to λ, except for the magnitude of $\dot{\lambda}$ appearing in Eq. (3.56).[14]

We said that the adiabatic approximation should be valid if the Hamiltonian varies "slowly enough," and we are now in a position to quantify this. Let ΔE be a typical spacing between energy levels and λ_0 be the inverse of a typical $\langle m|\partial_\lambda n\rangle$ for a neighboring energy level. We can think of λ_0 as the scale of λ over which the state $|n\rangle$ varies significantly. Then for an order-of-magnitude estimate, we can identify $\hbar\dot{\lambda}/\lambda_0\Delta E$ as the small dimensionless parameter describing the "slowness" of adiabatic evolution. Basically the evolution is adiabatic if the rate of variation of λ is slow compared to $\Delta E/\hbar$, the characteristic frequency of quantum evolution in the system.

An interesting feature of adiabatic perturbation theory is the fact that, leaving aside the phase information encoded in the Berry phase, the time-evolving wave function has only a short-term memory of the history of the path. Keeping terms to first order in $\dot{\lambda}$, for example, the state at time t depends only on $H_{\lambda(t')}$ for times t' at, and infinitesimally prior to, the current time t. This memory gets pushed back a little further as higher-order terms are included, but the overall picture is that the state vector rapidly "forgets" what happened earlier in its evolution along the path.

If we now use Eq. (3.56) when taking the expectation value of the current operator at order $\dot{\lambda}$, we can check that it does correctly describe the charge transport during the adiabatic evolution. This is carried through in Ex. 3.10.

Finally, we note that a common context for the application of adiabatic evolution is that of a system with "fast" and "slow" variables. The canonical example is that of electrons in molecules and solids, where the electron is many orders of magnitude lighter than the nuclei. From the viewpoint of the time-evolving electron system, it is often an excellent approximation to treat the nuclear coordinates $\mathbf{R}_j$ as classical variables that evolve slowly along a prescribed path. However, there is also a back-reaction on the system of nuclei, so that in a quantum treatment they experience a "gauge potential" $\mathbf{A}_j$ arising from the electron system as it adiabatically follows the nuclear one. This gauge potential is nothing other than the Berry connection $A_{j\mu} = \langle\psi_{\{\mathbf{R}\}}|i\partial\psi_{\{\mathbf{R}\}}/\partial R_{j\mu}\rangle$, where $|\psi_{\{\mathbf{R}\}}\rangle$ is the electronic ground state at a fixed set of $\mathbf{R}_j$. This is an important part of the theory of Berry phases as applied to molecular physics (see, for example, Bohm et al., 1992).

---

[14] See Thouless (1983). A similar formulation was derived shortly thereafter by Niu and Thouless (1984) in terms of the one-particle density matrix instead of the wave function; see, for example, their Eq. (2.10) .

Exercises

3.8 A beam of electrons with spin aligned along $+\hat{\mathbf{z}}$ is sent through an arrangement of mirrors as shown:

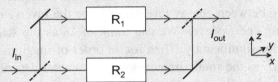

The dashed mirrors are half-silvered, such that identical beams with intensity $I_{in}/2$ and spin $+\hat{\mathbf{z}}$ enter regions R_1 and R_2. In the absence of these regions, full constructive interference would occur at the upper right with $I_{out} = I_{in}$. The electrons entering R_1 experience a constant magnetic field $\mathbf{B} = B_0\hat{\mathbf{z}}$, while those entering R_2 see a field with the slow spatial variation $\mathbf{B} = B_0\hat{\mathbf{n}}(x)$. Note that the magnitude is constant, but the direction $\hat{\mathbf{n}}$ slowly varies with x; $\hat{\mathbf{n}}$ starts out along $+\hat{\mathbf{z}}$ at the left edge and rotates along three consecutive great-circle arcs, first to $+\hat{\mathbf{x}}$, then to $+\hat{\mathbf{y}}$, then back to $+\hat{\mathbf{z}}$ at the point of exit from $\mathbf{R}_2$.

(a) What is the intensity I_{out} relative to I_{in}?

(b) Can we simplify the problem by letting $\mathbf{B}$ remain zero in region R_1? Why or why not?

3.9 Fill in the steps leading from Eq. (3.56) to Eq. (3.57) and Eq. (3.58).

3.10 Here we compute the current induced by an adiabatic change of the Hamiltonian and check that it correctly predicts the change in the electric dipole moment.

(a) Using Eq. (3.56), simplified to $|\psi(t)\rangle = e^{i\alpha}(|n\rangle + \dot{\lambda}\,|\delta n\rangle)$, show that the induced change in some arbitrary operator $\mathcal{O}$ is $\langle\mathcal{O}\rangle = 2\dot{\lambda}\,\mathrm{Re}\,\langle n|\mathcal{O}|\delta n\rangle$.

(b) Defining the current operator $\mathcal{J} = -e\mathbf{v}$ in terms of the velocity operator $\mathbf{v}$ and using Eq. (3.61), show that

$$\langle\mathcal{J}\rangle = -2e\hbar\dot{\lambda}\,\mathrm{Im}\sum_{m\neq n}\frac{\langle n|\mathbf{v}|m\rangle\langle m|\partial_\lambda n\rangle}{E_n - E_m}.$$

(c) Using Eq. (2.47), show that this becomes $\langle\mathcal{J}\rangle = -2e\dot{\lambda}\,\mathrm{Re}\,\langle n|\mathbf{r}|\partial_\lambda n\rangle$. *Hint:* Note that $\langle n|\mathbf{r}|n\rangle\langle n|\partial_\lambda n\rangle$ is pure imaginary (why?).

(d) Noting that $\langle\mathcal{J}\rangle$ has the interpretation of $d\mathbf{d}/dt$, where $\mathbf{d} = -e\mathbf{r}$ is the dipole operator, and canceling the dt, show that this becomes $\partial_\lambda\langle\mathbf{d}\rangle = -2e\,\mathrm{Re}\,\langle n|\mathbf{r}|\partial_\lambda n\rangle = -e\partial_\lambda\langle n|\mathbf{r}|n\rangle$, which is self-evident. This shows that the calculation of the adiabatically induced current does correctly predict the change in electric dipole of the system.

3.4 Berryology of the Brillouin Zone

Up to this point in this chapter, we have considered Berry phases, connections, and curvatures defined for some $|u_\lambda\rangle$ in a generic parameter space $\lambda = (\lambda_1, \lambda_2, \ldots)$. We now turn to the main theme of this book, where we specialize to the case that these parameters are the wavevector components k_j labeling Bloch states $|\psi_{nk}\rangle$ of band n in the Brillouin zone (BZ), as described in Section 2.1.3.

We assume for the moment that band n is *isolated*, meaning, that it does not touch bands $n \pm 1$ anywhere in the BZ. This is a significant restriction, as degeneracies between bands at high-symmetry points in the BZ are common in crystalline materials. When they occur, they typically introduce a nonanalytic dependence of $|\psi_{nk}\rangle$ on $\mathbf{k}$, which is problematic for the definitions of the Berry connection and curvature. We will lift this restriction in Section 3.6, but for now it allows us to assume that such singularities are not present for band n, and the Berry formalism should apply.

However, we immediately encounter an important subtlety. Should we define the Berry phase and curvature in terms of the Bloch functions $|\psi_{nk}\rangle$, or their cell-periodic versions $|u_{nk}\rangle$? The answer is that we *must* use the latter. To see this, consider the case of a discretized Berry phase for a 1D crystal of lattice constant a. If we were to substitute the $|u_j\rangle$ of Eq. (3.1) by the $|\psi_{nk}\rangle$, we would need to compute inner products $\langle \psi_{nk} | \psi_{n,k+b} \rangle$, which take a form like

$$\int_{-\infty}^{\infty} \psi_{nk}^*(x) \psi_{n,k+b}(x)\, dx = \int_{-\infty}^{\infty} e^{ibx} u_{nk}^*(x) u_{n,k+b}(x)\, dx \qquad (3.63)$$

for some small but finite k-space separation b. However, the product of u factors on the right-hand side of Eq. (3.63) is periodic with the unit cell, so that the phase factor e^{ibx} will average to zero when the integral is carried over all x, and the result vanishes. Integrating Eq. (3.63) over one unit cell would also be problematic, since the result would depend on the cell location (e.g., $[0, a]$ versus $[-a/2, a/2]$). Instead, the expression

$$\langle u_{nk} | u_{n,k+b} \rangle = \int_0^a u_{nk}^*(x) u_{n,k+b}(x)\, dx \qquad (3.64)$$

is finite and well behaved. (We adopt a single-unit-cell normalization convention for u functions: $\langle u_{nk} | u_{nk} \rangle = \int_0^a |u_{nk}(x)|^2\, dx = 1$).

The essential observation is that all of the $|u_{nk}\rangle$ have the same simple periodic boundary conditions, forcing $u_{nk}(x)$ at $x = 0$ and $x = a$ to be equal. Thus, all $|u_{nk}\rangle$ belong to the same Hilbert space. As a result, inner products between vectors at different k, or derivatives with respect to k, are well defined. This would not be the case if the formalism were based on the $|\psi_{nk}\rangle$ vectors. Note, however, that the

k dependence reappears in a different guise, by recalling that the $|u_{nk}\rangle$ are now solutions of a k-dependent Hamiltonian H_k as given by Eq. (2.43).

It is now straightforward to take the formalism of Section 3.1 over to the case of Bloch functions in the BZ. Returning to 2D or 3D, a Berry phase associated with band n takes the form

$$\phi_n = \oint \mathbf{A}_n(\mathbf{k}) \cdot d\mathbf{k} \tag{3.65}$$

where the Berry connection is

$$A_{n\mu}(\mathbf{k}) = \langle u_{nk}|i\partial_\mu u_{nk}\rangle \tag{3.66}$$

with $\partial_\mu = \partial/\partial k_\mu$, where μ is a Cartesian index. This can be written equivalently in vector form as $\mathbf{A}_n(\mathbf{k}) = \langle u_{nk}|i\nabla_\mathbf{k} u_{nk}\rangle$. Similarly, the Berry curvature is

$$\Omega_{n,\mu\nu}(\mathbf{k}) = \partial_\mu A_{n\nu}(\mathbf{k}) - \partial_\nu A_{n\mu}(\mathbf{k}) = -2\,\mathrm{Im}\,\langle \partial_\mu u_{nk}|\partial_\nu u_{nk}\rangle, \tag{3.67}$$

which in 3D can be reexpressed in pseudovector form (see p. 89) as $\boldsymbol{\Omega}_n(\mathbf{k})$. In 2D, the BZ can be regarded as a torus, and the Chern number of an entire band is

$$C_n = \frac{1}{2\pi} \int_{\mathrm{BZ}} \Omega_{n,xy}\, d^2k. \tag{3.68}$$

As before, we have a gauge freedom to transform the Bloch functions as

$$|\tilde{u}_{nk}\rangle = e^{-i\beta(\mathbf{k})}|u_{nk}\rangle \tag{3.69}$$

where $\beta(\mathbf{k})$ is some real function of $\mathbf{k}$. The Berry connection

$$\widetilde{\mathbf{A}}_n(\mathbf{k}) = \mathbf{A}_n(\mathbf{k}) + \nabla_\mathbf{k}\beta(\mathbf{k}) \tag{3.70}$$

is gauge-dependent; the Berry curvature $\Omega_{n,\mu\nu}(\mathbf{k})$ and Chern number C_n are fully gauge-invariant; and the Berry phase ϕ_n is invariant modulo 2π.

It is equally straightforward to develop the theory in terms of the reduced wavevector $\boldsymbol{\kappa}$ of Eqs. (2.27–2.28). In this case the derivative $\partial_\mu = \partial/\partial k_\mu$ with Cartesian reciprocal-space index μ is replaced by the derivative $\partial_j = \partial/\partial\kappa_j$, where j runs over the reciprocal-lattice directions. The Berry connections and curvatures appear as A_{nj} and $\Omega_{n,ij}$ in this formalism. The Chern number on the 2D BZ can be computed either from Eq. (3.68) or from $C_n = (1/2\pi)\int_{\mathrm{BZ}} \Omega_{n,12}\, d^2\kappa$, with the same result being obtained with either method. This follows intuitively from the fact that both correspond to a continuum limit of the same discretized system, but see also Ex. 3.12 for an analytic proof.

In the discretized version of this theory, which is well suited for use in practical calculations, we have to take inner products of the form $\langle u_{nk}|u_{n,k+b}\rangle$ between neighboring points $\mathbf{k}$ and $\mathbf{k} + \mathbf{b}$ in the BZ. This is quite unlike what we usually encounter when computing the expectation values of observables, Eq. (2.46), where the same wavevector **k** appears in both the bra and the ket. If the Berry phase and curvature are to have any physical significance, it will have to be in the context of a paradigm rather different from that of ordinary observables and their expectation values.

You may be wondering whether the Berry phases, connections, and curvatures defined previously are actually nonzero in the crystals of interest. This question requires better framing for the case of the Berry phase, where it depends on choice of path, and for the Berry connection, where it depends on the gauge choice. Nevertheless, the Berry curvature $\Omega_n(\mathbf{k})$ for band n is a uniquely defined function in the BZ. It is fairly straightforward to show that:

1. If the crystal has inversion symmetry, then $\Omega_n(\mathbf{k}) = \Omega_n(-\mathbf{k})$.
2. If the crystal has TR symmetry, then $\Omega_n(\mathbf{k}) = -\Omega_n(-\mathbf{k})$. Quantities involving an integral of Ω_n over the BZ will vanish.
3. It follows that if the crystal has both inversion and TR symmetry (or, for that matter, only the $\mathcal{I} * \mathcal{T}$ operator is a symmetry), then $\Omega_n(\mathbf{k}) = 0$ identically.
4. If the crystal has some other spatial or magnetic symmetries, as described by the magnetic point group, then additional relations are imposed on $\Omega_n(\mathbf{k})$. For example, a simple 3-fold axis imposes the corresponding 3-fold rotational symmetry on the $\Omega_n(\mathbf{k})$ field.

Items 2 and 3 suggest that the Berry curvature will be of most interest in magnetic systems, which have spontaneously broken TR symmetry. We shall see in Chapters 5 and 6 that this is indeed the case. Certainly the Berry curvature vanishes everywhere in the BZ for many nonmagnetic centrosymmetric materials, such as Si in the diamond structure, Cu in the face-centered cubic (fcc) structure, and Bi_2Se_3, which is a van der Waals–bonded layered structure. Nevertheless, it turns out that Berry-phase concepts also play a central role in aspects of these materials, as will be discussed in Chapters 4 and 5.

Since we have mentioned magnetic materials, we should briefly return to the subject of Section 2.1.2 and clarify the treatment of spin and spin-orbit coupling.

- If spin-orbit coupling is *not* included in the treatment, then the systems of spin-up and spin-down electrons can be treated independently. In this case the potentials appearing in Eq. (2.18) are real, and each spin system acts like a system of scalar electrons with unbroken TR symmetry. If in addition the system is *nonmagnetic*, then spin up and spin down behave identically; we typically label bands by a single integer n, including a factor of 2 in sums and integrals such as in Eq. (2.46) to account for spin degeneracy. In a *magnetic* system, we need to add a spin label s to the Bloch functions $|u_{ns\mathbf{k}}\rangle$ and carry the same label on Berry-related quantities such as ϕ_{ns}, $\mathbf{A}_{ns}$, and Ω_{ns}.

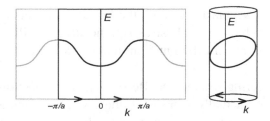

Figure 3.12 At left, conventional view of a 1D band structure, with the first BZ highlighted in black and the extended-zone scheme shown in gray. At right, a more topologically natural view in which the BZ is wrapped onto a circle and the band structure is plotted on a cylinder.

- If spin-orbit coupling *is included*, then the number of band labels n is doubled, and the Bloch functions $|u_{n\mathbf{k}}\rangle$ are spinor wave functions in the sense of Eq. (2.19). The Berry-related quantities carry only the label n, but the inner products in Eqs. (3.66–3.67) are taken between spinor wave functions.

We shall soon have to ask what physical interpretation can be given to the Berry phases and curvatures associated with the Bloch states. Before we do, let us discuss on which kinds of *paths* the Berry phases might be defined. For a 1D crystal (of lattice constant a), there is really only one closed path of interest – namely, the one that circulates around the BZ. Recall from the discussion on pp. 45–46 that the BZ of a 1D system is best viewed as a closed loop or unit circle. This is illustrated in Fig. 3.12, where the panel at left shows the conventional view in which an energy band $E_n(k)$ is plotted as a periodic function of k with period $2\pi/a$. However, this picture is misleading, since a state at k is *one and the same* as a state at $k + 2\pi/a$. In other words, these two wavevectors are duplicate labels for the same state. The panel at right shows a more unconventional, yet more natural, way of thinking about an energy band, in which the wavevector axis is wrapped into a circle and the energy is plotted on the surface of the resulting cylinder. Then a possible object of interest is the Berry phase

$$\phi_n = \oint_{\text{BZ}} A_n(k)\, dk = \oint_{\text{BZ}} \langle u_{nk} | i\partial_k u_{nk} \rangle \tag{3.71}$$

defined on this loop. The notation $\oint_{\text{BZ}}$ indicates an integral taken around the loop formed by the 1D BZ.

Such a Berry phase was first discussed by Zak (1989) and is sometimes called a "Zak phase." In the introductory paragraph of this paper, Zak listed some of the areas in which the concept of the Berry phase was making an impact in atomic, molecular, and nuclear physics, and then wrote:

It seems, however, that one important and natural system for the appearance of Berry's phase was left out. We have in mind the motion of an electron in a periodic solid.

This farsighted observation by Zak set the stage for many of the later developments that are discussed in this book. At the time, Zak was mainly concerned with symmetry properties and the relation of the Berry phase to the "band center," which we now identify with a Wannier center (see Section 3.5). It was to take a few years before the connection to the theory of electric polarization would be made, as will be discussed in the next chapter.

An important subtlety arises when it comes to computing this Berry phase. For the Berry phase to be well defined, we need $|u_{nk}\rangle$ to be a smooth function of k everywhere on the loop. If we represent the loop by letting k range from 0 to $2\pi/a$, for example,[15] then we also have to ensure smoothness across the artificial boundary point where k crosses from $2\pi/a$ back to 0. That is, we must insist that

$$\psi_{n,k=2\pi/a}(x) = \psi_{n,k=0}(x).\qquad(3.72)$$

Thus, the Bloch functions at the two ends of the interval $[0, 2\pi/a]$ must be equal not just up to a phase, but with the same phase. In an extended-zone context, where quantities such as E_{nk} are regarded as periodic functions of k, Eq. (3.72) is just the *periodic gauge condition* of Eq. (2.35).

But we insisted that the Berry connection should be defined not in terms of the Bloch functions $|\psi_{nk}\rangle$, but rather in terms of their cell-periodic partners $|u_{nk}\rangle$. Using Eq. (2.40), the condition of Eq. (3.72) translates to the condition

$$u_{n,k=2\pi/a}(x) = e^{-2\pi ix/a}u_{n,k=0}(x),\qquad(3.73)$$

introduced earlier as Eq. (2.42). So the vectors $|u_{n,k=2\pi/a}\rangle$ and $|u_{n,k=0}\rangle$ are not equal! Moreover, they differ by more than a global phase, since the phase factor in Eq. (3.73) is x-dependent. The extra $e^{-2\pi ix/a}$ phase is an essential feature of the cell-periodic Bloch functions, and one that we have to learn to live with.

To compute the Berry phase in practice, we discretize the BZ into N equal intervals and compute $|u_{nk_i}\rangle$ for $i = 0,\ldots,N-1$ with $k_i = 2\pi i/N$. This typically involves calling a matrix diagonalization routine that returns eigenvectors whose phase is not under our control. Thus, if we call this routine to compute $|u_{nk_0}\rangle$ and $|u_{nk_N}\rangle$ independently, we cannot assume that they will obey Eq. (3.73). Instead, we use Eq. (3.73) to construct $|u_{nk_N}\rangle$ from $|u_{nk_0}\rangle$, with the correct phase relation. That is, we compute the Berry phase of band n as

$$\phi_n = -\text{Im} \ln \left[\langle u_{nk_0}|u_{nk_1}\rangle \langle u_{nk_1}|u_{nk_2}\rangle \cdots \langle u_{nk_{N-1}}|e^{-2\pi ix/a}|u_{nk_0}\rangle \right].\qquad(3.74)$$

---

[15] Similar issues arise for any other choice of BZ, such as from $-\pi/a$ to π/a.

It is important not to forget the phase factor in the last term when completing the loop – that is, when k wraps from $k = 2\pi/a$ back to $k = 0$.

A simple PYTHTB program that computes the Berry phase around the 1D BZ for the alternating site chain model of Fig. 2.6, for the same parameter set as the band structure calculated in Appendix D.4, is given in Appendix D.7. The first part of the program computes the eigenvectors evec[0], ..., evec[60] for a set of nk = 61 points (including both end points) covering the 1D BZ. Near the end of chain_alt_bp.py we find the Berry phase calculated as follows:

```
prod=1.+0.j
for i in range(1,nk-1):                    # <evec_0|evec_1>...<evec_58|evec_59>
   prod*=np.vdot(evec[i-1],evec[i])        # a*=b means a=a*b

# now compute the phase factors needed for last inner product
orb=np.array([0.0,0.5])               # relative coordinates of orbitals
phase=np.exp((-2.j)*np.pi*orb)        # construct phase factors
evec_last=phase*evec[0]              # evec[60] constructed from evec[0]
prod*=np.vdot(evec[-2],evec_last)    # include <evec_59|evec_last>

print "Berry phase is %7.3f"% (-np.angle(prod))
\index{Berry phase}%
```

This portion of code implements the calculation of Eq. (3.74), including the insertion of the phase factor in the last inner product. Recall from Eq. (3.73) that the computed evec[60] will *not be* the same as evec[0], so it would be incorrect to compute the last inner product as <evec[59]|evec[0]>. Instead, evec[60] will be equal, up to an overall phase, to evec_last, which is constructed from evec[0]. Since we cannot be sure of this phase, it is essential to compute the final inner product as <evec[59]|evec_last> and not as <evec[59]|evec[60]>. The printed output reports "Berry phase is 2.217" which is about 0.71π.

At the very end of chain_alt_bp.py in Appendix D.7, the Berry phase is computed again using advanced features of the PYTHTB package that are provided for this purpose. An array of eigensolutions is initialized on a grid of k-points and then filled by calls to wf_array and solve_on_grid, and the Berry phase is computed by the berry_phase function applied to the array. As you run the program you can confirm that the results are the same.

Should we be surprised that the Berry phase is nonzero? In Ex. 3.4 we found that the Berry phase for the cyclic evolution of the ground state of the triatomic molecule vanished unless TR symmetry was broken by the introduction of complex hoppings. However, we *did* get a nontrivial result for the Berry phase here, even though the Hamiltonian is real. This is because the effective H_k of Eq. (2.43), and its eigenstates $|u_{nk}\rangle$, are complex at arbitrary k.

It is interesting to investigate the dependence of the Berry phase on the parameters of the model. This topic is explored in Ex. 3.15, where you may discover

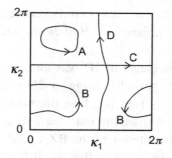

Figure 3.13 Sketch of four possible closed paths in the BZ of a 2D crystal. Paths A and B are trivially closed, while paths C and D wrap by reciprocal-lattice vectors $\mathbf{b}_1$ and $\mathbf{b}_2$, respectively.

that a cyclic evolution of the Hamiltonian can sometimes shift the Berry phase by $\pm 2\pi$. Such behavior will be discussed again in Section 4.2.3.

In two dimensions, there is more freedom in the choice of path for the Berry phase. Some examples are shown in Fig. 3.13. This figure is drawn in terms of the reduced wavevectors κ_j of Eq. (2.31), each of which runs between 0 and 2π. Paths A and B are simple closed paths. Remember that the left and right edges of the BZ are identified, as are the top and bottom, with the BZ regarded as a closed torus. From this point of view, there is no intrinsic difference between a path like A that lies entirely inside the square BZ shown in the figure, and one like B that traverses the artificial boundary at $\kappa_1 = 2\pi$.

In contrast, paths C and D wind around the BZ in a nontrivial way, in analogy to the 1D path described earlier. Path C returns to itself after κ_1 has been incremented by 2π. This corresponds to translating $\mathbf{k}$ by $\mathbf{b}_1$, so when crossing the artificial boundary at $\kappa_1 = 2\pi$ an extra phase factor

$$\langle u_{n\mathbf{k}_{N-1}} | e^{-i\mathbf{b}_1 \cdot \mathbf{r}} | u_{n\mathbf{k}_0} \rangle \tag{3.75}$$

has to be included, in analogy with Eq. (3.74). Similar considerations apply to path D, where a phase factor $e^{-i\mathbf{b}_2 \cdot \mathbf{r}}$ is needed instead.

In 2D, we can also define a Chern number C_n associated with each band n via

$$\oint_{\text{BZ}} \Omega_n \, dS = 2\pi C_n . \tag{3.76}$$

As in Eq. (3.71), the notation $\oint_{\text{BZ}}$ denotes an integral over the BZ regarded as a closed manifold, but now it is a surface integral over a 2D manifold.[16] If the

---

[16] In practice, the Chern number is most easily computed by discretizing the 2D BZ on an $N \times N$ mesh $\mathbf{k}_{i_1 i_2} = (i_1/N)\mathbf{b}_1 + (i_2/N)\mathbf{b}_2$. One then computes the eigenvectors on the $(N+1)^2$ mesh points as i_1 and i_2 each runs over $0, \ldots, N$; computes the Berry phase around each of the N^2 plaquettes; and sums

Chern index does not vanish, such a system is referred to as a *Chern insulator*, or equivalently, a *quantum anomalous Hall insulator*, as will be discussed in Section 5.1.

The concept of a periodic gauge in 1D was introduced on p. 46 and has been used several times since. A periodic gauge in 2D is defined as one for which $|\psi_{n,\mathbf{k}+\mathbf{b}_1}\rangle = |\psi_{n\mathbf{k}}\rangle$ and $|\psi_{n,\mathbf{k}+\mathbf{b}_2}\rangle = |\psi_{n\mathbf{k}}\rangle$ on the boundaries of the conventional BZ, so that it can be wrapped smoothly onto the 2-torus. A periodic gauge that is also smoothly defined on the interior of the BZ would, therefore, be smooth and continuous everywhere on the 2-torus. But recall from the discussion on p. 93 that it is *impossible* to construct a smooth and periodic gauge when the Chern index C_n is nonzero. Such a topological obstruction causes no problem for the calculation of particular Berry phases, such as those along paths A–D, or of the Chern number, which is an integral of a gauge-invariant quantity. But it will have other consequences, especially regarding the ability to construct Wannier functions, as we shall discuss in the next section.

For a 3D crystal, the BZ forms a 3-torus. Concerning Berry phases, we can consider simple loops that close without winding around the 3-torus, or ones that wind in the $\mathbf{b}_1$, $\mathbf{b}_2$, or $\mathbf{b}_3$ direction (e.g., like path D in Fig. 3.13 but in 3D). Concerning Chern indices, we can compute these on any closed 2D manifold lying in the 3D BZ. For example, the Fermi surface of a metal is a suitable closed surface on which a Chern index can be computed. But we can also consider 2D submanifolds that extend across the 3D BZ. For example, consider the Chern index $C_{n1}(\kappa_1)$ defined for band n on the κ_2–κ_3 "plane" lying at some fixed κ_1. This "plane" is really a 2-torus when we recall that the "edges" at $\kappa_2 = 0$ and 2π can be regarded as being seamlessly glued together, and similarly for κ_3. But if band n is isolated (i.e., it does not touch band $n-1$ or $n+1$ anywhere in the 3D BZ), then all properties of band n on this plane must evolve smoothly as κ_1 is varied. In particular, $C_{n1}(\kappa_1)$ must be a continuous function of κ_1. But it is also an integer-valued function, and a continuous integer-valued function must be completely constant. As a consequence, we are free to compute it at any κ_1 of our choice (say, $\kappa_1 = 0$) and to drop the κ_1 argument, writing it simply as C_{n1}. Similarly, defining C_{n2} and C_{n3} to be the Chern indices for κ_3–κ_1 and κ_1–κ_2 planes, respectively, we conclude that any fully isolated band n in a 3D crystal is characterized by a triplet of integer Chern indices (C_{n1}, C_{n2}, C_{n3}).

So far, all of this is very abstract. What, if any, physical interpretation can be attached to the geometric Berry-phase quantities described to this point? This is the subject that will concern us in subsequent chapters. Before we finally turn to

---

these values. The mesh chosen must be fine enough that each plaquette Berry phase is small compared to π. Following the argument given in Section 3.2.2, this sum must be an integer multiple of 2π, from which it is straightforward to extract the Chern index C_n.

this part of the story, however, it is useful to cover two more details of a somewhat mathematical nature. These are the construction of the Wannier representation and the multiband treatment, which are discussed in the two remaining sections of this chapter.

Exercises

3.11 Here we explore how the Berry phase of a band in a 1D insulator changes when the entire crystal is translated to the right by a distance x_0 (or equivalently, when the origin is shifted by $-x_0$). The new Bloch functions $|\tilde{\psi}_{nk}\rangle$ are related to the old ones $|\psi_{nk}\rangle$ through $\tilde{\psi}_{nk}(x) = \psi_{nk}(x - x_0)$. Find the corresponding relationship between $\tilde{u}_{nk}$ and u_{nk}, and use this to show that the Berry phase computed with this new choice of origin is $\tilde{\phi}_n = \phi_n + 2\pi x_0/a$.

3.12 We want to show that $\int_{BZ} \Omega \, d^2k = \int_{BZ} \overline{\Omega} \, d^2\kappa$, where the first integral is in the $\mathbf{k} = (k_x, k_y)$ space with $\Omega = \Omega_{xy}$, where $\Omega_{\mu\nu} = -2\text{Im} \langle \partial_\mu u_{nk} | \partial_\nu u_{nk} \rangle = \epsilon_{\mu\nu}\Omega$; the second is in the $\boldsymbol{\kappa} = (\kappa_1, \kappa_2)$ space with $\overline{\Omega} = \overline{\Omega}_{12}$, where $\overline{\Omega}_{ij} = -2\text{Im} \langle \partial_i u_{n\kappa} | \partial_j u_{n\kappa} \rangle = \epsilon_{ij}\overline{\Omega}$; and ϵ_{ij} is the 2×2 antisymmetric tensor. Assume the two frames are related by $k_\mu = B_{\mu j}\kappa_j$ (implied sum notation here and below).

(a) Show that $d^2k = \det(B) \, d^2\kappa$.

(b) Using the chain rule $\partial_j = B_{\mu j}\partial_\mu$, show that $\overline{\Omega}_{ij} = \epsilon_{ij}B_{\mu i}B_{\nu j}\Omega_{\mu\nu}$.

(c) Show that $\epsilon_{ij}B_{\mu i}B_{\nu j} = \epsilon_{\mu\nu} \det(B)$ and complete the proof.

3.13 Demonstrate the two claims made on p. 105:

(a) The presence of inversion symmetry implies $\Omega_n(\mathbf{k}) = \Omega_n(-\mathbf{k})$.

(b) The presence of time-reversal symmetry implies $\Omega_n(\mathbf{k}) = -\Omega_n(-\mathbf{k})$.

3.14 Let's see whether inversion symmetry restricts the allowed values of the Berry phase in a 1D insulator.

(a) Choose several parameter sets for the alternating site model of Fig. 2.6 such that the system is inversion-symmetric about a site, or about a midbond position, or neither. Explain your choice.

(b) Use one of the methods in `chain_alt_bp` of Appendix D.7 to compute the Berry phase for these cases. What do you observe?

(c) In one of these cases, an inversion center is present, but not at the origin. Try shifting the setting of the orbital locations so that inversion lies at the origin for this case. What is the Berry phase?

3.15 Reconsider the alternating site model discussed on p. 108, but now we want to follow the evolution of the Berry phase as we modify the parameters. Treat t as a constant and consider the 2D parameter space (Δ, δ). Consider *two closed paths* of your own choosing in this parameter space, one that encircles the origin $\Delta = \delta = 0$ and one that does not. Make a plot of ϕ versus progress along this path. (You may wish to follow the example at the end of `chain_alt_bp.py` of using the `wf_array` method, which has the added advantage that the call to the `wf_array.berry_phase` function avoids unwanted 2π jumps of ϕ during the evolution along the path). What is special about the origin in the two-parameter space? Do you observe an interesting cyclic change of ϕ for one of the cases?

3.5 Wannier Functions

3.5.1 Introduction

If we have an isolated band $E_n(\mathbf{k})$, meaning one that never touches the band below or above it, then we have a right to expect that $E_n(\mathbf{k})$ is a smooth and periodic function of $\mathbf{k}$ in 3D reciprocal space. It is then natural to consider its Fourier transform to real space, defined by

$$E_{n\mathbf{R}} = \frac{V_{\text{cell}}}{(2\pi)^3} \int_{\text{BZ}} e^{-i\mathbf{k}\cdot\mathbf{R}} E_{n\mathbf{k}} \, d^3k, \qquad (3.77\text{a})$$

$$\updownarrow \text{ FT}$$

$$E_{n\mathbf{k}} = \sum_{\mathbf{R}} e^{i\mathbf{k}\cdot\mathbf{R}} E_{n\mathbf{R}}. \qquad (3.77\text{b})$$

The second equation is the inverse transform; the consistency between this pair of equations is associated with the two orthogonality identities

$$\int_{\text{BZ}} e^{i\mathbf{k}\cdot(\mathbf{R}-\mathbf{R}')} \, d^3k = \frac{(2\pi)^3}{V_{\text{cell}}} \delta_{\mathbf{R},\mathbf{R}'}, \qquad (3.78)$$

which is relatively intuitive,[17] and

$$\sum_{\mathbf{R}} e^{i(\mathbf{k}-\mathbf{k}')\cdot\mathbf{R}} = \frac{(2\pi)^3}{V_{\text{cell}}} \delta^3(\mathbf{k} - \mathbf{k}'), \qquad (3.79)$$

which is less so. These and other Fourier transform conventions are summarized in Appendix A.[18] Insofar as $E_n(\mathbf{k})$ is smooth in $\mathbf{k}$-space, we can expect $E_{n\mathbf{R}}$ to be large only for a few lattice vectors $\mathbf{R}$ near the origin, and to decay rapidly with increasing $|\mathbf{R}|$.

---

[17] The phases cancel on the left unless $\mathbf{R} = \mathbf{R}'$, in which case the left side is the BZ volume.

[18] The wavevectors $\mathbf{k}$ on the right side of Eq. (3.79) should be interpreted as living on the 3-torus; that is, $\mathbf{k} - \mathbf{k}'$ can be replaced by $\mathbf{k} - \mathbf{k}' + \mathbf{G}$ for any reciprocal lattice vector $\mathbf{G}$. Also note that there is some freedom in the choice of prefactors in Eqs. (3.77a) and (3.77b), as discussed in Appendix A.3; our choice conveniently assigns the same energy units to $E_{n\mathbf{R}}$ as to $E_{n\mathbf{k}}$.

Now suppose we can choose a smooth and periodic gauge for the Bloch functions $|\psi_{n\mathbf{k}}\rangle$ associated with this band. Having done so, we should be able to Fourier transform these in a similar way, by defining

$$|w_{n\mathbf{R}}\rangle = \frac{V_{\text{cell}}}{(2\pi)^3} \int_{\text{BZ}} e^{-i\mathbf{k}\cdot\mathbf{R}} |\psi_{n\mathbf{k}}\rangle \, d^3k, \tag{3.80a}$$

$$\Updownarrow \text{ FT}$$

$$|\psi_{n\mathbf{k}}\rangle = \sum_{\mathbf{R}} e^{i\mathbf{k}\cdot\mathbf{R}} |w_{n\mathbf{R}}\rangle. \tag{3.80b}$$

The Fourier-transform partners to the Bloch functions defined in Eq. (3.80a) are known as the *Wannier functions* associated with band n; Eq. (3.80b) provides the inverse transform back from Wannier to Bloch functions. Again, the idea is that as long as $\psi_{n\mathbf{k}}(\mathbf{r})$ is a smooth function of $\mathbf{k}$, then $w_{n\mathbf{R}}(\mathbf{r})$ decays rapidly with $|\mathbf{R}|$ for a given $\mathbf{r}$. Actually, it turns out that each Wannier function $w_{n\mathbf{R}}(\mathbf{r})$ is a localized function centered near $\mathbf{R}$, so it is more natural to describe the situation by saying that $w_{n\mathbf{R}}(\mathbf{r})$ decays rapidly with $|\mathbf{r} - \mathbf{R}|$ for a given $\mathbf{R}$. Moreover, since the Fourier transform expressed by Eq. (3.80) is really just a special case of a unitary transformation, we can view the Bloch and Wannier functions as providing two different basis sets describing the same manifold of states associated with the electron band in question.

A word of warning is in order concerning the inner-product and normalization conventions to be used here. Recall that the true Bloch function $|\psi_{\mathbf{k}}\rangle$ is related to the cell-periodic function $|u_{\mathbf{k}}\rangle$ by $\psi_{\mathbf{k}}(\mathbf{r}) = e^{i\mathbf{k}\cdot\mathbf{r}}u_{\mathbf{k}}(\mathbf{r})$. Also let $|\chi_{\mathbf{k}}\rangle$ and $|v_{\mathbf{k}}\rangle$ be another pair related in the same way. For the cell-periodic functions we adopt the inner-product convention

$$\langle u_{\mathbf{k}}|v_{\mathbf{k}'}\rangle \equiv \int_{V_{\text{cell}}} u_{\mathbf{k}}^*(\mathbf{r}) v_{\mathbf{k}'}(\mathbf{r}) \, d^3r, \tag{3.81}$$

since the "natural domain" of such a function is a single unit cell. Moreover, we normalize $\psi_{n\mathbf{k}}(\mathbf{r})$ and $u_{n\mathbf{k}}(\mathbf{r})$ such that

$$\int_{V_{\text{cell}}} |\psi_{n\mathbf{k}}(\mathbf{r})|^2 = \int_{V_{\text{cell}}} |u_{n\mathbf{k}}(\mathbf{r})|^2 = \langle u_{n\mathbf{k}}|u_{n\mathbf{k}}\rangle = 1. \tag{3.82}$$

However, we use a very different inner-product convention for the Bloch functions themselves, namely

$$\langle \psi_{\mathbf{k}}|\chi_{\mathbf{k}'}\rangle \equiv \int \psi_{\mathbf{k}}^*(\mathbf{r}) \chi_{\mathbf{k}'}(\mathbf{r}) \, d^3r \tag{3.83}$$

where the integral is over *all space*. This is a natural choice for the Bloch functions, which describe physical electrons that are delocalized throughout the crystal. It follows that $\langle \psi_{nk} | \psi_{nk} \rangle$ is not unity, but is instead infinite, within our conventions. In the special case of $\mathbf{k} = \mathbf{k}'$, we may occasionally use the notation

$$\langle \psi_{\mathbf{k}} | \chi_{\mathbf{k}} \rangle_{\text{cell}} \equiv \int_{V_{\text{cell}}} \psi_{\mathbf{k}}^*(\mathbf{r}) \chi_{\mathbf{k}}(\mathbf{r}) \, d^3r = \langle u_{\mathbf{k}} | v_{\mathbf{k}} \rangle \tag{3.84}$$

to limit the integral to a single unit cell.

A useful formula, somewhat analogous to Eq. (3.79), is

$$\langle \psi_{\mathbf{k}} | \chi_{\mathbf{k}'} \rangle = \frac{(2\pi)^3}{V_{\text{cell}}} \langle u_{\mathbf{k}} | v_{\mathbf{k}'} \rangle \, \delta^3(\mathbf{k} - \mathbf{k}') \,. \tag{3.85}$$

This formula is derived in Appendix A as Eq. (A.13). From this, it follows that the Wannier functions obey the orthonormality condition $\langle w_{n\mathbf{R}} | w_{n'\mathbf{R}'} \rangle = \delta_{n,n'} \, \delta_{\mathbf{R},\mathbf{R}'}$, as will be shown in Ex. 3.16.

3.5.2 Properties of the Wannier Functions

In standard solid state physics texts, it is demonstrated that the Wannier functions have the following interesting and useful properties:

1. As hinted earlier, they are *localized* functions in real space. That is,

    $$|w_{n\mathbf{R}}(\mathbf{r})| \to 0 \quad \text{as} \quad |\mathbf{r} - \mathbf{R}| \quad \text{gets large.} \tag{3.86}$$

    We can think of $|w_{n\mathbf{R}}\rangle$ as being peaked in cell $\mathbf{R}$, even if its tails extend into neighboring unit cells.

2. The Wannier functions are translational images of one another:

    $$w_{n\mathbf{R}}(\mathbf{r}) = w_{n0}(\mathbf{r} - \mathbf{R}) \,. \tag{3.87}$$

    More formally, $|n\mathbf{R}\rangle = T_{\mathbf{R}} |n0\rangle$ where, as in Section 2.1.3, $T_{\mathbf{R}}$ is the operator that translates the system by lattice vector $\mathbf{R}$.

3. The Wannier functions form an orthonormal set:

    $$\langle w_{n\mathbf{R}} | w_{n\mathbf{R}'} \rangle = \delta_{\mathbf{R}\mathbf{R}'} \,. \tag{3.88}$$

4. The Wannier functions span the same subspace of the Hilbert space as is spanned by the Bloch functions from which they are constructed. Defining $\mathcal{P}_n$ to be the projection operator onto band n, this property can be expressed as

    $$\mathcal{P}_n = \frac{V_{\text{cell}}}{(2\pi)^3} \int_{\text{BZ}} |\psi_{n\mathbf{k}}\rangle\langle\psi_{n\mathbf{k}}| \, d^3k = \sum_{\mathbf{R}} |w_{n\mathbf{R}}\rangle\langle w_{n\mathbf{R}}| \,. \tag{3.89}$$

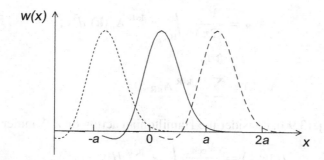

Figure 3.14 Sketch of three adjacent Wannier functions $w_{nR}(x)$ for band n in a 1D crystal of lattice constant a. The Wannier center assigned to the home unit cell $R=0$ is shown as the full curve; dotted and dashed curves represent those in cells at $-a$ and a, respectively. The Wannier functions are mutually orthonormal at the same time that they are translational images of one another.

From this, it also follows that the total charge density $\rho_n(\mathbf{r})$ in band n,

$$\rho_n(\mathbf{r}) = -e \, \langle \mathbf{r}|\mathcal{P}_n|\mathbf{r}\rangle = -e \, \frac{V_{\text{cell}}}{(2\pi)^3} \int_{\text{BZ}} |\psi_{n\mathbf{k}}(\mathbf{r})|^2 \, d^3k = -e \sum_{\mathbf{R}} |w_{n\mathbf{R}}(\mathbf{r})|^2 \,,$$

is the same when computed in either representation.

The first three of these properties are illustrated in Fig. 3.14, where a possible set of Wannier functions are sketched for a 1D crystal. Each one is exponentially localized and normalized, and the neighboring Wannier functions are periodic images of one another. Moreover, the Wannier functions are shown as having a negative lobe so that $\langle w_{n0}|w_{na}\rangle$ can plausibly vanish as a result of cancellation between contributions of opposite sign in the integral over x.

It is also possible to prove two remarkable properties about matrix elements of operators between Wannier functions:

5. The Hamiltonian matrix elements between Wannier functions are band-diagonal, and these diagonal elements are nothing other than the coefficients $E_{n\mathbf{R}}$ in the Fourier expansion of the band energy:

$$\langle w_{n0}|H|w_{n\mathbf{R}}\rangle = E_{n\mathbf{R}} \,. \tag{3.90}$$

6. The position matrix elements between Wannier functions are

$$\langle w_{n0}|\mathbf{r}|w_{n\mathbf{R}}\rangle = \mathbf{A}_{n\mathbf{R}} \,. \tag{3.91}$$

In this last equation, the $\mathbf{A}_{n\mathbf{R}}$ are the Fourier transform coefficients of the Berry connection $\mathbf{A}_n(\mathbf{k})$ defined in Eq. (3.66). That is, they are related in complete analogy with Eq. (3.77) or (3.80) by

$$\mathbf{A}_{n\mathbf{R}} = \frac{V_{\text{cell}}}{(2\pi)^3} \int_{\text{BZ}} e^{-i\mathbf{k}\cdot\mathbf{R}} \, \mathbf{A}_n(\mathbf{k}) \, d^3k, \tag{3.92a}$$

$$\Updownarrow \text{ FT}$$

$$\mathbf{A}_n(\mathbf{k}) = \sum_{\mathbf{R}} e^{i\mathbf{k}\cdot\mathbf{R}} \, \mathbf{A}_{n\mathbf{R}}. \tag{3.92b}$$

To derive Eq. (3.90), consider the Hamiltonian acting on a Wannier function,

$$H \, |w_{n\mathbf{R}}\rangle = \frac{V_{\text{cell}}}{(2\pi)^3} \int_{\text{BZ}} e^{-i\mathbf{k}\cdot\mathbf{R}} \, H|\psi_{n\mathbf{k}}\rangle \, d^3k, \tag{3.93}$$

and then multiply on the left by $\langle w_{n\mathbf{0}}|$. Since $H|\psi_{n\mathbf{k}}\rangle$ has the form of a Bloch function of wavevector $\mathbf{k}$,[19] we can use Eq. (3.85) to obtain

$$\langle w_{n\mathbf{0}}|H|w_{n\mathbf{R}}\rangle = \frac{V_{\text{cell}}}{(2\pi)^3} \int_{\text{BZ}} e^{-i\mathbf{k}\cdot\mathbf{R}} \, \langle u_{n\mathbf{k}}|H_{\mathbf{k}}|u_{n\mathbf{k}}\rangle \, d^3k, \tag{3.94}$$

which reduces to Eq. (3.90) using $E_{n\mathbf{k}} = \langle u_{n\mathbf{k}}|H_{\mathbf{k}}|u_{n\mathbf{k}}\rangle$ with Eq. (3.77). We shall defer the derivation of Eq. (3.91) for the moment, returning to it on p. 117.

Equation (3.90) is a remarkable result, as it implies that the Wannier functions provide an *exact* tight-binding representation of the dispersion $E_{n\mathbf{k}}$ of band n. That is, we construct a tight-binding model consisting of one orbital per cell having site energy $E_{n\mathbf{0}}$ and hoppings $E_{n\mathbf{R}}$ to its neighbors located at relative cell $\mathbf{R}$. In this context, Eq. (2.75) is exactly Eq. (3.77b), so that this Wannier-based tight-binding model reproduces the exact band dispersion. Since the Wannier functions are localized, the hopping matrix elements fall off quickly with distance, so that only a few hoppings typically have to be retained. This method of reproducing high-accuracy band structures from a Wannierized tight-binding representations is often referred to as *Wannier interpolation*.[20]

Equation (3.91) is in some ways even more remarkable. It says that matrix elements of the position operator between Wannier functions have a form that is highly reminiscent of the Berry-phase formalism. An especially important quantity is the center of charge of a Wannier function,[21] defined as the diagonal position-operator matrix element

$$\bar{\mathbf{r}}_n = \langle w_{n\mathbf{0}}|\mathbf{r}|w_{n\mathbf{0}}\rangle. \tag{3.95}$$

---

[19] Actually it is just $E_{n\mathbf{k}}|\psi_{n\mathbf{k}}\rangle$ but we keep it in this form for the moment to make an analogy later with Eq. (3.100).

[20] The Wannier interpolation approach can also be generalized to compute group velocities, effective masses, Berry connections and curvatures, and other quantities of interest. See, for example, Yates et al. (2007).

[21] This is analogous to the center of mass of an object, but defined based on charge rather than mass.

From Eq. (3.91) this is $\mathbf{A}_{n0}$, which Eq. (3.92a) tells us is just the BZ average of the Berry connection $\mathbf{A}_n(\mathbf{k})$:

$$\bar{\mathbf{r}}_n = \frac{V_{\text{cell}}}{(2\pi)^3} \int_{\text{BZ}} \mathbf{A}_n(\mathbf{k})\, d^3k$$

$$= \frac{V_{\text{cell}}}{(2\pi)^3} \int_{\text{BZ}} \langle u_{n\mathbf{k}} | i \nabla_{\mathbf{k}} u_{n\mathbf{k}} \rangle\, d^3k. \tag{3.96}$$

In 1D this becomes $\bar{x}_n = (a/2\pi) \int_0^{2\pi/a} \langle u_{nk} | i\partial_k u_{nk} \rangle\, dk$ or

$$\bar{x}_n = a \frac{\phi_n}{2\pi}. \tag{3.97}$$

Thus the location of the Wannier center is just proportional, by a scale factor $a/2\pi$, to the Berry phase! In other words, a Berry phase evolving from 0 to 2π would just correspond to a Wannier center evolving from $x=0$ to $x=a$.

This connection between Wannier centers and Berry phases is of fundamental importance, and lies at the heart of the developments to be presented in Chapter 4. It allows us to understand some of the behaviors that were observed for 1D insulators in the exercises at the end of the previous section. In particular, a translation of the crystal by x_0 was found in Ex. 3.11 to result in a change of the Berry phase by $2\pi x_0/a$, which corresponds to a shift of $\bar{x}_n$ by x_0 as expected. If the origin lies at a center of symmetry, then we anticipate that $\bar{x}_n$ should also lie at a center of symmetry ($\bar{x}_n = 0$ or $a/2$) with a corresponding constraint on the Berry phase ($\phi_n = 0$ or π), just as in Ex. 3.14. If the Hamiltonian of a 1D insulator is adiabatically carried around a cyclic loop in parameter space as in Ex. 3.15, ϕ_n may sometimes evolve by 2π, corresponding to a shift of $\bar{x}$ by a lattice constant. In such a case, the lattice of Wannier centers ($\bar{x}_n + la$ for integer l) returns to itself, but with a lattice-constant shift. Figure 1.8 from Chapter 1 can be taken to illustrate just such a case of "adiabatic charge transport" if the negative point charges in the figure are identified with the Wannier centers. As we shall see in Chapter 4, the possibility of such a shift is central to the theory of electric polarization in insulators.

We now return to the derivation of Eq. (3.96), focusing first on the 1D version in Eq. (3.97). The action of the position operator on a Wannier function can be obtained from

$$(x - R)|w_{nR}\rangle = \frac{a}{2\pi} \int_0^{2\pi/a} (x - R)\, e^{ik(x-R)}\, |u_{nk}\rangle\, dk$$

$$= \frac{a}{2\pi} \int_0^{2\pi/a} \left(-i\partial_k e^{ik(x-R)} \right) |u_{nk}\rangle\, dk$$

$$= \frac{a}{2\pi} \int_0^{2\pi/a} e^{ik(x-R)} \left(i\partial_k |u_{nk}\rangle \right)\, dk. \tag{3.98}$$

We have applied an integration by parts in taking the last step in Eq. (3.98), making use of the fact that $e^{ik(x-R)} u_{nk}(x) = \psi_{nk}(x - R)$ has the same value at $k = 0$ and $2\pi/a$. Generalizing this result to 3D and moving $\mathbf{R}$ to the other side, we find that the analogy to Eq. (3.93) is

$$\mathbf{r}\,|w_{n\mathbf{R}}\rangle = \frac{V_{\text{cell}}}{(2\pi)^3} \int_{\text{BZ}} e^{-i\mathbf{k}\cdot\mathbf{R}} \left[e^{i\mathbf{k}\cdot\mathbf{r}}(\mathbf{R} + i\nabla_{\mathbf{k}})\,|u_{nk}\rangle \right] d^3k \,. \tag{3.99}$$

The object in brackets on the right-hand side has the form of a Bloch function of wavevector $\mathbf{k}$, so we can again multiply on the left by $\langle w_{n0}|$ and use Eq. (3.85) to obtain

$$\langle w_{n0}|\mathbf{r}|w_{n\mathbf{R}}\rangle = \frac{V_{\text{cell}}}{(2\pi)^3} \int_{\text{BZ}} e^{-i\mathbf{k}\cdot\mathbf{R}} \langle u_{nk}|\mathbf{R} + i\nabla_{\mathbf{k}}|u_{nk}\rangle \, d^3k \,. \tag{3.100}$$

The term involving $\mathbf{R}$ yields $\mathbf{R}\,\delta_{0,\mathbf{R}} = 0$ and can be discarded, yielding the important result

$$\langle w_{n0}|\mathbf{r}|w_{n\mathbf{R}}\rangle = \frac{V_{\text{cell}}}{(2\pi)^3} \int_{\text{BZ}} e^{-i\mathbf{k}\cdot\mathbf{R}} \langle u_{nk}|i\nabla_{\mathbf{k}}u_{nk}\rangle \, d^3k \,. \tag{3.101}$$

We recognize the right-hand side as $\mathbf{A}_{n\mathbf{R}}$ from Eq. (3.92a), giving Eq. (3.91) as claimed.

To summarize, we have arrived at a view in which an electron band in a solid is represented by a lattice of Wannier functions, each a periodic replica of a representative one in the home unit cell, as sketched for a 2D crystal in Fig. 3.15. The position $\bar{\mathbf{r}}_n$ of the home-cell Wannier center, shown as the lower-left-hand cross in the figure, is given by Eq. (3.95) or, equivalently, Eq. (3.96). Three periodic replicas and their corresponding Wannier centers are also shown. The Hamiltonian matrix elements between Wannier functions, defined in Eq. (3.90) and obtainable

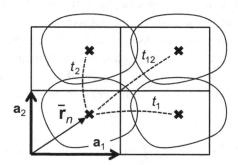

Figure 3.15 Sketch of four of the infinite lattices of Wannier functions (irregular blobs) for the single band n in 2D with lattice vectors $\mathbf{a}_1$ and $\mathbf{a}_2$. Crosses mark the Wannier centers located at $\bar{\mathbf{r}}_n + \mathbf{R}$; dashed lines indicate the nearest-neighbor hoppings $t_1 = \langle w_{n0}|H|w_{n\mathbf{a}_1}\rangle$, $t_2 = \langle w_{n0}|H|w_{n\mathbf{a}_2}\rangle$ and $t_{12} = \langle w_{n0}|H|w_{n,\mathbf{a}_1+\mathbf{a}_2}\rangle$ of the corresponding tight-binding model.

via Eq. (3.94), are indicated as the dashed hoppings in the figure. In the notation of Eq. (3.90), the hoppings t_1, t_2, and t_{12} correspond to $E_{n,\mathbf{a}_1}$, $E_{n,\mathbf{a}_2}$, and $E_{n,\mathbf{a}_1+\mathbf{a}_2}$, respectively, while E_{n0} (not shown) is the on-site energy of the Wannier function. If these are the dominant hoppings, this simple tight-binding model may be expected to provide a good approximation to the full band structure, but in any case the model is guaranteed to become more and more accurate as farther-neighbor hoppings are added.

3.5.3 Gauge Freedom

We emphasized earlier that the phases of the Bloch functions are not unique; they can be twisted by a gauge change $e^{-i\beta(\mathbf{k})}$ as in Eq. (3.69). What effect does this have on Eqs. (3.90) and (3.91)? Because Eq. (3.94) is diagonal in $\mathbf{k}$, the phase factors $e^{i\beta(\mathbf{k})}$ cancel out, and the matrix elements $E_{n\mathbf{R}}$ are *exactly unchanged* by the gauge transformation. On one level, this appears surprising, since the Wannier functions themselves may change shape, or become somewhat more or less localized, as a result of the gauge change. On another level, we showed that the $E_{n\mathbf{R}}$ are really just the Fourier-transform coefficients of the band structure $E_{n\mathbf{k}}$; these are unique, and so must be gauge-invariant.

Before discussing the effect on the real-space matrix elements in Eq. (3.91), it is worthwhile to review the distinction between *progressive* and *radical* gauge transformations that was introduced in Section 3.1.2. In 1D, the idea is that each gauge transformation is characterized by a winding number m, which is the integer appearing in Eq. (3.19); the transformation is "progressive" if $m = 0$ and "radical" otherwise. In 3D $\mathbf{k}$-space, the gauge twist $e^{i\beta(\mathbf{k})}$ is characterized by a triplet of integers that we denote by (n_1, n_2, n_3), such that $\beta(\mathbf{k} + \mathbf{b}_j) = \beta(\mathbf{k}) + 2\pi n_j$, where $\mathbf{b}_j$ is a primitive reciprocal lattice vector as defined in Eq. (3.29). Letting $\mathbf{R} = n_1\mathbf{a}_1 + n_2\mathbf{a}_2 + n_3\mathbf{a}_3$ be the real-space lattice vector defined by this triplet of integers, we find that each gauge transformation is characterized by a lattice vector $\mathbf{R}$ such that

$$\beta(\mathbf{k} + \mathbf{G}) = \beta(\mathbf{k}) + \mathbf{G} \cdot \mathbf{R}. \tag{3.102}$$

This condition is necessary and sufficient to ensure that $e^{-i\beta(\mathbf{k})}$ is invariant under a translation by $\mathbf{G}$; that is, the transformation preserves the periodic gauge property. A 3D gauge transformation is said to be "progressive" if $\mathbf{R} = 0$ and "radical" otherwise.

From the preceding discussion, it is clear that any 3D gauge transformation can be decomposed into two steps: a radical part $\beta(\mathbf{k}) = \mathbf{k} \cdot \mathbf{R}$ for some $\mathbf{R}$, which we denote as a "shift," and a progressive part that obeys $\beta(\mathbf{k} + \mathbf{G}) = \beta(\mathbf{k})$. The effect of the shift is just to transform $\mathbf{A}_{n\mathbf{k}}$ via Eq. (3.70) into $\tilde{\mathbf{A}}_{n\mathbf{k}} = \mathbf{A}_{n\mathbf{k}} + \mathbf{R}$ so that $\tilde{\bar{\mathbf{r}}}_n = \bar{\mathbf{r}}_n + \mathbf{R}$; in other words, the Wannier center simply shifts by a lattice vector. In fact, what has happened is that the Wannier functions themselves have all shifted by a lattice vector, $|\tilde{w}_{n\mathbf{R}'}\rangle = |w_{n,\mathbf{R}'+\mathbf{R}}\rangle$. This is really just a relabeling of the original Wannier functions by shifting all of them by the same lattice vector $\mathbf{R}$ or, equivalently, a change in the choice of the Wannier function assigned to the home unit cell $\mathbf{R}=0$.

The effect of the progressive part can easily be worked out via Eq. (3.70), and we find that

$$\langle \tilde{w}_{n0}|\mathbf{r}|\tilde{w}_{n\mathbf{R}}\rangle = \langle w_{n0}|\mathbf{r}|w_{n\mathbf{R}}\rangle + \frac{V_{\text{cell}}}{(2\pi)^3} \int_{\text{BZ}} e^{-i\mathbf{k}\cdot\mathbf{R}} \nabla_{\mathbf{k}}\beta(\mathbf{k}) \, d^3k$$

$$= \langle w_{n0}|\mathbf{r}|w_{n\mathbf{R}}\rangle + i\mathbf{R}\frac{V_{\text{cell}}}{(2\pi)^3} \int_{\text{BZ}} e^{-i\mathbf{k}\cdot\mathbf{R}} \beta(\mathbf{k}) \, d^3k \qquad (3.103)$$

where an integration by parts has been applied.[22] For the diagonal element ($\mathbf{R}=0$), the second term vanishes and we obtain the important result

$$\tilde{\bar{\mathbf{r}}}_n = \bar{\mathbf{r}}_n. \qquad (3.104)$$

This implies that the *centers of charge of the Wannier functions are gauge-invariant* in the progressive case. More generally, $\tilde{\bar{\mathbf{r}}}_n = \bar{\mathbf{r}}_n + \mathbf{R}$, where $\mathbf{R}$ characterizes the shift part of the transformation, but it remains true that the *lattice of Wannier centers* remains invariant. In the 1D case, this is nothing other than the gauge invariance of the Berry phase (modulo 2π) in Eq. (3.97).

While all choices of gauge lead to the same energy-band coefficients $E_{n\mathbf{R}}$ and the same Wannier centers $\bar{\mathbf{r}}_n$, it is not true that all Wannier functions are created equal. For practical calculations, it is often preferable to find a set of *maximally localized Wannier functions* (MLWFs) that have as little spread as possible. The construction of such functions for a single band in 1D is the subject of Ex. 3.17, and we shall return to the general case of multiple bands in 3D in Section 3.6.1.

To summarize what we have learned so far, the band subspace associated with band n can be thought of equally well as spanned by the Bloch states $|\psi_{n\mathbf{k}}\rangle$ for $\mathbf{k}$ running over the BZ, or by a periodic lattice of localized Wannier functions $|w_{n\mathbf{R}}\rangle$ constructed from the $|\psi_{n\mathbf{k}}\rangle$ via a Fourier transform. Because of the gauge freedom in the phases of the $|\psi_{n\mathbf{k}}\rangle$, the Wannier functions are not unique, but the location of the Wannier center $\bar{\mathbf{r}}_n$ in the home unit cell is unique up to a lattice vector. This gauge-invariance property provides a strong hint that the Wannier charge centers may be related to some physically measurable property, and we shall see in Chapter 4 that this is indeed the case.

---

[22] The last term in Eq. (3.103) is just $i\mathbf{R}\,\beta(\mathbf{R})$, where $\beta(\mathbf{R})$ is the Fourier transform of $\beta(\mathbf{k})$.

Exercises

3.16 Using Eq. (3.85), prove the orthonormality condition $\langle w_{n\mathbf{R}}|w_{n'\mathbf{R}'}\rangle = \delta_{n,n'}\,\delta_{\mathbf{R},\mathbf{R}'}$ for the Wannier functions. This implies that each Wannier function is automatically orthogonal to all of its periodic images for the same band, as well as to any Wannier function belonging to a different band, regardless of the choice of gauge.

3.17 Here we explore the construction of maximally localized Wannier functions (MLWFs) for a single band in 1D.

(a) Show that the spatial second moment of a Wannier function for band n is given by

$$\langle w_n|x^2|w_n\rangle = \frac{a}{2\pi}\int_0^{2\pi/a}\langle\partial_k u_{nk}|\partial_k u_{nk}\rangle\,dk\,.$$

Here $|w_n\rangle = |w_{n0}\rangle$ is the Wannier function in the home unit cell.

(b) Using $1 = \mathcal{P}_{nk} + \mathcal{Q}_{nk}$, where $\mathcal{P}_{nk} = |u_{nk}\rangle\langle u_{nk}|$, show that

$$\langle w_n|x^2|w_n\rangle = \widetilde{\Omega}_n + \frac{a}{2\pi}\int_0^{2\pi/a} A_{nk}^2\,dk$$

where

$$\widetilde{\Omega}_n = \frac{a}{2\pi}\int_0^{2\pi/a}\langle\partial_k u_{nk}|\mathcal{Q}_{nk}|\partial_k u_{nk}\rangle\,dk\,.$$

(c) The quadratic spatial spread of the Wannier function, or mean square variation relative to its center, is

$$\Omega_n = \langle w_n|(x - \bar{x}_n)^2|w_n\rangle = \langle w_n|x^2|w_n\rangle - \bar{x}_n^2\,.$$

Show that this is also given by

$$\Omega_n = \widetilde{\Omega}_n + \frac{a}{2\pi}\int_0^{2\pi/a}(A_{nk} - \bar{A}_n)^2\,dk$$

where we have used $\bar{A}_n$, the BZ average of A_{nk}, as a synonym for $\bar{x}_n$.

(d) Show that $\widetilde{\Omega}_n$ is gauge-invariant.

(e) Now consider the change of Ω_n under a progressive gauge change (i.e., one for which $\beta_{\lambda=1} = \beta_{\lambda=0}$). Since $\bar{A}_n$ is also invariant in this context, conclude that the minimizing gauge is one that makes $A_{nk} = \bar{A}_n$ independent of k. (This is just the "twisted parallel-transport gauge" discussed on p. 82.)

Note: The generalization of these results to 2D and 3D is nontrivial, because the Berry connection cannot be flattened to be constant in all Cartesian directions simultaneously. See pp. 128 and 135 for further discussion.

3.18 Consider a translation of the entire crystal by an arbitrary vector $\mathbf{r}_0$ (or equivalently, the result of moving the origin to $-\mathbf{r}_0$), such that $\tilde{\psi}_{n\mathbf{k}}(\mathbf{r}) = \psi_{n\mathbf{k}}(\mathbf{r} - \mathbf{r}_0)$. Show that the Wannier centers move to $\tilde{\bar{\mathbf{r}}}_n = \bar{\mathbf{r}}_n + \mathbf{r}_0$. (See also Ex. 3.11.)

3.19 An instructive model is the 1D s-p chain, defined as follows. The unit cell consists of one s and one p_x orbital, both lying at the origin. The parameters are the site energies E_s and E_p, and hoppings V_{ss}, V_{sp}, and V_{pp}, as defined in the sketch.

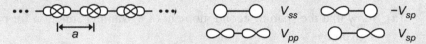

Write a PYTHTB program to solve this model as a function of two input parameters: ΔE, which controls the energy difference between E_s and E_p, and t, which scales the strengths of the hoppings according to $V_{ss} = -1.40t$, $V_{pp} = 3.24t$, and $V_{sp} = 1.84t$ [as suggested in the table at the back of Harrison (1989)]. You can use one or more of the sample programs in Appendix D as a model.

(a) Make plots of the band structure for fixed $\Delta E = 8\,\text{eV}$ with t taking on several values between 0.5 and 1.2 eV. What do you observe? Is there a critical value of t at which something special happens?

(b) Compute the Berry phase, and thus the position of the Wannier center, for the lower band, for several values of t both above and below the critical value. What do you find?

(c) Discuss. By considering the limit that t is very strong or very weak, can you describe the qualitative nature of the Wannier functions in the two cases? Is there any difference in the symmetry of the Hamiltonian above and below the critical value of t? How about in the symmetry of the Wannier function?

3.6 Multiband Formulation

To carry out the Wannier construction for band n as described in the preceding section, we had to assume that $|\psi_{n\mathbf{k}}\rangle$ was smooth and periodic over the entire BZ. This is usually not hard to arrange for an *isolated band*, meaning one that remains separated by a finite energy gap from the next-lowest and next-highest bands everywhere in the BZ. Unfortunately, that situation is actually rather uncommon. In most crystals, the occupied valence bands become degenerate at some high-symmetry points, and as a result the Bloch functions, defined as energy eigenstates, often have a singularity as a function of $\mathbf{k}$ in the vicinity of the degeneracy. This presents a problem not just for the formulation and construction of Wannier

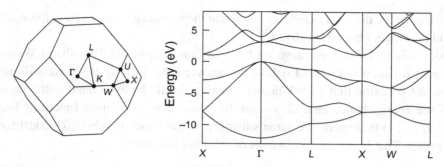

Figure 3.16 Band structure of silicon plotted along a path connecting several high-symmetry points in the fcc Brillouin zone shown at left. The four bands at negative energies are the occupied valence bands. Note degeneracies among bands at symmetry points and along some symmetry lines.

functions, but also for other quantities to be discussed later, such as the electric polarization.

Fortunately, there is an elegant solution to these problems. One can consider a group of bands that are glued together by degeneracies in this way as constituting a *composite group*, and develop methods for treating this group as a whole. This is the subject of the remainder of this chapter.

3.6.1 Multiband Wannier Functions

Consider, for example, the band structure of Si, shown in Fig. 3.16. As is conventional, the band structure is plotted along a series of segments connecting some chosen high-symmetry $\mathbf{k}$-points in the 3D BZ, which for the fcc lattice has the geometry shown at left in the figure.[23] There are four occupied valence bands covering the region from about $-15\,\mathrm{eV}$ to the valence-band maximum at $0\,\mathrm{eV}$. Bands 1 and 2 are degenerate at the zone-boundary point X, while bands 2–4 are degenerate at the BZ center Γ. If we tried to construct a Wannier function for band $n=2$ using Eq. (3.80a), we would encounter a serious problem in that $|\psi_{2\mathbf{k}}\rangle$ has a singularity for $\mathbf{k}$ at Γ. This occurs because $\lim_{\mathbf{k}\to 0}|\psi_{2\mathbf{k}}\rangle$ depends on the direction along which the limit is taken; the states turn out to have Si $3p_x$, $3p_y$, or $3p_z$ character when the approach is along the (100), (010), or (001) axis, respectively. A similar nonanalytic behavior occurs at the X point because of the crossing with the lowest band. In the context of a Fourier transform, a sharp structure in $\mathbf{k}$-space translates

---

[23] Si adopts the diamond structure in which the lattice is fcc, spanned by $\mathbf{a}_1=(0,a/2,a/2)$, $\mathbf{a}_2=(a/2,0,a/2)$, and $\mathbf{a}_3=(a/2,a/2,0)$, with two atoms per unit cell. Then in reciprocal space the origin is Γ; there are three X points at $2\pi/a(1,0,0)$, $2\pi/a(0,1,0)$, and $2\pi/a(0,0,1)$; there are four L points including at $\pi/a(1,1,1)$; and so on.

into a loss of localization in **r**-space, such that the resulting Wannier function $w_{n\mathbf{R}}(\mathbf{r})$ would no longer be well localized.[24]

Does this mean that we have no hope of constructing well-localized Wannier functions for such a system? Luckily, the answer is no, as long as we are willing to abandon the notion that each Wannier function should be associated with one and only one energy band. Instead, we ask for a set of four Wannier functions $|w_{n\mathbf{R}}\rangle$ ($n = 1, \ldots, 4$) that spans the same subspace as the Bloch bands $|\psi_{n\mathbf{k}}\rangle$ considered as a group. That is, we construct a set of Wannier functions

$$|w_{n\mathbf{R}}\rangle = \frac{V_{\text{cell}}}{(2\pi)^3} \int_{\text{BZ}} e^{-i\mathbf{k}\cdot\mathbf{R}} \, |\tilde{\psi}_{n\mathbf{k}}\rangle \, d^3k \qquad (3.105)$$

out of a set of Bloch-like functions $|\tilde{\psi}_{n\mathbf{k}}\rangle$ that are smooth functions of **k** everywhere in the BZ, and that are related to the true (energy-eigenstate) Bloch functions via a unitary transformation of the form

$$|\tilde{\psi}_{n\mathbf{k}}\rangle = \sum_{m=1}^{4} U_{mn}(\mathbf{k}) \, |\psi_{m\mathbf{k}}\rangle \, . \qquad (3.106)$$

Here $U_{mn}(\mathbf{k})$ is a manifold of 4×4 unitary matrices whose **k**-dependence near Γ and X has to be chosen in such a way as to "iron out" the nonanalytic behavior of $|\psi_{n\mathbf{k}}\rangle$, so that $|\tilde{\psi}_{n\mathbf{k}}\rangle$ is smooth everywhere. If this can be done, then the Wannier functions resulting from Eqs. (3.105–3.106) should be legitimate exponentially localized Wannier functions similar to those generated from single isolated bands.

In general, we define an *isolated group of J bands* to be a set of J consecutive energy bands that do not become degenerate with any lower or higher band anywhere in the BZ. In insulators, this is normally taken to coincide with the set of occupied valence bands, but other choices are possible. Noting that the $|\psi_{n\mathbf{k}}\rangle$ and $|u_{n\mathbf{k}}\rangle$ transform in the same way, we can rewrite Eq. (3.106) as

$$|\tilde{u}_{n\mathbf{k}}\rangle = \sum_{m=1}^{J} U_{mn}(\mathbf{k}) \, |u_{m\mathbf{k}}\rangle \, , \qquad (3.107)$$

with the same unitary rotation matrices U_{mn} appearing in both equations. We refer to Eqs. (3.106) and (3.107) as *multiband* or *non-Abelian*[25] gauge transformations. Equation (3.107) is the natural generalization of Eq. (3.69) to the multiband case, since the special case of diagonal matrices $U_{mn}(\mathbf{k}) = \delta_{mn} \, e^{-i\beta_n(\mathbf{k})}$ corresponds to the application of a simple phase twist $\beta_n(\mathbf{k})$ to each band individually.

---

[24] Typically, $w_{n\mathbf{R}}(\mathbf{r})$ acquires power-law tails, such that matrix elements of operators such as H and **r** are no longer well defined.

[25] The name refers to the fact that $J \times J$ matrices generally do not commute, and to the connection with non-Abelian gauge theories.

Note that the $|\tilde{\psi}_{n\mathbf{k}}\rangle$ are no longer eigenstates of H, since we have mixed different energy bands to construct them. Nevertheless, because the transformations in Eqs. (3.105) and (3.106) are both unitary, we can equally well use the $|\psi_{n\mathbf{k}}\rangle$, the $|\tilde{\psi}_{n\mathbf{k}}\rangle$, or the Wannier functions $|w_{n\mathbf{R}}\rangle$ constructed from the latter, as a representation of the occupied band subspace. Thus, total charge densities constructed from any one of these sets of functions must be identical, as are the expectation values of other operators when expressed as traces over the band subspace.

Returning to the question of smoothness of the $|\tilde{\psi}_{n\mathbf{k}}\rangle$, it is far from obvious whether it is always possible to apply an ironing-out procedure that will make these functions globally smooth and periodic, as they must be if we are going to construct well-localized Wannier functions from them. Clearly, the manifold $U_{mn}(\mathbf{k})$ must itself have nonanalytic behaviors that cancel out those of the underlying $|\psi_{m\mathbf{k}}\rangle$.

It is not hard to show that such an ironing-out procedure can be accomplished locally in any small region within the BZ, as follows. Choose some point $\mathbf{k}_0$ in the region of interest, and choose a set $|\tilde{u}_{n\mathbf{k}_0}\rangle$ of orthonormal functions spanning the band subspace of $H_{\mathbf{k}_0}$; these could be the eigenstates $|u_{n\mathbf{k}_0}\rangle$ themselves, or some unitary rotation of them. Then, for each nearby point $\mathbf{k}$, carry out a unitary rotation on the $|u_{n\mathbf{k}}\rangle$ such that the new set is optimally aligned with the $|\tilde{u}_{n\mathbf{k}_0}\rangle$. The meaning of "optimal alignment" and the means to accomplish it are described in Appendix B, but the basic idea is as follows. In the single-band case, optimal alignment (phase alignment) means that the inner product $\langle \tilde{u}_{n\mathbf{k}_0} | \tilde{u}_{n\mathbf{k}} \rangle$ should be *real* and *positive*. In the multiband case, the corresponding criterion is that the $J \times J$ matrix of inner products $\tilde{M}_{mn}^{(\mathbf{k}_0,\mathbf{k})} = \langle \tilde{u}_{m\mathbf{k}_0} | \tilde{u}_{n\mathbf{k}} \rangle$ should be *Hermitian* and *positive definite* (i.e., with all positive eigenvalues). As explained in Appendix B, this is accomplished by computing the singular value decomposition $M^{(\mathbf{k}_0,\mathbf{k})} = U\Sigma V^{\dagger}$ of the overlap matrix, and then applying the unitary rotation $VU^{\dagger}$ to the $|u_{n\mathbf{k}}\rangle$ to obtain a new set of states $|\tilde{u}_{n\mathbf{k}}\rangle$ that are optimally aligned with the $|\tilde{u}_{n\mathbf{k}_0}\rangle$. That is, these obey $\tilde{M}^{(\mathbf{k}_0,\mathbf{k})} = U\Sigma U^{\dagger}$, which is Hermitian and positive definite as advertised. The $|\tilde{u}_{n\mathbf{k}}\rangle$ will then be smooth functions of $\mathbf{k}$, assuming only that the region is small enough that the overlap matrix $M_{mn}^{(\mathbf{k}_0,\mathbf{k})}$ never becomes singular.

This shows that the singularities originally present in the $|\psi_{n\mathbf{k}}\rangle$ can always be ironed out locally, but the fact that it can be done globally, in such a way as to preserve a periodic gauge, is highly nontrivial. This fact was proved, under rather general conditions, by Brouder et al. (2007) via a deep mathematical analysis. On a practical level, the "projection method" that will be discussed shortly usually succeeds in finding a smooth and periodic set of $|\tilde{\psi}_{n\mathbf{k}}\rangle$ from which a set of well-localized Wannier functions can be constructed. If desired, the localization properties of these initial Wannier functions can be improved further by using iterative methods described by Marzari and Vanderbilt (1997) and Marzari et al. (2012).

A practical procedure that is frequently used to construct the multiband Wannier functions is the *projection method*, which proceeds as follows. Choose a set of J localized "trial functions" $|t_n\rangle$ whose locations and characters are similar to the expected Wannier functions. For example, for the four valence bands of Si or GaAs, one could choose Gaussians centered at the four midbond positions. The precise location and shape of these functions are typically not very important. Expand the set of functions to include the periodic images $|t_{n\mathbf{R}}\rangle$, and then construct Bloch-like functions $|\chi_{n\mathbf{k}}\rangle = \sum_{\mathbf{R}} e^{i\mathbf{k}\cdot\mathbf{R}} |t_{n\mathbf{R}}\rangle$ in analogy with Eq. (3.80b). These $|\chi_{n\mathbf{k}}\rangle$ are smooth and periodic throughout the BZ. While they are not orthonormal, we can make them so by carrying out a symmetric orthonormalization at each $\mathbf{k}$ to obtain a revised set $|\chi'_{n\mathbf{k}}\rangle$. That is, we compute $S_{\mathbf{k},mn} = \langle\chi_{m\mathbf{k}}|\chi_{n\mathbf{k}}\rangle$ and construct $|\chi'_{n\mathbf{k}}\rangle = \sum_m (S_{\mathbf{k}}^{-1/2})_{mn}|\chi_{m\mathbf{k}}\rangle$. Finally, we perform a $J \times J$ unitary rotation on the $|\psi_{n\mathbf{k}}\rangle$ so as to obtain a set of $|\tilde{\psi}_{n\mathbf{k}}\rangle$ that are optimally aligned with the $|\chi'_{n\mathbf{k}}\rangle$ using the singular value decomposition as described in Appendix B. These $|\tilde{\psi}_{n\mathbf{k}}\rangle$ will then be smooth and periodic functions of $\mathbf{k}$, and can be used to construct Wannier functions through Eq. (3.105).

This procedure is not fool-proof. For some choices of trial functions, it can happen that the vectors $|\chi_{n\mathbf{k}}\rangle$ are not linearly independent at some $\mathbf{k}$. In that case, $S_{\mathbf{k}}$ becomes singular, the symmetric orthonormalization cannot be performed, and the procedure fails. For ordinary insulators, this problem can usually be fixed by a modified choice of trial functions. In certain kinds of topological insulators, however, there can be a topological obstruction to the construction of Wannier functions; in this case, a singularity in $S_{\mathbf{k}}$ is guaranteed to occur somewhere in the BZ. This will be discussed in some detail in Sections 5.1.5 and 5.2.4.

Once again, the Wannier functions are nonunique, since many choices of trial functions are possible. More generally, many different multiband gauge transformations $U_{mn}(\mathbf{k})$ can successfully iron out the nonanalytic behavior of the Bloch energy eigenstates. If fact, as soon as we have one such gauge transformation, we can always construct another one by following it with a second multiband gauge transformation having the form of Eq. (3.107), but this time with a $U(\mathbf{k})$ that itself is smooth and periodic everywhere in the BZ. As a result, the Wannier functions may change shape, becoming somewhat more or less localized.

In spite of this broader gauge dependence, the six properties of single-band Wannier functions enumerated in Section 3.5.2 all have their counterparts in the multiband case. Equations (3.86) and (3.87) remain true, while Eq. (3.88) is generalized to a multiband orthonormality condition $\langle w_{m\mathbf{R}}|w_{n\mathbf{R}'}\rangle = \delta_{mn}\delta_{\mathbf{R}\mathbf{R}'}$. As for Eq. (3.89), the total band projection $\mathcal{P} = \sum_n \mathcal{P}_n$ is now invariant.

The fact that the Wannier functions provide an exact tight-binding–like representation remains true, with the modification that instead of just computing the $E_{n\mathbf{R}}$ as in Eq. (3.90), we have to construct a $J \times J$ *Hamiltonian matrix* $H_{mn}(\mathbf{R}) = \langle w_{m0}|H|w_{n\mathbf{R}}\rangle$. This is then Fourier transformed in the manner of Eq. (3.92b) to obtain $H_{mn}(\mathbf{k})$:

$$H_{mn}(\mathbf{k}) = \sum_{\mathbf{R}} e^{i\mathbf{k}\cdot\mathbf{R}} H_{mn}(\mathbf{R}) . \tag{3.108}$$

From this, it is easily shown that $H_{mn}(\mathbf{k}) = \langle \tilde{\psi}_{n\mathbf{k}}|H|\tilde{\psi}_{n\mathbf{k}}\rangle$. The eigenvalues $E_{n\mathbf{k}}$ are then the eigenvalues of the matrix secular equation $\det[H(\mathbf{k}) - E_{n\mathbf{k}}] = 0$.

For the single-band case, we found a relation between the position operator matrix elements in the Wannier representation and the Berry connection in the form of Eq. (3.91). To generalize this relation, we first define the multiband Berry connection as

$$A_{mn,\mu}(\mathbf{k}) = \langle \tilde{u}_{m\mathbf{k}}|i\partial_{\mu}\tilde{u}_{n\mathbf{k}}\rangle , \tag{3.109}$$

where $\mathbf{A}$ is now a $J \times J$ matrix (and a Cartesian 3-vector) at each $\mathbf{k}$. The generalized versions of Eqs. (3.91–3.92) are

$$\langle w_{m0}|\mathbf{r}|w_{n\mathbf{R}}\rangle = \mathbf{A}_{mn,\mathbf{R}} \tag{3.110}$$

and

$$\mathbf{A}_{mn,\mathbf{R}} = \frac{V_{\text{cell}}}{(2\pi)^3} \int_{\text{BZ}} e^{-i\mathbf{k}\cdot\mathbf{R}} \mathbf{A}_{mn}(\mathbf{k}) \, d^3k , \tag{3.111a}$$

$$\updownarrow \text{ FT}$$

$$\mathbf{A}_{mn}(\mathbf{k}) = \sum_{\mathbf{R}} e^{i\mathbf{k}\cdot\mathbf{R}} \mathbf{A}_{mn,\mathbf{R}} . \tag{3.111b}$$

Once more, this expresses a Fourier relation between the Wannier position matrix elements and the Berry connection in $\mathbf{k}$ space.

It is again of special interest to focus on the diagonal elements

$$\bar{\mathbf{r}}_n = \frac{V_{\text{cell}}}{(2\pi)^3} \int_{\text{BZ}} e^{-i\mathbf{k}\cdot\mathbf{R}} \mathbf{A}_{nn}(\mathbf{k}) \, d^3k . \tag{3.112}$$

Under a multiband gauge transformation, these may shift in position. Even so, as we shall see in Ex. 3.23, the sum of Wannier-center vectors

$$\bar{\mathbf{r}}_{\text{tot}} = \sum_{n=1}^{J} \bar{\mathbf{r}}_n \tag{3.113}$$

remains multiband-gauge-invariant, modulo a lattice vector, just as for the single-band Wannier center as expressed in Eq. (3.104).[26] This fact will prove important later on.

If the Wannier functions are not unique, how can we calculate them and plot them in practice for materials of interest? An answer to this question was provided by Marzari and Vanderbilt (1997), who suggested focusing on a set of maximally localized Wannier functions (MLWFs) chosen so as to minimize the sum of quadratic spreads of the Wannier functions in one unit cell,[27]

$$
\Omega_{\text{spread}} = \sum_{n=1}^{J} \left[\langle w_{n0} | r^2 | w_{n0} \rangle - |\bar{\mathbf{r}}_n|^2 \right]. \tag{3.114}
$$

The quantity in brackets represents the mean square variation of the Wannier electron density away from its mean position $\bar{\mathbf{r}}_n = \langle w_{n0} | \mathbf{r} | w_{n0} \rangle$. We encountered the problem of constructing MLWFs for a single band in 1D in Ex. 3.17, and we shall return to it shortly for the multiband case in Section 3.6.3. A general construction procedure for multiple bands in 3D was given by Marzari and Vanderbilt (1997); such a construction is by now a standard procedure in the computational electronic structure community, as reviewed by Marzari et al. (2012) and implemented in the open-source WANNIER90 code package (Mostofi et al., 2008, 2014).

As an example, Fig. 3.17 shows a MLWF constructed for Si. The minimization of Eq. (3.114) results in four equivalent Wannier functions, each of which is located on one of the four nearest-neighbor Si–Si bonds in the unit cell, and each one having the character of a σ-bond orbital, namely, a bonding combination of two inward-directed sp^3 hybrids from the two neighboring atoms. We emphasize that there is no one-to-one correspondence between the four Wannier functions in the unit cell and the four bands in Fig. 3.16; instead, the four Wannier functions taken together span the same space as is spanned by the four lowest bands in Fig. 3.16.

It is satisfying to see that the MLWfs correspond to our usual understanding of the chemical bonding in Si. This turns out to be a rather general feature of the MLWF construction (Marzari and Vanderbilt, 1997; Marzari et al., 2012), and is responsible in part for its popularity. For the purposes of this book, however, the MLWF criterion is not particularly important in the 3D context. When speaking of Wannier functions in the multiband case, we shall require only that these are reasonably well localized, such that Hamiltonian and position matrix elements

---

[26] On p. 119 we introduced the distinction between radical and progressive gauge transformations in the single-band case, depending on whether $e^{i\beta(\mathbf{k})}$ winds by a nonzero multiple of 2π as $\mathbf{k}$ loops around the BZ. For the multiband gauge transformation of Eq. (3.107), the corresponding question is whether $\det[U(\mathbf{k})]$ has a corresponding winding. For a progressive gauge transformation, there is no such winding and $\bar{\mathbf{r}}_{\text{tot}}$ is fully gauge-invariant.

[27] Despite a similarity in notation, Ω_{spread} is not to be confused with a Berry curvature.

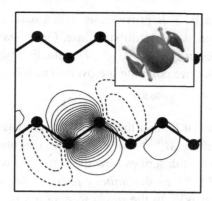

Figure 3.17 One of the four maximally localized Wannier functions in Si, shown as an amplitude contour plot on a (110) plane containing the chains of Si atoms (filled circles connected by bonds). Inset: 3D view showing isosurfaces of the Wannier function at a positive (central region) and a negative (two end caps) contour level.

between Wannier functions are well defined. (By contrast, the MLWFs do play a special role in 1D; see Sec. 3.6.3.)

3.6.2 Multiband Parallel Transport

In Section 3.1.1 we introduced a notion of parallel transport for the case of a single state $|u_\lambda\rangle$ for discretized λ. When this concept was applied to the case of a single band transported along a path in **k**-space in the BZ of a 3D crystal, in which the phase of the first state $|\bar{u}_{n\mathbf{k}_0}\rangle$ has already been chosen, the procedure was to choose the phase of the next state $|\bar{u}_{n\mathbf{k}_1}\rangle$ such that $\langle\bar{u}_{n\mathbf{k}_0}|\bar{u}_{n\mathbf{k}_1}\rangle$ is real and positive, and then similarly for $\langle\bar{u}_{n\mathbf{k}_1}|\bar{u}_{n\mathbf{k}_2}\rangle$ and all subsequent points. The criterion that $\langle\bar{u}_{n\mathbf{k}_i}|\bar{u}_{n\mathbf{k}_{i+1}}\rangle$ should be real and positive is equivalent to requiring that this inner product should be as close to unity as possible, which intuitively corresponds to a notion of "optimal alignment" of the states. In the case that this string of points is a closed loop, the physical states must be identical at the final and initial points, but the phases do not necessarily match. Translating Eq. (3.21) to this context, the Berry phase of the loop is then just the phase mismatch

$$\phi_n = -\text{Im} \ln \langle\bar{\psi}_{n\mathbf{k}_N}|\bar{\psi}_{n\mathbf{k}_0}\rangle = -\text{Im} \ln \langle\bar{u}_{n\mathbf{k}_N}|e^{-i\mathbf{b}_j\cdot\mathbf{r}}|\bar{u}_{n\mathbf{k}_0}\rangle, \tag{3.115}$$

where the $e^{-i\mathbf{b}_j\cdot\mathbf{r}}$ factor is needed only if the loop winds around the BZ by a primitive reciprocal lattice vector $\mathbf{b}_j$.

We have already seen on p. 125 that there is a natural multiband generalization of the concept of optimal phase alignment, based on the singular value decomposition as described in Appendix B. This provides us with a natural generalization of the notion of parallel transport to the multiband case. Once again, we start with some set of J orthonormal states $|\bar{u}_{n\mathbf{k}_0}\rangle$ $(n = 1, \ldots, J)$ at the first point $\mathbf{k}_0$ along the chain. Then, for the next point $\mathbf{k}_1$, we compute the overlap matrix

$$M_{mn}^{(\mathbf{k}_0, \mathbf{k}_1)} = \langle \bar{u}_{m\mathbf{k}_0} | u_{n\mathbf{k}_1} \rangle \tag{3.116}$$

between the chosen states at $\mathbf{k}_0$ and the initial states at $\mathbf{k}_1$. Subjecting this to the singular value decomposition $M^{(\mathbf{k}_0, \mathbf{k}_1)} = V \Sigma W^\dagger$, we can obtain the unitary matrix $\mathcal{M}^{(\mathbf{k}_0, \mathbf{k}_1)} = V W^\dagger$, which best describes how the states got rotated, as in Eq. (B.8). We can regard each $\mathcal{M}^{(\mathbf{k}_0, \mathbf{k}_1)}$ as the *unitary part* of the corresponding $M^{(\mathbf{k}_0, \mathbf{k}_1)}$. We then apply its inverse $W V^\dagger$ to the $|u_{n\mathbf{k}_1}\rangle$ to obtain a new set of $|\bar{u}_{n\mathbf{k}_1}\rangle$ that are optimally aligned to the $|\bar{u}_{n\mathbf{k}_0}\rangle$. We repeat the procedure to choose states $|\bar{u}_{n\mathbf{k}_2}\rangle$ that are optimally aligned with the $|\bar{u}_{n\mathbf{k}_1}\rangle$, and so on, providing a parallel-transport gauge $|\bar{u}_{n\mathbf{k}_0}\rangle, \ldots, |\bar{u}_{n\mathbf{k}_N}\rangle$ for the states along the chain.

Let us return to the case that the chain of N $\mathbf{k}$-points forms a loop such that $\mathbf{k}_0$ and $\mathbf{k}_N$ represent the same Hamiltonian. After the sequential parallel-transport procedure is applied around this loop, the states at $\mathbf{k}_N$ will have acquired an overall global unitary rotation characterized by the $J \times J$ unitary matrix[28]

$$\mathfrak{U}_{mn} = \langle \bar{u}_{m\mathbf{k}_N} | \bar{u}_{n\mathbf{k}_0} \rangle . \tag{3.117}$$

We now define the *total Berry phase* associated with this path to be

$$\phi_{\text{tot}} = -\text{Im} \ln \det \mathfrak{U}. \tag{3.118}$$

Equations (3.117) and (3.118) constitute the multiband generalization of Eq. (3.8).

We can do more and extract a set of individual Berry phases associated with the multiband parallel transport. Since $\mathfrak{U}$ is unitary, its eigenvalues λ_m are unimodular; that is, $\lambda_m = e^{-i\phi_m}$ for real ϕ_m. Thus, Eq. (3.118) can also be written as

$$\phi_{\text{tot}} = -\text{Im} \ln \prod_{m=1}^{J} \lambda_m = -\sum_{m=1}^{J} \text{Im} \ln \lambda_m = \sum_{m=1}^{J} \phi_m . \tag{3.119}$$

The individual phases ϕ_m, which are known as *multiband Berry phases* or *Wilson loop eigenvalues* (Wilson, 1974), correspond to the multiband generalization of the single-band Berry phase of Eq. (3.8).

Note that we can also determine the total Berry phase ϕ, or the individual multiband Berry phases ϕ_m, without explicitly constructing the parallel-transport gauge at all. We start with the states $|u_{n\mathbf{k}_0}\rangle, \ldots, |u_{n\mathbf{k}_{N-1}}\rangle$ in any arbitrary gauge, and

---

[28] If the path involves a winding by reciprocal lattice vector $\mathbf{b}_j$, then as in Eq. (3.115), a factor of of $e^{-i\mathbf{b}_j \cdot \mathbf{r}}$ has to be inserted into the inner product in Eq. (3.117).

for each neighboring pair of points, we construct the overlap matrix $M_{mn}^{(\mathbf{k}_i,\mathbf{k}_{i+1})} = \langle u_{m\mathbf{k}_i} | u_{n\mathbf{k}_{i+1}} \rangle$. We then replace each $M^{(\mathbf{k}_i,\mathbf{k}_{i+1})}$ by its unitary part $\mathcal{M}^{(\mathbf{k}_i,\mathbf{k}_{i+1})}$ as described earlier [see also Eq. (B.8)], and compute the matrix product

$$\mathfrak{U} = \prod_{i=0}^{N-1} \mathcal{M}^{(\mathbf{k}_i,\mathbf{k}_{i+1})}. \tag{3.120}$$

This is easily shown to be identical to the $\mathfrak{U}$ of Eq. (3.117), and the ϕ_m are extracted from its eigenvalues as described earlier. The total Berry phase is just the sum of these eigenvalues, and can also be written as

$$\phi_{\text{tot}} = -\text{Im} \ln \det \prod_{i=0}^{N-1} \mathcal{M}^{(\mathbf{k}_i,\mathbf{k}_{i+1})}. \tag{3.121}$$

Often ϕ_{tot} is approximated in practice by

$$\phi_{\text{tot}} = -\text{Im} \ln \det \prod_{i=0}^{N-1} M^{(\mathbf{k}_i,\mathbf{k}_{i+1})} \tag{3.122}$$

since this gives the same result in the limit of increasing mesh density and is easier to calculate in practice.

3.6.3 Maximally Localized Wannier Functions in 1D

One way to attach a meaning to these multiband Berry phases is to recognize that they correspond to the centers of the MLWFs in 1D. To see this, consider what happens if we use the eigenvectors of $\mathfrak{U}$ to rotate the states $|\bar{u}_{nk}\rangle$ into new states $|\tilde{u}_{nk}\rangle$. That is, let V be the $J \times J$ unitary matrix that diagonalizes $\mathfrak{U}$, $(V^\dagger \mathfrak{U} V)_{mn} = \delta_{mn} e^{-i\phi_m}$, and construct

$$|\tilde{u}_{nk}\rangle = \sum_n V_{mn} |\bar{u}_{mk}\rangle. \tag{3.123}$$

Since this is a global unitary rotation applied to the states at all k, the $|\tilde{u}_{nk}\rangle$ again constitute a parallel-transport gauge, but now arranged such that each state $|\tilde{u}_{nk}\rangle$ returns precisely to itself at the end of the loop, having acquired the Berry phase ϕ_n. The phase mismatch at the end of the loop can be ironed out separately for each n by smoothly twisting the gauge in the sense of Eq. (3.10):

$$|\tilde{u}_{nk_j}\rangle = e^{-i\phi_n(j/N)} |\bar{u}_{nk_j}\rangle. \tag{3.124}$$

This corresponds to the construction of a *twisted parallel-transport gauge* in the multiband case, in which the phase evolution of the manifold of states for each n is optimally smooth.

We can then use these $|\bar{u}_{nk_j}\rangle$ to construct a set of especially well-localized Wannier functions $|w_{nR}\rangle$ via the usual Fourier-transform procedure of Eq. (3.105), which for a discrete mesh in 1D becomes

$$w_{n0}(x) = \frac{1}{N} \sum_{j=0}^{N-1} e^{ik_j x} \, \tilde{u}_{nk_j}(x) \tag{3.125}$$

for the J Wannier functions in the home unit cell. Taken together with their periodic images $|w_{nR}\rangle$ for other lattice vectors R, these form an orthonormal set whose centers are given by $\bar{x}_{nR} = R + \bar{x}_n$, where for the home cell

$$\bar{x}_n = \frac{\phi_n}{2\pi} a . \tag{3.126}$$

Because we have constructed these Wannier functions out of a Bloch manifold that is maximally smooth in k, namely those of a twisted parallel-transport gauge, we expect them to be optimally localized in real space in some sense.

In fact, they are precisely the MLWFs introduced on p. 128, specialized to 1D, and $\bar{x}_n$ are their maximally localized Wannier centers (see also Ex. 3.17 for the single-band case). To demonstrate this, we will first show that these diagonalize the position operator,

$$\langle w_{m0}|x|w_{nR}\rangle = \bar{x}_n \, \delta_{mn} \, \delta_{R0} , \tag{3.127}$$

and then explain why this corresponds to maximal localization.

To arrive at Eq. (3.127), first note that a parallel-transport gauge makes $A_{mn}(k) = \langle u_{mk}|i\partial_\mu u_{nk}\rangle$ vanish. This follows from the discussion on p. 125 and in Appendix B by noting that in a parallel-transport gauge, two points k and $k + \delta k$ on the path have an inner product matrix $M_{mn} = (V\Sigma V^\dagger)_{mn}$. As $\delta k \rightarrow 0$, the diagonal elements making up Σ approach unity to order $(\delta k)^2$ or higher (a first-order variation is impossible since the singular values cannot exceed unity). Thus, to first order in δ_k, we have $M = VV^\dagger$, which is the unit matrix. This does not mean that the states at $k+\delta k$ are the same as those at k, but rather that the difference has no projection inside the occupied space, which is the same as saying that $\langle u_{mk}|\partial_\mu u_{nk}\rangle$ must vanish. The twisted parallel-transport gauge adds an extra k-dependent phase $e^{-i\phi_n ka/2\pi}$ so that the Berry connection becomes just

$$A_{mn}(k) = \frac{\phi_n}{2\pi} a \, \delta_{mn} . \tag{3.128}$$

Since this is independent of k, only the $R = 0$ term of the Fourier transform survives. Using Eq. (3.110) we have that $A_{mn}(R) = \bar{x}_n \, \delta_{mn} \, \delta_{R0}$, which translates into Eq. (3.127).

Having explained that the twisted parallel-transport gauge corresponds to a choice of Wannier functions that diagonalizes the x operator, we still need to explain why this corresponds to the maximal localization. It is easiest to do this for a *finite* chain cut from the infinite system, for which we can define Wannier-like localized functions in a way that parallels the definition of true Wannier functions in the corresponding extended bulk system. Suppose, for example, that we cut a segment consisting of M unit cells from an infinite insulating chain having J occupied bands. Let N be the number of occupied states in this chain segment (typically $\simeq MJ$). From the N eigenstates $|\psi_i\rangle$, we construct the $N \times N$ "position matrix"

$$X_{ij} = \langle \psi_i | x | \psi_j \rangle \qquad (3.129)$$

and obtain its eigenvalues $\bar{x}_i$. From its eigenvectors $\xi_i(j)$, we construct the finite-chain "Wannier functions"[29] $|w_i\rangle = \sum_j \xi_i(j) |\psi_j\rangle$ such that

$$\langle w_i | x | w_j \rangle = \bar{x}_i \, \delta_{ij}. \qquad (3.130)$$

Relative to the original set of $|\psi_i\rangle$, or for that matter any other set of $|\varphi_i\rangle$ related to them by an $N \times N$ unitary rotation, we claim that these are as localized as possible in the sense of minimizing the total spread

$$\Omega_{\text{spread}} = \sum_{i=1}^{N} \left[\langle \varphi_i | x^2 | \varphi_i \rangle - \langle \varphi_i | x | \varphi_i \rangle^2 \right]. \qquad (3.131)$$

To see this, note that the first term, which is essentially just a trace over the occupied subspace of the x^2 operator, is invariant to the choice of unitary rotation and, therefore, can be dropped. It follows that it is enough to *maximize* the sum of squares $\sum_i \langle \varphi_i | x | \varphi_i \rangle^2 = \sum_i X_{ii}^2$ of the orbital centers. However, the sum

$$\sum_{ij} |X_{ij}|^2 = \sum_i X_{ii}^2 + \sum_{i \neq j} |X_{ij}|^2$$

is also an invariant (being the matrix norm $||X||^2 = \text{Tr}\,[X^\dagger X]$), so we can minimize the spread by *minimizing* the sum of off-diagonal elements $\sum_{i \neq j} |X_{ij}|^2$. But this is exactly what the previously described construction of the Wannier functions $|w_i\rangle$ accomplished, by making all of the off-diagonal elements of X vanish.

By analogy, we would like to argue that the true bulk Wannier functions obeying Eq. (3.127) are also maximally localized for an infinite chain. Unfortunately, the argument provided earlier for finite systems cannot be taken over directly to the

---

[29] Localized Wannier-like functions in finite systems are sometimes called "generalized Wannier functions" or "localized molecular orbitals" (see p. 136) to distinguish them from true periodic-system Wannier functions. We will not be so careful, however, and will use the terms "Wannier functions" or "MLWFs" to include the finite-system ones described here.

infinite chain, since traces of infinite-dimensional matrices are generally ill defined. For a proper demonstration, the reader is referred to the article by Marzari and Vanderbilt (1997). Here, we content ourselves with pointing out the similarities of the band projection operator $\mathcal{P}$, which takes the form $\mathcal{P} = \sum_i |\psi_i\rangle\langle\psi_i| = \sum_i |w_i\rangle\langle w_i|$ for finite systems and the form $\mathcal{P} = (a/2\pi)\sum_n \int_{\text{BZ}} |\psi_{nk}\rangle\langle\psi_{nk}| \, dk = \sum_{nR} |w_{nR}\rangle\langle w_{nR}|$ for extended ones. In both cases, the construction of maximally localized Wannier centers and Wannier functions corresponds to finding the eigenvalues and eigenvectors of the projected position operator $\mathcal{P}x\mathcal{P}$ (Kivelson, 1982). However, deep in the interior of a finite chain, this projection operator should become equal to the bulk one; that is, $P(x, x') = \langle x|\mathcal{P}|x'\rangle$ should be the same for the finite and extended system for (x, x') deep in the chain. Thus, we can expect the Wannier functions and their centers to match once we are sufficiently far from the surface of the finite system.

We can illustrate this equivalence using the PYTHTB package applied to a 1D tight-binding model with three alternating site energies, as was introduced in Fig. 1.4 in Chapter 1. The model has basis states located at 0, $\frac{1}{3}$, and $\frac{2}{3}$ in a 1D crystal of unit lattice constant, with site energies Δ_0, Δ_1, and Δ_2, respectively, and hoppings t between nearest neighbors. The model represents a kind of sliding charge-density wave system, where the location of the peaks and troughs of the site-energy variation is parametrized by λ as

$$\Delta_j = V_0 \cos(2\pi j/3 - \lambda). \qquad (3.132)$$

A PYTHTB implementation of this model is given in the program chain_3_site.py in Appendix D.8.[30] For some arbitrary choice of parameters, the program uses Berry-phase methods to compute the MLWF centers for the infinite chain, and then uses position-matrix diagonalization to find the corresponding centers of a 30-site chain cut from the bulk, for some particular value of λ. The results are compared for three cases: first treating band 0 individually, then band 1 individually, and then bands 0 and 1 together as an occupied group. The results are summarized in the printed output from the PYTHTB code:

```
Wannier centers of bands 0 and 1:
   Individual Wannier centers:  0.3188 0.9092
   Multiband  Wannier centers:  0.3419 0.8860

Finite-chain eigenenergies associated with
Band 0: -3.15 -3.12 -3.08 -3.02 -2.96 -2.88 -2.81 -2.75 -2.69 -2.66
Band 1: -0.32 -0.25 -0.14  0.01  0.17  0.34  0.51  0.66  0.77  1.16
```

---

[30] This is the same model treated in the example programs 3site_cycle.py and 3site_cycle_fin.py in the PYTHTB package.

```
Finite-chain Wannier centers associated with band 0:
   0.3329 1.3193 2.3188 3.3188 4.3188 5.3188 6.3188 7.3188 8.3184 9.3073
Compare with bulk:
   0.3188 0.3188 0.3188 0.3188 0.3188 0.3188 0.3188 0.3188 0.3188 0.3188

Finite-chain Wannier centers associated with band 1:
   0.0697 0.9225 1.9106 2.9093 3.9092 4.9092 5.9093 6.9100 7.9155 8.9548
Compare with bulk:
   0.9092 0.9092 0.9092 0.9092 0.9092 0.9092 0.9092 0.9092 0.9092 0.9092

First 10 finite-chain Wannier centers associated with bands 0 and 1:
   0.0195 0.3627 0.8962 1.3436 1.8871 2.3421 2.8861 3.3419 3.8860 4.3419
Compare with bulk:
   0.8860 0.3419 0.8860 0.3419 0.8860 0.3419 0.8860 0.3419 0.8860 0.3419
```

In each case it is clear that the MLWF centers of the finite system converge rapidly (in fact, exponentially) to the corresponding bulk positions as one enters deep into the interior of the chain. By making simple modifications to the PYTHTB code provided (see Ex. 3.22), it is easy to check that while the positions of the MLWF centers closest to the surface are sensitive to perturbations applied to the Hamiltonian at the chain end, these influences disappear deep in the chain, where the finite-system MLWF centers always converge on the bulk values.

We can recap this section as follows. In the bulk, if each of the J occupied bands is isolated (i.e., nowhere degenerate with the band below or above), then we can treat each separately using the methods of Section 3.5.2, where we found that the Wannier center for band n is just given through Eq. (3.97) by the Berry phase of that band. Alternatively, we can treat all J bands as a composite group – indeed, we have to do so if they are connected by degeneracies. In this case we again obtain J MLWF centers $\bar{x}_n$ through the identical Eq. (3.126); the only difference is that these Wannier functions no longer have a one-to-one relation to individual energy bands, but rather are extracted from the entire set of bands treated as a group. The ϕ_n in Eq. (3.126) are now the multiband Berry phases obtained from the eigenvalues of the global unitary evolution matrix $\mathfrak{U}$ in Eq. (3.117), which is typically computed from Eq. (3.120) in practice, and the unit-cell sum of the $\bar{x}_n$ is an invariant (modulo a lattice vector). We have also seen that the MLWF centers obtained from $\mathfrak{U}$ correspond to the eigenvalues of the band-projected position operator $\mathcal{P}x\mathcal{P}$. Moreover, a corresponding diagonalization of $\mathcal{P}x\mathcal{P}$ for a large but finite system provides a corresponding set of MLWF centers, obtained in practice as the eigenvalues of the position matrix of Eq. (3.129). We have seen that these MLWF centers are expected to match the bulk ones deep in the system interior.

In 2D and 3D, the construction of MLWFs is not so straightforward. The essential problem is that $\mathcal{P}x\mathcal{P}$, $\mathcal{P}y\mathcal{P}$, and $\mathcal{P}z\mathcal{P}$ do not commute, so it is impossible to diagonalize them simultaneously. Practical methods seek a compromise such that all three are as nearly codiagonal as possible. Such methods were introduced by Marzari and Vanderbilt (1997) and are reviewed by Marzari et al. (2012).

For extended periodic systems, the construction is done starting from the Bloch functions in reciprocal space, and yields a set of Wannier functions in the home unit cell that can be replicated into all the neighboring cells by lattice translations. For finite 3D (or 2D) systems, the procedure starts from the real-space eigenstates and seeks to make all three (or the first two) of the matrices X_{ij}, Y_{ij} and Z_{ij} defined following Eq. (3.129) as codiagonal as possible. This results in well-localized Wannier-like functions analogous to the 1D ones discussed on p. 133, but living in 3D. Localized functions of this kind have played an important part in the chemistry literature, where they go under the name of "localized molecular orbitals" (Boys, 1960, 1966; Edmiston and Ruedenberg, 1963; Foster and Boys, 1960a, 1960b).

3.6.4 Multiband Berry Connections and Curvatures

In the single-band case, our considerations of Berry phases and Berry connections also led to the concept of Berry curvatures and associated topological properties such as Chern numbers. The same can be done for the multiband case, although we clearly need to work in two or more dimensions for a Berry curvature to be defined. We therefore return now to 3D, and see how the entire formalism of Berry connections and curvatures introduced earlier in this chapter can be generalized to the multiband case, where objects that were previously scalars are now replaced by $J \times J$ matrices. The resulting formalism is often referred to as *non-Abelian* because the matrices do not generally commute. The formalism can be developed in any space of parameters λ_j as in the early parts of this chapter, but here we shall use the notation of Bloch states of a 3D crystal for consistency with the preceding discussion.

We once again assume that some definite choice of a smooth and periodic gauge has been made for the Bloch-like states $|\psi_{nk}\rangle$ and their cell-periodic versions $|u_{nk}\rangle$. These are the quantities that were written as $|\tilde{\psi}_{nk}\rangle$ and $|\tilde{u}_{nk}\rangle$ in Section 3.6.1, but we now dispense with the tilde notation for the sake of brevity. They are unitarily related to the true (energy-eigenstate) Bloch functions of the band group in question. In 3D, the multiband Berry connection matrix was defined in Eq. (3.109), which in our current notation is

$$A_{mn,\mu}(\mathbf{k}) = \langle u_{mk}|i\partial_\mu u_{nk}\rangle,\qquad(3.133)$$

and the Berry curvature matrix[31] is again defined as its curl,

$$\Omega_{mn,\mu\nu}(\mathbf{k}) = \partial_\mu A_{mn,\nu}(\mathbf{k}) - \partial_\nu A_{mn,\mu}(\mathbf{k})$$
$$= i\langle\partial_\mu u_{mk}|\partial_\nu u_{nk}\rangle - i\langle\partial_\nu u_{mk}|\partial_\mu u_{nk}\rangle.\qquad(3.134)$$

---

[31] This is the noncovariant Berry curvature matrix. See Eq. (3.147) or (3.149) for the definition of the gauge-covariant Berry curvature matrix.

As before, we sometimes hide the Cartesian indices by writing these as vectors $\mathbf{A}_{mn}$ and pseudovectors $\mathbf{\Omega}_{mn}$.

It is also useful to define the matrix traces

$$A_\mu^{tr}(\mathbf{k}) = \sum_n A_{nn,\mu}(\mathbf{k}) \tag{3.135}$$

and

$$\Omega_{\mu\nu}^{tr}(\mathbf{k}) = \sum_n \Omega_{nn,\mu\nu}(\mathbf{k}) \tag{3.136}$$

where the sum is over the J bands in the group. These traced quantities behave very much like the single-band Berry quantities of Section 3.4. In particular, for any closed path C in $\mathbf{k}$-space we can define a total Berry phase

$$\phi_{tot} = \oint_C \mathbf{A}^{tr}(\mathbf{k}) \cdot d\mathbf{k} = \int_S \mathbf{\Omega}^{tr}(\mathbf{k}) \cdot d\mathbf{S}, \tag{3.137}$$

which precisely corresponds to the discretized total Berry phase of Eqs. (3.118–3.119). In Eq. (3.137), Stokes' theorem has been invoked to convert the line integral into a surface integral over a patch of surface S whose boundary is C. In the special case that the individual bands are nowhere degenerate and the $|u_{n\mathbf{k}}\rangle$ are Bloch energy eigenstates, we can identify the diagonal matrix elements $A_{nn\mu}$ and $\Omega_{nn,\mu\nu}$ in the current notation with $A_{n\mu}$ and $\Omega_{n,\mu\nu}$ of Section 3.4. The total Berry phase ϕ_{tot} around a closed path is then just the sum $\sum_n \phi_n$ of the individual-band Berry phases.

Let's see how these Berry connection and curvature matrices transform under the multiband gauge change of Eq. (3.107). It is useful first to introduce the notion of *gauge covariance*. Any J-component vector v that transforms according to

$$\tilde{v}_n = \sum_m U_{mn} v_m \tag{3.138}$$

is said to be a gauge-covariant vector, and any $J \times J$ matrix V that transforms according to

$$\tilde{V}_{mn} = (U^\dagger V U)_{mn}, \tag{3.139}$$

is referred to as a gauge-covariant matrix, where $J \times J$ matrix products are implied on the right-hand side. These are useful definitions since, for example, any product of gauge-covariant objects is also automatically gauge-covariant, and the trace of any product of gauge-covariant matrices is automatically gauge-invariant.

We now write Eq. (3.107) as

$$|\tilde{u}_{n\mathbf{k}}\rangle = \sum_m U_{mn} |u_{m\mathbf{k}}\rangle. \tag{3.140}$$

(We have dropped the explicit $\mathbf{k}$-dependence of U_{mn} and it is understood that all sums are over the J states.) The states $|\tilde{u}_{n\mathbf{k}}\rangle$ in Eq. (3.140) are clearly gauge-covariant vectors (regarded as J-component vectors), and a matrix of expectation values of an ordinary operator $\mathcal{O}$ between such states

$$\mathcal{O}_{mn} = \langle u_{m\mathbf{k}}|\hat{\mathcal{O}}|u_{n\mathbf{k}}\rangle \tag{3.141}$$

is also automatically gauge-covariant. This is easily checked by inserting Eq. (3.140) into $\tilde{\mathcal{O}}_{mn} = \langle \tilde{u}_{m\mathbf{k}}|\hat{\mathcal{O}}|\tilde{u}_{n\mathbf{k}}\rangle$.

Unfortunately, the vector

$$|\partial_\mu \tilde{u}_{n\mathbf{k}}\rangle = \sum_m U_{mn} |\partial_\mu u_{m\mathbf{k}}\rangle + \sum_m (\partial_\mu U_{mn}) |u_{m\mathbf{k}}\rangle \tag{3.142}$$

is not gauge-covariant, in view of the appearance of the second term on the right side of the expression. By the same token, the Berry connection matrix

$$\tilde{A}_{\mu,mn} = (U^\dagger A_\mu U)_{mn} + (U^\dagger i\partial_\mu U)_{mn} \tag{3.143}$$

is not gauge-covariant, again because of the extra term on the right side. A similar extra term appears in the transformed Berry curvature tensor $\tilde{\Omega}_{\mu\nu,mn}$. For some purposes this is inconvenient. For example, any physical observable should be gauge-invariant, and it turns out to be easy to construct invariants out of gauge-covariant matrices.[32] We should not be surprised that the Berry connection is not gauge-covariant here, since it was not gauge-invariant in the single-band case. However, the Berry curvature *was* gauge-invariant in the single-band case, and it is unsatisfactory that we have lost the corresponding property here.

This situation can be repaired as follows. First, at each $\mathbf{k}$ define the projection operator onto the group of J bands in question,

$$\mathcal{P}_\mathbf{k} = \sum_n |u_{n\mathbf{k}}\rangle\langle u_{n\mathbf{k}}|, \tag{3.144}$$

and its complement $\mathcal{Q}_\mathbf{k} = 1 - \mathcal{P}_\mathbf{k}$. (In the case that the group is the full set of occupied states of an insulator, $\mathcal{P}_\mathbf{k}$ and $\mathcal{Q}_\mathbf{k}$ are just the projections onto valence and conduction manifolds, respectively.) Then we can define a "gauge-covariant derivative" of $|u_{n\mathbf{k}}\rangle$ as

$$|\breve{\partial}_\mu u_{n\mathbf{k}}\rangle = \mathcal{Q}_\mathbf{k} |\partial_\mu u_{n\mathbf{k}}\rangle, \tag{3.145}$$

which deserves this name because it satisfies

$$|\breve{\partial}_\mu \tilde{u}_{n\mathbf{k}}\rangle = \sum_m U_{mn} |\breve{\partial}_\mu u_{m\mathbf{k}}\rangle \tag{3.146}$$

---

[32] For example, the trace of any gauge-covariant matrix is automatically gauge-invariant. More generally, if $A_1 \ldots A_M$ are gauge-covariant matrices, then the trace of their product is gauge-invariant.

since the $\mathcal{Q}_k$ term yields zero when acting on the $|u_{mk}\rangle$ in the last term of Eq. (3.142). There is no such thing as a gauge-covariant Berry connection tensor, since $\langle u_{nk}|\tilde{\partial}_\mu u_{mk}\rangle$ vanishes, but we can define a gauge-covariant Berry curvature tensor

$$\check{\Omega}_{mn,\mu\nu} = i\langle\tilde{\partial}_\mu u_{mk}|\tilde{\partial}_\nu u_{nk}\rangle - i\langle\tilde{\partial}_\nu u_{mk}|\tilde{\partial}_\mu u_{nk}\rangle$$

$$= i\langle\partial_\mu u_{mk}|\mathcal{Q}_k|\partial_\nu u_{nk}\rangle - i\langle\partial_\nu u_{mk}|\mathcal{Q}_k|\partial_\mu u_{nk}\rangle \qquad (3.147)$$

satisfying an equation like Eq. (3.139). We can write this covariant curvature more concisely using

$$|\tilde{\partial}_\mu u_{nk}\rangle = |\partial_\mu u_{nk}\rangle - \sum_m |u_{mk}\rangle\langle u_{mk}|\partial_\mu u_{nk}\rangle$$

$$= |\partial_\mu u_{nk}\rangle + i\sum_m A_{mn,\mu}|u_{mk}\rangle . \qquad (3.148)$$

It then follows that

$$\check{\Omega}_{mn,\mu\nu} = \Omega_{mn,\mu\nu} - i[A_\mu, A_\nu]_{mn} . \qquad (3.149)$$

Because of its gauge covariance as expressed in Eq. (3.139), $\check{\Omega}$ is a natural quantity to exploit when deriving expressions for physical quantities that ought to be gauge-invariant.[33] An example of such an expression will appear in Eq. (6.54) in Chapter 6. However, after tracing over occupied bands one finds that $\check{\Omega}_{\mu\nu}^{\text{tr}}$ and $\Omega_{\mu\nu}^{\text{tr}}$ are identical, with both yielding Eq. (3.136), since the trace of the commutator appearing in Eq. (3.149) vanishes.

Exercises

3.20 The *space group* is the group of all symmetry operations of a crystal; it is *nonsymmorphic* if some of these operations involve fractional translations. Consider a nonsymmorphic 2D insulator whose space group includes a glide mirror symmetry $\{M_y|\frac{1}{2}a\hat{x}\}$ – that is, a mirror symmetry M_y ($y \leftrightarrow -y$) followed by a translation of half a lattice vector along $\hat{x}$, in addition to primitive translations by $\mathbf{a}_1 = (a, 0)$ and $\mathbf{a}_2 = (0, b)$. Show that such a

---

[33] The gauge-covariant Berry curvature tensor $\check{\Omega}_{mn,\mu\nu}$ can be regarded as part of a more general curvature-metric tensor

$$\mathcal{F}_{mn,\mu\nu} = \langle\tilde{\partial}_\mu u_{mk}|\tilde{\partial}_\nu u_{nk}\rangle = \langle\partial_\mu u_{mk}|\mathcal{Q}_k|\partial_\nu u_{nk}\rangle$$

such that $\check{\Omega}_{mn,\mu\nu} = i\mathcal{F}_{mn,\mu\nu} - i\mathcal{F}_{nm,\mu\nu}^*$ is $2i$ times its anti-Hermitian part (for given $\mu\nu$). Its Hermitian part $g_{mn,\mu\nu} = (\mathcal{F}_{mn,\mu\nu} + \mathcal{F}_{nm,\mu\nu}^*)/2$ is known as the *metric tensor*. Usually these enter the physics in the form of $\Omega_{\mu\nu}^{\text{tr}}$ and $g_{\mu\nu}^{\text{tr}}$, i.e., after tracing over occupied bands. See, for example, Appendix C of Marzari and Vanderbilt (1997) and references therein for discussion. Note in particular that $g_{\mu\nu}^{\text{tr}}$ is closely related to the gauge-invariant part of the Wannier spread function as expressed through Eq. (C.12) in the Marzari and Vanderbilt article.

system cannot have a single isolated band. (*Hint:* Assume that $|w\rangle$ is a Wannier function for this band, and consider the result of the glide mirror applied to it; is this also a Wannier function? Conclude that the multiband Wannier functions should come in pairs related to each other by the glide mirror.)

3.21 Show that Eq. (3.117) and Eq. (3.120) describe the same unitary matrix $\mathfrak{U}$.

3.22 Test the correctness of the statements made at the end of Section 3.6.3 concerning the insensitivity of the deep-chain Wannier centers to details of the Hamiltonian near the chain end. Modify the PYTHTB code `chain_3_site.py` of Appendix D.8 by making additional site-energy shifts, or by changing the hopping matrix elements, or both, near the chain end. Show empirically that these changes affect the MLWF centers near the chain end, but the influence always decays rapidly as one goes deep in the chain.

3.23 We want to show that the vector sum $\bar{r}_{tot}$ of Wannier centers in Eq. (3.113) is gauge-invariant, modulo a lattice vector, under a multiband gauge transformation as in Eq. (3.107).

(a) Starting from Eq. (3.143), show that the traced Berry connection of Eq. (3.135) transforms as

$$\mathrm{Tr}\,[\tilde{A}_\mu(\mathbf{k})] = \mathrm{Tr}\,[A_\mu(\mathbf{k})] + \mathrm{Tr}\,[U^\dagger(\mathbf{k})i\partial_\mu U(\mathbf{k})].$$

(b) Show that this can be rewritten as

$$\mathrm{Tr}\,[\tilde{A}_\mu] = \mathrm{Tr}\,[A_\mu] + \partial_\mu\beta$$

where $\beta(\mathbf{k}) = -\mathrm{Im}\ln\det U(\mathbf{k})$. Note the similarity with Eq. (3.70). *Hint:* Recall that a trace can be evaluated in any representation. Here it is convenient to use the one that diagonalizes $U(\mathbf{k})$; that is, assume $U(\mathbf{k}) = V(\mathbf{k})B(\mathbf{k})V^\dagger(\mathbf{k})$, where V is unitary and B is unimodular diagonal, $B_{mn} = e^{-i\beta_n}\delta_{mn}$ for real β_n.

(c) Assuming that a global branch choice has been made for $\beta(\mathbf{k})$ such that it is smooth and periodic over the 3D BZ, prove the invariance of $\bar{r}_{tot}$, starting from Eq. (3.96). (For this part and the next, you can assume an orthorhombic $a \times b \times c$ unit cell and let $\mu = x$.)

(d) The gauge transformation in part (c) is multiband "progressive" in the sense of the discussion on p. 119. It is also possible for $\beta(\mathbf{k})$ to increase by $2\pi n_j$ as $\mathbf{k} \to \mathbf{k} + \mathbf{b}_j$, where n_j is an integer associated with primitive reciprocal lattice vector $\mathbf{b}_j$, and still be consistent with a $U(\mathbf{k})$ that is smooth and periodic over the 3D BZ. In this case of a "radical" gauge transformation, show that $\bar{r}_{tot}$ shifts by a lattice vector.

4

Electric Polarization

Clearly the electric polarization $\mathbf{P}$ of a crystalline insulator should be defined so as to carry the meaning of dipole moment per unit volume, but a precise formulation is surprisingly subtle. In Section 1.1 we briefly discussed the problems with a definition of $\mathbf{P}$ in terms of the (ionic plus electronic) charge density $\rho(\mathbf{r})$. Such problems will be spelled out more fully in Section 4.1. In fact, we shall see that in principle, even a perfect knowledge of $\rho(\mathbf{r})$ is not enough to determine $\mathbf{P}$. As was strongly hinted in Chapter 1, a suitable definition can instead be given in terms of Berry phases or, equivalently, in terms of Wannier centers. This formulation will be developed in Sections 4.2 and 4.3. In this way, a concept of bulk electric polarization is recovered, but at the expense of abandoning a completely unique value of $\mathbf{P}$. Instead, we have to accept a definition that determines $\mathbf{P}$ only modulo a quantum $e\mathbf{R}/V_{\text{cell}}$, as in Eq. (1.7). The interpretation of this quantum is discussed at some length in Section 4.4. The remaining sections of this chapter cover the relation of $\mathbf{P}$ to surface charges and the treatment of insulators in finite electric fields.

4.1 Statement of the Problem

Imagine that we have a perfect knowledge of the charge density $\rho(\mathbf{r})$ in some insulating crystal. In the context of Chapter 2, this would take a form like

$$\rho(\mathbf{r}) = e \sum_{\mathbf{R}\mu} Z_\mu \, \delta^3(\mathbf{r} - \mathbf{R} - \boldsymbol{\tau}_\mu) - \frac{e}{(2\pi)^3} \sum_n \int_{\text{BZ}} |\psi_{n\mathbf{k}}(\mathbf{r})|^2 \, d^3k \qquad (4.1)$$

where μ runs over atoms in the unit cell with atomic numbers Z_μ. The first term gives the nuclear contribution as a sum of delta functions at atom locations $\boldsymbol{\tau}_\mu$ in cell $\mathbf{R}$; the second term is the electronic contribution following from Eq. (2.39). It is natural to think that the electric polarization $\mathbf{P}$ can be expressed in terms of $\rho(\mathbf{r})$ in some way, since it expresses a dipolar property and dipoles are normally

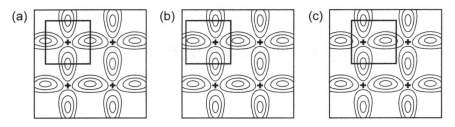

Figure 4.1 Model crystal with plus signs denoting delta-function nuclear charges and contours indicating electron clouds. The three panels indicate three different choices of unit cell boundaries (black squares), illustrating the dependence of Eq. (4.2) on unit cell location.

defined in terms of charge densities. Unfortunately, any effort to do so quickly runs into problems.

One tempting, but ultimately unsuccessful, approach is to identify **P** with

$$\mathbf{P}_{\text{dip}} = \frac{1}{V_{\text{cell}}} \int_{\text{cell}} \mathbf{r}\, \rho(\mathbf{r})\, d^3r \tag{4.2}$$

where the integral represents the dipole moment of the charge distribution inside one unit cell. This was introduced as Eq. (1.1) in Section 1.1, but as mentioned there, it does not give a unique definition because of a dependence on the choice of unit cell. This is illustrated for a model crystal in Fig. 4.1. The result of evaluating Eq. (4.2) is evidently $P_x = 0$ in Fig. 4.1(a), while the cells in Fig. 4.1(b–c) clearly contain dipoles pointing to the right and left ($P_x > 0$ and $P_x < 0$), respectively. If the crystal has an inversion or mirror symmetry, it may be possible to argue for a "most natural" choice of unit cell, but these are typically the cases where **P** vanishes anyway. Thus, we need an approach that works for cases with arbitrarily low symmetry. One clever suggestion could be to *average* over all possible unit cell locations, but unfortunately the result of this procedure is always exactly zero (see Ex. 4.1), which is hardly useful!

Another hint that this approach is not fruitful comes from evaluating Eq. (4.2) for the case of, say, a silicon crystal polarized by an external electric field in the context of density-functional theory (DFT) calculations. It turns out that, no matter how the unit cell boundaries are chosen, the field-induced polarization change computed according to Eq. (4.2) is an order of magnitude too small to be consistent with the known dielectric response of Si as characterized by its dielectric constant $\epsilon \simeq 12$ (Resta and Vanderbilt, 2007).

To understand why Eq. (4.2) fails, consider how **P** changes under a slow adiabatic change of the crystal Hamiltonian. We assume this Hamiltonian remains cell-periodic at all times, so that $\rho(\mathbf{r})$ and $\mathbf{j}(\mathbf{r})$ (the local microscopic induced current density) also remain cell-periodic. But recall that in Chapter 1 we argued for an intimate relation between polarization and current in the form of Eq. (1.2) or (1.9), which says that the change in polarization $\delta\mathbf{P}$ after some infinitesimal time δt must be

$$\delta\mathbf{P} = \mathbf{J}\,\delta t \tag{4.3}$$

where

$$\mathbf{J}(t) = \frac{1}{V_{\text{cell}}} \int_{V_{\text{cell}}} \mathbf{j}(\mathbf{r}, t)\, d^3 r \tag{4.4}$$

is the adiabatically induced macroscopic current density expressed as a unit-cell average. Equation (4.4) has a formal similarity to Eq. (4.2), but there is an all-important difference in that Eq. (4.4) does *not* depend on the location of the cell boundary. This follows because $\mathbf{j}(\mathbf{r}, t)$, unlike $\mathbf{r}\rho(r)$, remains a cell-periodic function of $\mathbf{r}$; the problematic position operator $\mathbf{r}$ is avoided, and the cell average is unambiguous.

Let's now see whether $\mathbf{P}_{\text{dip}}$ of Eq. (4.2) satisfies Eq. (4.3). We find

$$
\begin{aligned}
\delta\mathbf{P}_{\text{dip}} &= \frac{1}{V_{\text{cell}}} \int_{V_{\text{cell}}} \mathbf{r}\,\delta\rho(\mathbf{r})\, d^3 r \\
&= -\frac{\delta t}{V_{\text{cell}}} \int_{V_{\text{cell}}} \mathbf{r}\,[\nabla \cdot \mathbf{j}(\mathbf{r})]\, d^3 r \\
&= \frac{\delta t}{V_{\text{cell}}} \left[\int_{V_{\text{cell}}} \mathbf{j}(\mathbf{r})\, d^3 r - \int_{S_{\text{cell}}} \mathbf{r}\,(\mathbf{j} \cdot d\mathbf{S}) \right]
\end{aligned} \tag{4.5}
$$

where the divergence theorem $\dot{\rho} = -\nabla \cdot \mathbf{j}$ has been used in the form $\delta\rho(\mathbf{r}) = -\nabla \cdot \mathbf{j}(\mathbf{r})\,\delta t$ to express local charge conservation during the process. An integration by parts has been applied to get the last line, where the second term is a surface integral over the cell boundary S_{cell}. The first term in this last line is nothing other than the true $\delta\mathbf{P}$, so moving the surface term to the other side of the equation we obtain

$$\delta\mathbf{P} = \delta\mathbf{P}_{\text{dip}} + \frac{\delta t}{V_{\text{cell}}} \int_{S_{\text{cell}}} \mathbf{r}\,(\mathbf{j} \cdot d\mathbf{S}). \tag{4.6}$$

Clearly $\delta\mathbf{P} \neq \delta\mathbf{P}_{\text{dip}}$, so that $\mathbf{P}_{\text{dip}}$ does not satisfy Eq. (4.3) and is therefore not valid, unless it can be shown that the surface term in Eq. (4.6) vanishes. In general, though, there is no reason why it should. For a simple rectilinear $a \times b \times c$ unit cell, the x-component of this term is just $(\delta t/bc) \int j_x\, dy\, dz$, where the integral is over the right-side face of the unit cell. This extra term just measures *the current flowing through the face of the unit cell* during the adiabatic evolution. In the silicon example mentioned earlier, it is a much more important contribution than $\delta\mathbf{P}_{\text{dip}}$. In fact, there is no hope of extracting a knowledge of this current flow from the

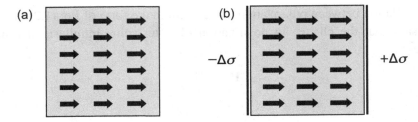

Figure 4.2 Illustration of the difficulty of defining polarization P in terms of the dipole moment of a finite sample divided by sample volume. Starting from (a), the addition of a charge density per unit area $\pm\Delta\sigma$ to the side surfaces changes the dipole moment by an amount proportional to sample volume (see text), so that P is not well defined by this procedure.

change of charge density $\delta\rho(\mathbf{r})$ alone.[1] This line of argument suggests that we must somehow go beyond a focus on charge densities, instead considering currents, so as to arrive at a suitable definition of electric polarization.

Of course, if we know the charge density $\rho(\mathbf{r})$ everywhere in space for some finite crystallite, including at its surface, we could try to adopt the "thermodynamic limit" approach and let $\mathbf{P}$ be defined as the dipole per unit volume of this large crystallite in the limit that its size L goes to infinity. This approach only works, however, if the surface contributions can be shown to vanish in the thermodynamic limit. Unfortunately, this is not the case here, as illustrated in Fig. 4.2. Suppose that a crystallite of size $L \times L \times L$ is prepared under two different sets of conditions, such that the second set leads to the appearance of an extra surface charge density $\Delta\sigma$ on the right-side face and $-\Delta\sigma$ on the left. This extra charge density will make a contribution of $(L^2\Delta\sigma)(L)$ to the total dipole moment of the sample (charges on the surfaces multiplied by their separation); dividing by volume L^3, we see that the contamination of $\mathbf{P}$ by the surface contribution does not vanish in the thermodynamic limit.

In a sense, the issue here is tied up with the definitions of "bound" and "free" charge. If we could distinguish between the two at the surface, we could define $\mathbf{P}$ as the dipole moment of bound charges divided by sample volume. Unfortunately, electrons do not come labeled as "bound" or "free," making this approach problematic. In any case, a suitable bulk definition of $\mathbf{P}$ should depend on bulk properties alone, so that an indirect definition in terms of surface properties is less than satisfactory.

---

[1] As an exception, note that if the crystal is composed of nonoverlapping molecular entities such that $\rho(\mathbf{r})$ and $\mathbf{j}(\mathbf{r})$ vanish on the cell boundary, as in Fig. 1.1(a), then the troublesome surface term in Eq. (4.6) vanishes. It is only in this ideal Clausius–Mossotti limit that $\mathbf{P}_{\text{dip}}$ can be accepted as a valid expression for the electric polarization.

If the information needed to compute $\mathbf{P}$ is not contained in the crystalline charge density $\rho(\mathbf{r})$, it seems very likely to be contained somehow in the underlying Bloch wave functions $\psi_{n\mathbf{k}}(\mathbf{r})$ from which $\rho(r)$ is constructed. For a finite system, such as a molecule or cluster, for which $\rho(\mathbf{r}) = -e \sum_j |\psi_j(\mathbf{r})|^2$, the electronic contribution to the dipole moment can be written in terms of expectation values of the position operator as $\int \mathbf{r} \, \rho(\mathbf{r}) \, d^3 r = -e \sum_j \langle \psi_j | \mathbf{r} | \psi_j \rangle$. Unfortunately, this kind of definition does not easily generalize to the crystalline case, where the expectation value $\langle \psi_{n\mathbf{k}} | \mathbf{r} | \psi_{n\mathbf{k}} \rangle = \int \mathbf{r} \, |\psi_{n\mathbf{k}}(\mathbf{r})|^2 \, d^3 r$ is ill defined (the integrand blows up at $|\mathbf{r}| \to \infty$). This suggests that, if we are to base a theory of electric polarization on the Bloch functions $\psi_{n\mathbf{k}}(\mathbf{r})$, it should perhaps depend not on probability densities $|\psi_{n\mathbf{k}}(\mathbf{r})|$, but rather on some phase property.

Another problem arises with all of the definitions proposed to this point. Recall that in Section 1.1 we argued, on the basis of several different physical arguments, that $\mathbf{P}$ should only be well defined modulo the quantum $e\mathbf{R}/V_{\text{cell}}$ of Eq. (1.7). Instead, all the preceding attempts at a definition lack this characteristic indeterminacy, and should be regarded with suspicion on this basis alone.

This is more or less how the situation stood in the computational electronic structure community in 1992. The problems with conventional attempts at a definition of electric polarization were understood, and various hints were pointing in the direction of phase properties as the basis of a proper definition. The "modern theory of polarization," based on the work of Resta (1992) and King-Smith and Vanderbilt (1993), was finally able to put the theory on a sound footing by making use of Berry-phase concepts, as we discuss next.

Exercises

4.1 Show that $\mathbf{P}_{\text{dip}}$ in Eq. (4.2) vanishes when averaged over all possible locations of the unit cell origin.

Hint: First show that the average is proportional to

$$\int_{V_{\text{cell}}} \int_{V_{\text{cell}}} \mathbf{r} \, \rho(\mathbf{r} - \mathbf{r}') \, d^3 r \, d^3 r'$$

where $\mathbf{r}'$ is the shift of origin of the cell. Then focus on the $\mathbf{r}'$ integral and argue that the result vanishes.

4.2 Berry-Phase Theory of Polarization

The essential problem with using a *static* dipole-moment–based approach to the definition of $\mathbf{P}$ is that attempts to average the static quantity $\mathbf{r}\rho(\mathbf{r})$ over a unit cell are doomed to failure, as this quantity is not a cell-periodic function.

The difficulty arises from the appearance of the unbounded position operator $\mathbf{r}$ in the integrand $\mathbf{r}\rho(\mathbf{r})$. In Chapter 1, however, we introduced a *dynamic* definition of the *change* in electric polarization, Eq. (1.9), for a system undergoing some adiabatic evolution that preserves the crystal periodicity, such as a sublattice displacement or the application of some external field. In this case $\Delta\mathbf{P}$ is given as the time integral of the current $\mathbf{J}(t)$ that flows during the interval between the initial and final states of the evolving system, as in Eq. (1.9), with $\mathbf{J}$ understood to be the macroscopic average of the microscopic current density $\mathbf{j}(\mathbf{r})$ as expressed in Eq. (4.4). Here we set out to build a theory of polarization on this current-based view of the problem.

4.2.1 First-Order Change in Polarization

In such a theory we consider a system evolving adiabatically according to Hamiltonian $H(\lambda)$ for some slowly time-varying $\lambda(t)$, with the current and polarization related by

$$\mathbf{J} = \frac{d\mathbf{P}}{dt} = \frac{d\mathbf{P}}{d\lambda}\frac{d\lambda}{dt} = \dot{\lambda}\,\partial_\lambda\mathbf{P} \tag{4.7}$$

where $\partial_\lambda = d/d\lambda$. The system described by $H(\lambda)$ is assumed to preserve the crystal periodicity and remain insulating for all λ. Our hope is to derive an expression for $\partial_\lambda\mathbf{P}$ that can be inserted into Eq. (1.13) to obtain the change in polarization as $\int(\partial_\lambda\mathbf{P})d\lambda$. We shall first give a heuristic derivation of an expression for $\partial_\lambda\mathbf{P}$ using the methods of ordinary perturbation theory summarized in Section 2.3, and then follow that derivation with a more careful line of argument using the adiabatic perturbation theory of Section 3.3.

The theory is most naturally formulated in terms of the cell-periodic functions obeying $H_\mathbf{k}|u_{n\mathbf{k}}\rangle = E_{n\mathbf{k}}|u_{n\mathbf{k}}\rangle$ with $H_\mathbf{k} = e^{-i\mathbf{k}\cdot\mathbf{r}}He^{i\mathbf{k}\cdot\mathbf{r}}$ as introduced in Section 2.1.5. Adapting the notation of Section 2.3 to this context, we define

$$Q_\mathbf{k} = \sum_m^{\text{unocc}} |u_{m\mathbf{k}}\rangle\langle u_{m\mathbf{k}}| = 1 - P_\mathbf{k} \tag{4.8}$$

where $P_\mathbf{k} = \sum_n^{\text{occ}} |u_{n\mathbf{k}}\rangle\langle u_{n\mathbf{k}}|$ and

$$T_{n\mathbf{k}} = \sum_m^{\text{unocc}} \frac{|u_{m\mathbf{k}}\rangle\langle u_{m\mathbf{k}}|}{E_{n\mathbf{k}} - E_{m\mathbf{k}}}. \tag{4.9}$$

Then the first-order change in an occupied-state wave function $|u_{n\mathbf{k}}\rangle$ with the perturbation $\partial_\lambda H_\mathbf{k}$ in Eq. (2.97) becomes

$$Q_\mathbf{k}\,|\partial_\lambda u_{n\mathbf{k}}\rangle = T_{n\mathbf{k}}\,(\partial_\lambda H_\mathbf{k})\,|u_{n\mathbf{k}}\rangle \tag{4.10}$$

and its contribution to the change in the expectation value of operator $\mathcal{O}$, Eq. (2.101), becomes

$$\partial_\lambda \langle \mathcal{O} \rangle_{nk} = 2\mathrm{Re} \, \langle u_{nk} | \mathcal{O} \mathcal{Q}_k | \partial_\lambda u_{nk} \rangle . \tag{4.11}$$

Summing over all the occupied bands $|u_{nk}\rangle$ and reversing the order of the matrix element, the total change is, in view of Eq. (2.36),

$$\partial_\lambda \langle \mathcal{O} \rangle = \frac{V_{\mathrm{cell}}}{(2\pi)^3} \sum_n^{\mathrm{occ}} \int_{\mathrm{BZ}} 2\mathrm{Re} \, \langle \partial_\lambda u_{nk} | \mathcal{Q}_k \mathcal{O} | u_{nk} \rangle \, d^3k . \tag{4.12}$$

Formally, the electric polarization should be dipole moment per unit volume, or $-e\langle \mathbf{r} \rangle / V_{\mathrm{cell}}$, so ideally we should like to replace $\mathcal{O}$ with $\mathbf{r}$ in Eq. (4.12). Is this permissible? From Eq. (4.8) we see that this would involve matrix elements of the form $\langle u_{mk} | \mathbf{r} | u_{nk} \rangle$ between *different* Bloch states $|u_{mk}\rangle$ (empty) and $|u_{nk}\rangle$ (occupied). We saw previously that *diagonal* matrix elements of the form $\langle u_{nk} | \mathbf{r} | u_{nk} \rangle$ are ill defined for the position operator, but we can argue that *off-diagonal* ones *are* well defined, using a trick involving the velocity operator.

Recall from Section 2.1.5 that the velocity operator is properly defined as

$$\mathbf{v} = \frac{-i}{\hbar} [\mathbf{r}, H], \tag{4.13}$$

from which it follows that $\mathbf{v_k} = e^{-i\mathbf{k}\cdot\mathbf{r}} \mathbf{v} e^{i\mathbf{k}\cdot\mathbf{r}}$ is

$$\mathbf{v_k} = (-i/\hbar)[\mathbf{r}, H_k] \tag{4.14}$$

(since $\mathbf{r}$ commutes with $e^{i\mathbf{k}\cdot\mathbf{r}}$). For a simple Hamiltonian of the form $H = p^2/2m + V(\mathbf{r})$ the velocity operator is just $\mathbf{v} = \mathbf{p}/m$, where m is the electron mass; this simplified expression often appears in the literature. However, we will adopt the proper expression in terms of the commutator, in part because it must be used in more complicated cases, as when spin-orbit coupling or external fields are present, as discussed in Section 2.1.2. More importantly, it is necessary to use this form in the next steps of our derivation. Starting from Eq. (4.14) and using the identity $\langle u_{mk} | [\mathbf{r}, H_k] | u_{nk} \rangle = (E_{nk} - E_{mk}) \langle u_{mk} | \mathbf{r} | u_{nk} \rangle$, we find that

$$\langle u_{mk} | \mathbf{r} | u_{nk} \rangle = i\hbar \frac{\langle u_{mk} | \mathbf{v_k} | u_{nk} \rangle}{E_{nk} - E_{mk}} . \tag{4.15}$$

We appear to have tamed the unruly position operator appearing on the left-hand side of this equation! After all, the matrix elements of the velocity operator, which is just $\mathbf{p}/m$ in simple cases, are perfectly well defined. However, it is crucial to keep in mind that Eq. (4.15) holds only when $|u_{nk}\rangle$ and $|u_{mk}\rangle$ are both eigenstates of H_k, not for any arbitrary states of Bloch form, and only when $m \neq n$. When $\mathbf{v} = \mathbf{p}/m$, this equation takes the familiar form[2] $\langle u_{mk}|\mathbf{r}|u_{nk}\rangle = (-i/m\omega_{mn,k})\langle u_{mk}|\mathbf{p}|u_{nk}\rangle$, where $\hbar\omega_{mn,k} = E_{mk} - E_{nk}$. This relation often appears in solid state physics texts, as for example in the discussion of the equivalence between expressions based on $\mathbf{r}$ and $\mathbf{p}$ for the optical dipole matrix element.

Multiplying both sides of Eq. (4.15) on the left by $|u_{mk}\rangle$ and summing over all unoccupied bands m then yields

$$\mathcal{Q}_\mathbf{k}\,\mathbf{r}\,|u_{nk}\rangle = i\hbar\,T_{n\mathbf{k}}\,\mathbf{v_k}\,|u_{nk}\rangle. \tag{4.16}$$

Inserting this in Eq. (4.12) with the heuristic identification $\mathbf{P} = -e\langle\mathbf{r}\rangle/V_{\text{cell}}$, we find the important result

$$\partial_\lambda\mathbf{P} = \frac{e\hbar}{(2\pi)^3}\sum_n^{\text{occ}}\int_{\text{BZ}} 2\text{Im}\,\langle\partial_\lambda u_{nk}|T_{n\mathbf{k}}\mathbf{v_k}|u_{nk}\rangle\,d^3k. \tag{4.17}$$

Using Eq. (4.10), this expression can also be written as

$$\partial_\lambda\mathbf{P} = \frac{e\hbar}{(2\pi)^3}\sum_n^{\text{occ}}\int_{\text{BZ}} 2\text{Im}\,\langle u_{nk}|(\partial_\lambda H_\mathbf{k})T_{n\mathbf{k}}^2\mathbf{v_k}|u_{nk}\rangle\,d^3k \tag{4.18}$$

involving energy denominators squared. These equations give the first-order change of polarization with respect to some perturbation λ.

Formulas like these formed the basis for the computation of polarization responses in solids that was developed in the 1980s (see, for example, Baroni and Resta, 1986; Baroni et al., 1987; Littlewood, 1980), drawing on the early formulation of Pick et al. (1970). These methods are nicely reviewed in Baroni et al. (2001) and in the pair of papers by Gonze (1997) and Gonze and Lee (1997). Two important examples of polarization responses that can be computed using these methods are the Born or dynamical effective charge tensor

$$Z_{\alpha,\tau\mu}^* = \frac{\partial P_\alpha}{\partial R_{\tau\mu}} \tag{4.19}$$

describing the polarization in direction α induced by a sublattice displacement of atom τ in direction μ, and the piezoelectric tensor

$$c_{\alpha,\mu\nu} = \frac{\partial P_\alpha}{\partial\eta_{\mu\nu}} \tag{4.20}$$

describing the corresponding derivative with respect to a strain $\eta_{\mu\nu}$.

Similar methods can be used to treat perturbations arising from a first-order applied electric field $\mathcal{E}$. Since the field enters the Hamiltonian in the form $e\mathcal{E}\cdot\mathbf{r} = e\mathcal{E}_\nu r_\nu$, this involves solving for the field derivative of $|u_{nk}\rangle$ using Eq. (4.10) in the form

---

[2] See Ex. 4.2.

$$Q_{\mathbf{k}}|\partial_\nu u_{n\mathbf{k}}\rangle = e\,T_{n\mathbf{k}}\,r_\nu\,|u_{n\mathbf{k}}\rangle$$
$$= i\hbar e\,T_{n\mathbf{k}}^2 v_{\mathbf{k},\nu}\,|u_{n\mathbf{k}}\rangle\,, \tag{4.21}$$

where $\partial_\nu = \partial/\partial\mathcal{E}_\nu$ and the problematic position operator on the right-hand side has been tamed using Eq. (4.16). Then it follows that the first-order perturbation of some operator $\mathcal{O}$ is given by[3]

$$\frac{\partial_\nu\langle\mathcal{O}\rangle}{V_{\text{cell}}} = \frac{e\hbar}{(2\pi)^3}\sum_n^{\text{occ}}\int_{\text{BZ}} -2\text{Im}\,\langle u_{n\mathbf{k}}|\mathcal{O}T_{n\mathbf{k}}^2 v_{\mathbf{k},\nu}|u_{n\mathbf{k}}\rangle\,d^3k\,. \tag{4.22}$$

This could be used, for example, to compute the change in magnetization in an applied $\mathcal{E}$-field by letting $\mathcal{O}$ be the magnetic dipole operator. It can also be used to compute the electric susceptibility, which is the change in the expectation value of $-e\mathbf{r}$ per unit cell. In that case, however, we have to tame the position operator one more time using the methods discussed earlier, arriving at

$$\chi_{\mu\nu} = \frac{\partial P_\mu}{\partial\mathcal{E}_\nu} = \frac{e^2\hbar^2}{(2\pi)^3}\sum_n^{\text{occ}}\int_{\text{BZ}} 2\text{Re}\,\langle u_{n\mathbf{k}}|v_{\mathbf{k},\mu}T_{n\mathbf{k}}^3 v_{\mathbf{k},\nu}|u_{n\mathbf{k}}\rangle\,d^3k\,. \tag{4.23}$$

In practice, the sums over unoccupied states are usually eliminated by replacing the formal Eqs. (4.10) and (4.21) with an iterative solution of a Sternheimer equation taking the form of Eq. (2.90).

Some readers may be concerned about whether the left-hand side of Eq. (4.15) is even well defined, or about whether the heuristic identification of $\mathbf{P}$ with $e\mathbf{r}/V_{\text{cell}}$ even makes sense in an infinite crystal. We can confirm that the results derived here are correct in a more convincing way by abandoning a purely static view of the perturbation, and returning to the adiabatic perturbation theory derived in Section 3.3. There we found that the instantaneous wave function at time t has a first-order correction proportional to $\dot\lambda$, as in Eq. (3.56). Translating Eq. (3.59) into the present context of a crystalline solid, the first-order-in-$\dot\lambda$ correction to each occupied Bloch function is given by

$$|\delta u_{n\mathbf{k}}\rangle = -i\hbar T_{n\mathbf{k}}|\partial_\lambda u_{n\mathbf{k}}\rangle\,. \tag{4.24}$$

The $|u_{n\mathbf{k}}\rangle$ are normalized to the unit cell of volume V_{cell}, so the induced response from a single Bloch state is

$$\langle\mathcal{O}\rangle_{n\mathbf{k}}/\dot\lambda = 2\text{Re}\,\langle u_{n\mathbf{k}}|\mathcal{O}|\delta u_{n\mathbf{k}}\rangle$$
$$= 2i\hbar\,\text{Im}\,\langle\partial_\lambda u_{n\mathbf{k}}|T_{n\mathbf{k}}\mathcal{O}|u_{n\mathbf{k}}\rangle\,, \tag{4.25}$$

---

[3] This equation is essentially identical to Eq. (4.18), as expected in view of the equality of mixed partials expressed in Eq. (2.106). The difference in signs occurs because Eq. (4.18) expresses the response of $\mathbf{P}$ to a perturbation $\lambda\mathcal{O}$, while Eq. (4.22) is the response of $\langle\mathcal{O}\rangle$ to a perturbation $-\mathcal{E}\cdot\mathbf{P}$.

in which we substituted Eq. (4.24) and reversed the matrix element to get the second line. Specializing $\mathcal{O}$ now to the current operator $\mathcal{J} = -e\mathbf{v}$ and summing over all occupied bands, the total induced current is[4]

$$\mathbf{J}/\dot{\lambda} = \frac{e\hbar}{(2\pi)^3} \sum_n \int_{\text{BZ}} 2\text{Im} \langle \partial_\lambda u_{n\mathbf{k}} | T_{n\mathbf{k}} \mathbf{v}_k | u_{n\mathbf{k}} \rangle \, d^3k . \tag{4.26}$$

The left-hand side is just $(d\mathbf{P}/dt)/(d\lambda/dt) = d\mathbf{P}/d\lambda = \partial_\lambda \mathbf{P}$, confirming the result presented earlier in Eq. (4.17) that was derived using a less rigorous argument. Similarly, the response to an $\mathcal{E}$-field perturbation in Eqs. (4.21–4.23) can be rederived by other methods, as by treating $\mathcal{E}$ as arising from a vector potential $\mathbf{A}(t) = -c\mathcal{E}t$ that is increasing linearly in time and again applying the methods of adiabatic perturbation theory.

4.2.2 Change of Polarization on an Adiabatic Path

At this point, we find ourselves in a strange situation. On the one hand, we have derived Eqs. (4.17–4.18) expressing the *derivatives* of the polarization with respect to various perturbations such as sublattice displacements or external fields. On the other hand, we still have no formal expression for the *polarization itself*. Usually one expects that the computation of the derivative of some object should be more difficult than the computation of the object itself. In the early 1990s, however, the electronic structure community found itself in the strange position of being able to calculate derivatives of $\mathbf{P}$, but not $\mathbf{P}$ itself. This paradoxical situation was resolved with the arrival of the "modern theory of polarization" based on Berry phases in the early 1990s. We are finally now ready to develop this theory.

To do so, we follow Resta (1992) and temporarily avoid the question of defining $\mathbf{P}$ itself. Instead, we focus on the *change* of $\mathbf{P}$ during an adiabatic evolution of the system from some initial configuration i to a final one f, as in Eq. (1.13), which we repeat here:

$$\Delta\mathbf{P}_{i\to f} = \int_i^f (\partial_\lambda \mathbf{P}) \, d\lambda . \tag{4.27}$$

Since we have well-defined formulas such as Eqs. (4.17) and (4.18) for $(\partial_\lambda \mathbf{P})$, it should be possible to calculate the polarization change in practice by numerically integrating Eq. (4.27) on a mesh of λ values connecting λ_i to λ_f. For example, to compute the spontaneous polarization $\mathbf{P}_s$ in a ferroelectric material such as $PbTiO_3$, one can let λ_i and λ_f refer to the centrosymmetric cubic structure and the ground-state ferroelectric structure, respectively, and let λ linearly interpolate

---

[4] Note that the definitions of $\mathcal{J}$ and $\mathbf{J} = \langle \mathcal{J} \rangle / V_{\text{cell}}$ differ by a factor of V_{cell}.

between them. Since we expect the polarization to vanish in the centrosymmetric initial configuration, we can then identify $\Delta\mathbf{P}$ with $\mathbf{P_s}$. In practice, choosing only a few mesh points in λ often provides adequate convergence.

In one sense, Resta's numerical integration algorithm solves the problem of computing $\mathbf{P}$ in polar insulators. However, it is less than satisfactory on two counts. On the practical side, the calculation is tedious, requiring the construction of a path connecting the system of interest to a nonpolar reference configuration, checking that the system remains insulating along this path, and then carrying out the somewhat heavy linear response calculation of the polarization derivative at each intermediate configuration. And on the philosophical side, it begs a number of questions: Is the construction of an intermediate path really necessary? If one were to choose two different paths leading from i to f, would Eq. (4.27) give the same answer, independent of path? If so, shouldn't it be possible to replace the formulation of Eq. (4.27) with one that requires only a pair of calculations, one at the initial configuration and one at the final configuration?

We discussed related questions in some detail in Section 1.1, where we gave arguments that the change in polarization should be the same modulo a quantum for two paths connecting the same states as in Fig. 1.5(a), or zero modulo a quantum for a closed path as in Fig. 1.5(b). Other arguments given later in Chapter 1 reinforced this hypothesis. But can we show mathematically that it should be so? To do this, we return to Eq. (4.17), leaving computational questions aside for the moment, and concentrating instead on formal manipulations that can convert Eq. (4.17) into a path-independent form.

We now make use of another trick involving the velocity operator, namely the simple identity

$$\partial_{\mathbf{k}} H_{\mathbf{k}} = \hbar v_{\mathbf{k}}, \qquad (4.28)$$

whose proof follows directly by taking the $\mathbf{k}$-derivative[5] of $H_{\mathbf{k}}$ as defined in Eq. (2.43) (see also Ex. 2.2). We then substitute this identity into Eq. (4.10), letting $\mathbf{k}$ play the role of the parameter λ in that equation. This allows us to write the first $\mathbf{k}$-derivative of the Bloch function $|u_{n\mathbf{k}}\rangle$ as

$$\mathcal{Q}_{\mathbf{k}} |\partial_{\mathbf{k}} u_{n\mathbf{k}}\rangle = \hbar T_{n\mathbf{k}} v_{\mathbf{k}} |u_{n\mathbf{k}}\rangle . \qquad (4.29)$$

Incidentally, combining this equation with Eq. (4.16) yields the remarkably simple relation $\mathcal{Q}_{\mathbf{k}} \mathbf{r} |u_{n\mathbf{k}}\rangle = i \mathcal{Q}_{\mathbf{k}} |\partial_{\mathbf{k}} u_{n\mathbf{k}}\rangle$, suggesting the heuristic replacement

$$\mathbf{r} \longrightarrow i\partial_{\mathbf{k}} \qquad (4.30)$$

---

[5] We use the notations $\partial_{\mathbf{k}}$ and $\nabla_{\mathbf{k}}$ interchangeably. The form $\partial_{\mathbf{k}}$ is used here for consistency with the notation ∂_{λ} used for other perturbations.

(compare $\mathbf{p} \to -i\hbar\partial_r$ in quantum mechanics), which can sometimes be used as a shortcut in derivations. However, our goal here is to use Eq. (4.29) to eliminate $\mathbf{v_k}$ from Eq. (4.17). Remarkably, this also eliminates the sum over unoccupied states entirely, and Eq. (4.17) becomes

$$\partial_\lambda \mathbf{P} = \frac{e}{(2\pi)^3} \sum_n^{\text{occ}} \int_{\text{BZ}} 2\text{Im} \langle \partial_\lambda u_{n\mathbf{k}} | \mathcal{Q}_\mathbf{k} | \partial_\mathbf{k} u_{n\mathbf{k}} \rangle \, d^3 k. \tag{4.31}$$

It is not hard to show that $\sum_n^{\text{occ}} \langle \partial_\lambda u_{n\mathbf{k}} | \mathcal{P}_\mathbf{k} | \partial_\mathbf{k} u_{n\mathbf{k}} \rangle$ is pure real (see Ex. 4.4). Thus, using $\mathcal{Q}_\mathbf{k} = 1 - \mathcal{P}_\mathbf{k}$, this expression can be written more concisely as

$$\partial_\lambda \mathbf{P} = \frac{e}{(2\pi)^3} \sum_n^{\text{occ}} \int_{\text{BZ}} 2\text{Im} \langle \partial_\lambda u_{n\mathbf{k}} | \partial_\mathbf{k} u_{n\mathbf{k}} \rangle \, d^3 k. \tag{4.32}$$

The disappearance of the sum over unoccupied states coming from the $T_{n\mathbf{k}}$ operator is very encouraging. After all, we expect the electric polarization to be a ground-state property, and as such, we expect it to depend only on the occupied Bloch states $|u_{n\mathbf{k}}\rangle$. This is manifestly true of ordinary observables, for which the expectation value per unit cell is

$$\langle \mathcal{O} \rangle_{\text{cell}} = \frac{1}{(2\pi)^3} \sum_n^{\text{occ}} \int_{\text{BZ}} \langle u_{n\mathbf{k}} | \mathcal{O} | u_{n\mathbf{k}} \rangle \, d^3 k. \tag{4.33}$$

Equation (4.32) is not of this form, but having lost the problematic sum over unoccupied states, it is much closer to Eq. (4.33) than was Eq. (4.17).

There is another remarkable aspect of Eq. (4.32): Its integrand takes the form of a Berry curvature! Compare Eq. (4.32), for example, with Eq. (3.31) in Section 3.2.1. To develop this connection, let's restrict our attention for the moment to the case of a single occupied band n in a 1D insulator of lattice constant a. Translating Eq. (4.32) to this context, the contribution P_n from this band obeys

$$\partial_\lambda P_n = \frac{e}{2\pi} \int_0^{2\pi/a} 2\text{Im} \langle \partial_\lambda u_{n\mathbf{k}} | \partial_\mathbf{k} u_{n\mathbf{k}} \rangle \, dk. \tag{4.34}$$

It is then natural to regard (λ, k) as defining a 2D parameter space in which we define Berry connections $A_n^{(\lambda)} = i\langle u_{n\mathbf{k}} | \partial_\lambda u_{n\mathbf{k}} \rangle$ and $A_n^{(k)} = i\langle u_{n\mathbf{k}} | \partial_\mathbf{k} u_{n\mathbf{k}} \rangle$ and the Berry curvature $\Omega_n^{(\lambda k)} = -2\text{Im} \langle \partial_\lambda u_{n\mathbf{k}} | \partial_\mathbf{k} u_{n\mathbf{k}} \rangle$. Then Eq. (4.34) reduces to the marvelously simple form

$$\partial_\lambda P_n = \frac{-e}{2\pi} \int_0^{2\pi/a} \Omega_n^{(\lambda k)} \, dk. \tag{4.35}$$

That is, the first-order change in polarization is just proportional to the Brillouin zone (BZ) integral of the Berry curvature in (λ, k) space!

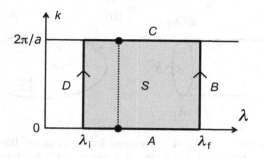

Figure 4.3 Region of integration of Berry curvature $\Omega_n^{(\lambda k)}$ in Eq. (4.36).

The next step is to insert this identity into Eq. (4.27) to compute the contribution to the change in polarization in going from λ_i to λ_f, giving

$$\Delta P_{n,i \to f} = \frac{-e}{2\pi} \iint_S \Omega_n^{(\lambda k)} \, d\lambda \, dk = \frac{-e}{2\pi} \Phi_{S,n}^{(\lambda k)}, \tag{4.36}$$

where S is the rectangular region shown in Fig. 4.3. In other words, ΔP_n is just proportional to $\Phi_{S,n}^{(\lambda k)}$, the Berry flux of band n passing through this rectangle! Recall that the Berry curvature $\Omega_n^{(\lambda k)}$ is gauge-invariant, so the expression for ΔP in Eq. (4.36) is completely independent of the choice of gauge. Thus, we now have a physically meaningful expression for the change of polarization along an adiabatic path. Even so, it would be still more convenient to derive an expression that does not depend on the details of the path.

To do so, we make use of the fact that k is really a cyclic variable. Following the discussion in Section 3.4, we insist on choosing a *periodic gauge* in the k-direction. As specified by Eq. (3.72), this means that for any given λ, such as the one indicated by the dotted vertical line in Fig. 4.3, $|\psi_{nk}\rangle$ is identical, in phase as well as in character, at the top and bottom points indicated by the dots.

Having identified segments A and C in this way, we have effectively wrapped surface S into a cylinder, as shown in Fig. 4.4(a), and we can apply all of the results developed in Section 3.2.3. In particular, Eq. (3.41) becomes

$$\Phi_{S,n}^{(\lambda k)} = \phi_n(\lambda_f) - \phi_n(\lambda_i) \tag{4.37}$$

where the right-hand side represents the total change in the k-loop Berry phase as λ goes from λ_i to λ_f (i.e., the difference between the Berry phases computed on segments B and D in Fig. 4.3). This also follows directly from Stokes' theorem: The integral of $\Omega_n^{(\lambda k)}$ on this cylinder is given by the Berry phase on the boundary, which in this case is composed of the two disconnected loops at the ends. After taking account of the sense of circulation on these loops, Stokes' theorem immediately yields Eq. (4.37).

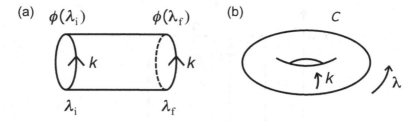

Figure 4.4 (a) Region S of Fig. 4.3, wrapped into a cylinder. Berry phases ϕ at boundary loops are indicated. (b) Same, but wrapped further into a torus in the case of a closed loop in λ space. The Chern number C of the surface is indicated.

Plugging Eq. (4.37) into Eq. (4.36) and restoring the sum over occupied bands, we then find the central result

$$\Delta P_{i \to f} = \frac{-e}{2\pi} \sum_n^{occ} [\phi_n(\lambda_f) - \phi_n(\lambda_i)] . \tag{4.38}$$

With this equation we have not only eliminated the undesirable sum over unoccupied states and obtained a gauge-invariant expression, but also expressed the change of polarization only in terms of calculations that can be carried out independently at the initial and final configurations λ_i and λ_f, with no detailed knowledge of the path connecting them. In fact, Eq. (4.38) can be recast simply as

$$\Delta P_{i \to f} = P(\lambda_f) - P(\lambda_i) \tag{4.39}$$

where the polarization of a 1D insulator is defined to be

$$P = \frac{-e}{2\pi} \sum_n^{occ} \phi_n . \tag{4.40}$$

Equations (4.39) and (4.40) express the central result of what is often referred to as the "Berry phase theory" or "modern theory" of polarization.

Various alternative forms are possible. Equation (4.40) can be rewritten in terms of the Berry connection $A_n(k) = i\langle u_{nk} | \partial_k u_{nk} \rangle$ as

$$P = \frac{-e}{2\pi} \sum_n^{occ} \int_0^{2\pi/a} A_n(k) \, dk . \tag{4.41}$$

As written, these formulas apply only when each band n is isolated. By using the methods of Section 3.6.2, however, we can treat the entire group of occupied bands as a unit and write

$$P = \frac{-e}{2\pi} \phi = \frac{-e}{2\pi} \int_0^{2\pi/a} A^{tr}(k) \, dk \tag{4.42}$$

where ϕ is ϕ_{tot} of Eq. (3.118) and $A^{\text{tr}}(k)$ is given by Eq. (3.135). When the bands are isolated, $\phi = \sum_n^{\text{occ}} \phi_n$ and $A^{\text{tr}}(k) = \sum_n^{\text{occ}} A_n(k)$, but the forms given in Eq. (4.42) are robust even in the presence of band degeneracies and crossings within the manifold of occupied states. In view of Eq. (3.97) as derived in Section 3.5.2, it is clear that Eq. (4.42) can also be written in terms of Wannier center positions; this viewpoint will be developed in Section 4.3.3. When generalized to 3D (see Section 4.3), Eqs. (4.39–4.41) become

$$\Delta \mathbf{P}_{i \to f} = \mathbf{P}(\lambda_f) - \mathbf{P}(\lambda_i) \tag{4.43}$$

where

$$\mathbf{P} = \frac{-e}{(2\pi)^3} \sum_n^{\text{occ}} \int_{\text{BZ}} \mathbf{A}_n(\mathbf{k}) \, d^3k \tag{4.44}$$

for isolated bands or

$$\mathbf{P} = \frac{-e}{(2\pi)^3} \int_{\text{BZ}} \mathbf{A}^{\text{tr}}(\mathbf{k}) \, d^3k \tag{4.45}$$

when treating the occupied bands as a unified group. All of these expressions are written assuming spinor bands; a factor of 2 should be inserted as usual whenever a spin-degenerate band convention is adopted.

4.2.3 *Quantized Adiabatic Charge Transport in 1D*

A special case arises when the parametric variation is over a closed loop, so that λ_f and λ_i describe identical 1D Hamiltonians. Then we can "glue together" the two ends of the cylinder in Fig. 4.4(a) to get the torus shown in Fig. 4.4(b). Since the torus is a closed manifold, the Chern theorem of Eq. (3.36) applies to it, and one concludes that $\iint_S \Omega_n^{(\lambda k)} \, d\lambda \, dk$ must be 2π times an integer C.

We encountered precisely this kind of situation in Section 3.2.3, where we saw that the flow of the Berry phase ϕ^ν over a cycle in μ (c.f. Fig. 3.10) is just 2π times the Chern number C associated with the Berry flux on the 2D (μ, ν) manifold. In the present context, μ and ν play the role of λ and k, respectively, and $\phi = \phi^{(k)}$. Moreover, we are assuming N occupied bands, each with its own Berry phase $\phi_n(\lambda)$ and a Chern number given by

$$2\pi C_n = \phi_n(\lambda_f) - \phi_n(\lambda_i) \quad \text{(continuous evolution)}. \tag{4.46}$$

The total change in polarization is then

$$\Delta P = -e \sum_n^{\text{occ}} C_n = -eC \tag{4.47}$$

where C is the total Chern number of the occupied band subspace along this path. The physical interpretation is, of course, that exactly C electrons have been pumped one unit cell to the right along the chain during the cyclic adiabatic evolution. This is precisely the quantization of adiabatic charge transport that was anticipated in Chapter 1.

This interpretation is reinforced when we recall that the Wannier centers and Berry phases are simply related via Eq. (3.97): $\bar{x}_n = a\phi_n/2\pi$, where a is the lattice constant. That is, $\bar{x}_n$ shifts to the right by C_n lattice constants as λ is carried around the loop. If we associate a charge $-e$ with each Wannier function, then indeed this amounts to a net transport of $C = \sum_n^{\text{occ}} C_n$ electrons by a unit cell, consistent with Eq. (4.47). Figure 1.8(b) of Section 1.1.4 illustrates this kind of situation, in which the relevant 1D direction is the vertical one (along z) in the figure, and the negative point charges shown there are now reinterpreted as being located at the Wannier centers of some occupied band.

This phenomenon of adiabatic charge transport is easily illustrated using the PYTHTB package.[6] In Section 3.6.3 we introduced the program `chain_3_site.py` of Appendix D.8 to treat the three-site model shown in Fig. 1.4(d–f). That program computes the Wannier center positions (or equivalently, the Berry phases) for one particular value of the parameter λ that describes the progress of the sliding charge-density wave according to Eq. (1.11) or Eq. (3.132). We now generalize this treatment in program `chain_3_cycle.py` of Appendix D.9, which follows the Wannier center of the bottom band of this model as λ is carried over a cyclic variation by 2π. The output of `chain_3_cycle.py` is reproduced here as Fig. 4.5. The crystal is taken to have a lattice constant of $a = 1$, corresponding to three sites in Fig. 1.4(d–f). Clearly the Wannier center advances by one lattice constant as λ cycles by 2π.[7] As discussed in Section 3.2.3, this implies a Chern number of $C = 1$. The situation is very similar to that shown in Fig. 3.11(a–b), except that there the 2D (α, φ) manifold was built from two parameters describing a finite system, while here one of those parameters has been replaced by the wavevector k of the chain.

This example evidently confirms the hypothetical picture introduced in Section 1.1.2, where it was argued that an adiabatic cyclic variation of the Hamiltonian of a 1D insulator can act as a quantum charge pump, transporting exactly an integer number of electrons to the right during the course of the cycle. It also underlines the essential nonuniqueness of the definition of polarization, which has to reflect the possibility of such a cycle. We will return to this line of argument, and generalize the picture from 1D to 3D, in Section 4.3.1.

---

[6] See also Ex. 3.15, which provides an example using the alternating site model of Appendix D.7.
[7] Actually, this model has an additional symmetry, in that advancing λ by $2\pi/3$ has the effect of converting the Hamiltonian of Fig. 1.4(d) into that of Fig. 1.4(f) – that is, translating the system to the right by $a/3$. This is the reason for the three-step appearance of the curve in Fig. 4.5.

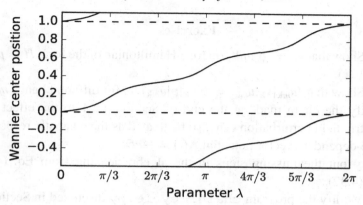

Figure 4.5 Evolution of the Wannier center positions of the lowest band of the three-site model described by Eq. (3.132) as parameter δ is carried around an adiabatic cycle from 0 to 2π. The plot is the output of program `chain_3_cycle.py` in Appendix D.9.

4.2.4 Historical Development

Before closing this section, it may be of interest to briefly recap the historical development of this theory. From the perspective of the computational electronic structure community, the first critical step was taken by Resta (1992), who focused attention on polarization differences as opposed to some absolute concept of polarization, and suggested that such differences could be calculated via Eq. (4.27) with $dP/d\lambda$ computed using Eq. (4.17) or (4.18). The remainder of the derivation leading to Eq. (4.44) then appeared (in a condensed form) in King-Smith and Vanderbilt (1993). However, there were many important antecedents, stretching back to the foundational work of Blount (1962), who developed methods very early on for incorporating the position operator into the formalism of the Bloch representation. A centrally important precedent was the paper of Thouless (1983), in which he first derived formulas for the adiabatically induced current and its time integral and outlined their topological implications. Thouless essentially carried out the previously discussed derivation, albeit using somewhat different language (notably, electric polarization was not mentioned) and with a focus on the quantization of charge transport for closed loops. This was around the same time as some of the foundational papers of Michael Berry were published (e.g., Berry, 1984). Another crucial precedent was the work of Zak (1989), focusing on the role of Berry phases for energy bands in solids. Nevertheless, it was not until after the papers of Resta (1992) and King-Smith and Vanderbilt (1993) (which drew heavily on the 1983 Thouless paper) that the importance of these developments came to be understood in the computational electronic structure community.

Exercises

4.2 (a) Show that $\mathbf{v_k} = (\mathbf{p}+\hbar\mathbf{k})/m$ for a Hamiltonian of the form $H = p^2/2m + V(\mathbf{r})$.

    (b) Show that $\langle u_{mk}|\mathbf{v_k}|u_{mk}\rangle = \langle u_{mk}|\mathbf{p}|u_{mk}\rangle/m$ for different bands $m \neq n$.

4.3 Justify the claim made at the end of Section 4.2.1 to the effect that an electric field perturbation can also be treated as the adiabatic response to a time-dependent vector potential $\mathbf{A}(t) = -c\mathcal{E}t$.

4.4 Carry out the missing steps in the algebra leading from Eq. (4.31) to Eq. (4.32).

4.5 (a) Modify the program `chain_3_cycle.py` discussed in Section 4.2.3 to plot the Wannier center positions of each of the three bands separately, thereby obtaining the cyclic Chern number for each band. Before generating the plots, ask yourself what you expect. The Wannier center of the lowest-energy band moves to the right during the cycle. Do you expect the same for the highest-energy band, or do you expect it to move in the reverse direction? Any guess about the middle band?

    (b) Now use the multiband features of PYTHTB to treat the lowest two bands as a group, and plot the two Wannier centers *vs.* λ. Again, ask yourself what you expect before observing the results of the calculation.

    Note: The Chern number of a set of bands treated as a group must be the sum of the Chern numbers of the individual bands (why?). Also, if you treat all three bands as a group, you must get $C = 0$. (Why? You may want to set up the calculation to see what happens.) These results may help you understand some of the observed behaviors.

4.3 Discussion

The arguments of the previous section imply that the polarization **P** cannot be expressed as the expectation value of a quantum operator, as is the case for most observable properties of quantum systems. Instead, it is related to geometric Berry phases and Berry curvatures of the Bloch wave functions in reciprocal space. Since this is a somewhat unfamiliar situation, it is worth discussing several aspects of the theory and clarifying some of its subtleties.

4.3.1 The Quantum of Polarization

Let us briefly review the discussion of the "quantum of polarization" that appeared in Section 1.1.2, returning to the case of a 1D insulator. The expression $\Delta P_{i\to f}$ appearing on the left-hand side of Eq. (4.39) represents the time-integrated current that flows for a *particular* adiabatic path connecting the initial and final states. It may be different for two different paths, such as the two shown in Fig. 1.5(a), but if so the ΔP values can differ only by the quantum, an integer multiple of e.

The expression $P(\lambda_f) - P(\lambda_i)$ appearing on the right-hand side of Eq. (4.39) can be interpreted in two different ways, depending on the context. If we have information only on the initial and final states, and not on the path connecting them, then each Berry phase ϕ_n entering Eq. (4.40) should be regarded as defined only modulo 2π, so that $P(\lambda_i)$ and $P(\lambda_f)$ are each known only modulo e. In this case it is safer to rewrite Eq. (4.39) with the '=' sign replaced by the ':=' notation introduced in Chapter 1 [e.g., Eqs. (1.6) and (1.15)]; this indicates that the definite quantity on the left-hand side is equal to the right-hand side up to a quantum (here e). This is illustrated in Fig. 4.6(a), where $\phi = \sum_n^{occ} \phi_n$ is plotted versus λ. If we have no knowledge of the path, then we will typically make some arbitrary branch choice such as $\phi \in [0, 2\pi]$ as shown by the shaded region, and can conclude only that $\Delta P_{i \to f} = -e[(\phi_f - \phi_i)/2\pi + N]$ for some unknown integer N.

In contrast, if we have full information about the path and compute the Berry phase $\phi(\lambda)$ for each point along the path, then Eq. (4.39) becomes a true equality, assuming that we insist that $\phi(\lambda)$ vary smoothly along the path as in Fig. 4.6(b), avoiding any 2π discontinuities like the one visible in Fig. 4.6(a). In this case we can conclude that $\Delta P_{i \to f} = -e(\widetilde{\phi}_f - \phi_i)/2\pi$ with no uncertainty modulo the quantum.

Most commonly we will be working in the former context, where we are interested in P for some particular configuration; in this case we suppress the λ-dependence and just write P as in Eq. (4.40) or (4.41). There are several equivalent ways of thinking about the fact that P is only well defined modulo the quantum e. One option is to assign it some specific value on the real axis – say,

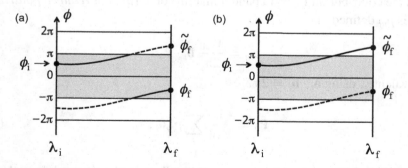

Figure 4.6 (a) Berry phase is assigned to branch choice $\phi \in [-\pi, \pi]$ (shaded region) independent of λ. (b) Branch choice is chosen in such a way that ϕ varies smoothly with λ. Solid and dashed curves differ by the quantum of 2π.

$0.2\,e$ – while understanding that either $-0.8\,e$ or $1.2\,e$ would be an equivalent specification. A second option is to wrap the real axis onto a circle of circumference e, and specify the value of P as a point on this circle. A third alternative is to think of P as a *lattice-valued* quantity; that is, P is associated with the lattice of values $\{\ldots, -1.8\,e, -0.8\,e, 0.2\,e, 1.2\,e, \ldots\}$.

Quantities of this kind, which are only defined modulo a quantum, do not arise very frequently, but they are not entirely without precedent. To give one example, suppose you are standing in front of two mechanical clocks, each of which ticks once a second, one softly and one loudly, and you want to describe the relative phase of the ticks. Let's call this the asynchronicity A, defined as the time delay from a soft tick to a loud one. Specifications that $A = 0.2\,$s or $A = 1.2\,$s or even $A = -0.8\,$s are obviously equivalent, and using the lattice-valued approach one could also say $A = \{\ldots, -0.8\,\text{s}, 0.2\,\text{s}, 1.2\,\text{s}, \ldots\}$. The polarization and the asynchronicity are two examples of physical quantities that are naturally defined only up to a quantum.

We now explore how to generalize this Berry-phase formulation to the case of a 3D crystal. The expression for P in terms of the Berry connection $A(k)$ in Eq. (4.41) was already generalized to 3D in Eqs. (4.44) and (4.45), but it is usually better to avoid working with gauge-dependent quantities such A, and the "modulo a quantum" nature of P is not obvious in this framework. Instead, here we generalize the approach in which $\partial_\lambda P$ was expressed in terms of a Berry curvature in Eq. (4.35), leading to a Berry-phase expression in Eq. (4.37). To do this concisely, it is useful to work in terms of the three reduced wavevector coordinates κ_j, each running from 0 to 2π, introduced in Eq. (2.31). Then

$$\frac{\partial f}{\partial \kappa_j} = \frac{\partial f}{\partial \mathbf{k}} \cdot \frac{\partial \mathbf{k}}{\partial \kappa_j} = \frac{\mathbf{b}_j}{2\pi} \cdot \partial_{\mathbf{k}} f$$

for any reciprocal-space function $f(\mathbf{k})$ defined over the 3D BZ.

Our derivation now starts from Eq. (4.32) for $\partial_\lambda \mathbf{P}$. It is convenient at this point to express the contribution $\mathbf{P}_n$ from band n in terms of a triplet of *reduced polarization* variables p_{nj} defined via

$$p_{nj} = \frac{V_{\text{cell}}}{2\pi} \mathbf{b}_j \cdot \mathbf{P}_n \tag{4.48}$$

or equivalently, using $\mathbf{a}_i \cdot \mathbf{b}_j = 2\pi \delta_{ij}$,

$$\mathbf{P}_n = \frac{1}{V_{\text{cell}}} \sum_j p_{nj} \mathbf{a}_j. \tag{4.49}$$

The p_{nj}, which have units of charge and are well defined only modulo e, describe the components of the electric polarization in each of the the three lattice directions. Similar relations hold between the occupied-band sums $p_j = \sum_n p_{nj}$ and $\mathbf{P} = \sum_n \mathbf{P}_n$. With this notation, Eq. (4.32) becomes

$$\frac{2\pi}{V_{\text{cell}}} \partial_\lambda p_{nj} = \frac{e}{(2\pi)^2} \int_{\text{BZ}} 2\text{Im} \langle \partial_\lambda u_{n\mathbf{k}} | \partial_{\kappa_j} u_{n\mathbf{k}} \rangle \, d^3k \qquad (4.50)$$

where we have used $\mathbf{b}_j \cdot \partial_\mathbf{k} = 2\pi \partial_{\kappa_j}$, which follows from Eq. (2.31). Except for a minus sign, the integrand on the right is nothing other than the Berry curvature $\Omega_n^{(\lambda \kappa_j)}$ of $|u_{n\mathbf{k}}(\lambda)\rangle$ with respect to variables λ and κ_j. Noting that the volume elements are related by $d^3\kappa = V_{\text{cell}} \, d^3k$, Eq. (4.50) becomes

$$\partial_\lambda p_{nj} = \frac{-e}{(2\pi)^3} \int_{\text{BZ}} \Omega_n^{(\lambda \kappa_j)} \, d^3k \,. \qquad (4.51)$$

Here 'BZ' now refers to the Brillouin zone of volume $(2\pi)^3$ in κ-space, so that the right-hand side is just $-e$ times the average of $\Omega_n^{(\lambda \kappa_j)}$ over the BZ.

To relate this to Berry phases, let's specialize to $j = 1$ for the moment and write

$$\partial_\lambda p_{n1} = \frac{-e}{(2\pi)^3} \iint \left[\int_0^{2\pi} \Omega_n^{(\lambda \kappa_1)} \, d\kappa_1 \right] d\kappa_2 d\kappa_3 \,. \qquad (4.52)$$

The object inside the square brackets is then perfectly analogous to Eq. (4.35), except that we are now working in reduced coordinates and all quantities carry an extra (κ_2, κ_3) dependence. Following the same logic that led from there to Eq. (4.40), we find

$$p_{n1} = \frac{-e}{(2\pi)^3} \iint \phi_n^{(\kappa_1)}(\kappa_2, \kappa_3) \, d\kappa_2 \, d\kappa_3 \qquad (4.53)$$

where

$$\phi_n^{(\kappa_1)}(\kappa_2, \kappa_3) = \int_0^{2\pi} i \langle u_{n\kappa} | \partial_{\kappa_1} u_{n\kappa} \rangle \, d\kappa_1 \qquad (4.54)$$

is the Berry phase of band n computing along a string of $\mathbf{k}$-vectors in the BZ extending parallel to $\mathbf{b}_1$. More generally, we find that

$$p_{nj} = \frac{-e}{2\pi} \bar{\phi}_n^{(\kappa_j)} \qquad (4.55)$$

where the bar indicates an average over the 2D BZ in the directions complementary to j. Plugging into Eq. (4.49) and restoring the sum over bands, we obtain the important result

$$\mathbf{P} = \frac{-e}{V_{\text{cell}}} \sum_n^{\text{occ}} \sum_j \frac{\bar{\phi}_{nj}}{2\pi} \mathbf{a}_j \qquad (4.56)$$

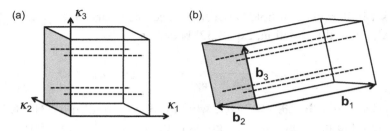

Figure 4.7 (a) Strings of **k**-points along which Berry phases are computed to obtain $\bar{\phi}_{n1}$, shown in the reduced BZ of variables $(\kappa_1, \kappa_2, \kappa_3)$, each of which runs between 0 and 2π. (b) Same, but shown in the parallelogram unit cell in the usual reciprocal space (k_x, k_y, k_z), illustrated for the case of a crystal of low symmetry.

for the Berry-phase polarization in 3D. This becomes just

$$P = \frac{-e}{V_{\text{cell}}} \sum_j \frac{\bar{\phi}_j}{2\pi} a_j \tag{4.57}$$

after defining $\bar{\phi}_j = \sum_n^{\text{occ}} \bar{\phi}_{nj}$ to include the sum over the bands. [When treating all occupied bands as a unified group, one should replace the sum over individual $\bar{\phi}_{nj}$ by a total Berry phase defined in the manner of Eq. (3.118).]

This situation is illustrated in Fig. 4.7. The **k**-point strings along which the Berry phases ϕ_{n1} need to be computed are shown in Fig. 4.7(a) in the κ-space BZ, which is a cube of side 2π. These Berry phases could equivalently be computed along the strings parallel to reciprocal lattice vector $\mathbf{b}_1$ in the true **k**-space parallelogram BZ in Fig. 4.7(b). Note that Eq. (4.53) instructs us to average $\phi_{n1}(\kappa_2, \kappa_3)$ over the (κ_2, κ_3) plane; this is what we are calling $\bar{\phi}_{n1}$. Similarly, $\bar{\phi}_{n2}$ is the average of $\phi_{n2}(\kappa_1, \kappa_3)$ over the (κ_1, κ_3) plane, where the Berry phases are now taken for strings along κ_2 (or parallel to $\mathbf{b}_2$), and similarly for $\bar{\phi}_{n3}$. Thus, three separate calculations need to be set up, each of which determines one component of **P** (although symmetry often reduces the number of components to just one or two, simplifying the calculation).

The quantum of indeterminacy in the definition of **P** is now clearly evident in Eq. (4.56). Each Berry phase is insensitive to a change $\bar{\phi}_j \rightarrow \bar{\phi}_j + 2\pi m_j$, which amounts to shifting **P** by $(-e/V_{\text{cell}})\mathbf{R}$, where **R** is the real-space lattice vector $m_1\mathbf{a}_1 + m_2\mathbf{a}_2 + m_3\mathbf{a}_3$. This is exactly the indeterminacy that was anticipated in Eq. (1.7).

There is a subtlety in the definition of the average Berry phase $\bar{\phi}_j$ that should not be overlooked. For $\bar{\phi}_1$ to make sense, we need to ensure that $\phi_1(\kappa_2, \kappa_3)$ is a smooth function of κ_2 and κ_3 with no 2π jumps. The difficulty is illustrated in Fig. 4.8, where we imagine that the 2D (κ_2, κ_3) face of the BZ has been sampled with a 3×3 sampling mesh, and the nine resulting Berry phases are shown as a

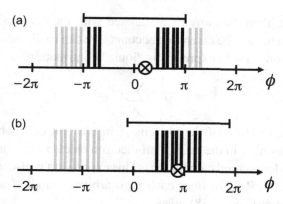

Figure 4.8 Histograms of Berry phases computed for parallel **k**-point strings on a sampling of the orthogonal face of the 2D BZ (see Fig. 4.7).

histogram. It would be a mistake to make the branch choice $\phi \in [-\pi, \pi]$ for each phase individually, as in Fig. 4.8(a); the resulting average $\bar{\phi}$, shown as the circled cross, would be meaningless. If the branch choice is made arbitrarily for one **k**-point string and then the others are chosen such that $|\Delta\phi| \ll 2\pi$, then a clustered group of Berry phases is correctly identified as in Fig. 4.8(b), and the average $\bar{\phi}$ is suitable for use in Eq. (4.56).

It is important to note that the indeterminacy of **P** modulo a quantum is rarely a problem in practice. In virtually all dielectric materials, the change $\Delta \mathbf{P}_{i \to f}$ that can be induced by applying a moderate electric field (e.g., whose strength is insufficient to cause dielectric breakdown) is a small fraction of the quantum $e\mathbf{R}/V_{cell}$, and the same is true for most ferroelectric materials upon reversal of the spontaneous polarization by an external field. In this case the appropriate algorithm is to make an arbitrary branch choice when computing $\bar{\phi}_j$ at the initial λ_i, and then choose the branch for $\bar{\phi}_j$ at the final λ_f such that the difference $\Delta\bar{\phi}_j$ between these is small compared to π. In some unusual cases, as for polarization reversal in strongly polarized ferroelectric materials such as $PbTiO_3$ or $BiFeO_3$, it may be that $\Delta \mathbf{P}_{i \to f}$ is not small compared to the quantum. In this case the best practical solution is to compute **P** on some number of intermediate λ values as well, choosing the branch choices such that $\Delta\bar{\phi}_j$ is small compared to π for each neighboring pair of λ values (or, if the $\Delta\bar{\phi}_j$ are still too large, repeating the procedure with finer λ meshes until they are not).

4.3.2 Ionic Contribution and Origin Dependence

Strictly speaking, the polarization **P** given by Eq. (4.44) or (4.56) is only the electronic contribution. If the atoms are moving during the adiabatic evolution parametrized by λ, then the currents associated with the classical motions of the charged nuclei also have to be taken into account. We can easily adapt the formalism developed in Section 1.1.4 to express this "ionic" contribution

$$\mathbf{P}_{\text{ion}} = \frac{1}{V_{\text{cell}}} \sum_\mu e Z_\mu \, \boldsymbol{\tau}_\mu . \tag{4.58}$$

Here $\boldsymbol{\tau}_\mu$ gives the location of the nucleus μ in the unit cell, while Z_μ specifies either the atomic number in the case of all-electron calculations, or the core charge in the case of pseudopotential calculations. This expression is also well defined only modulo the quantum $e\mathbf{R}/V_{\text{cell}}$, since each $\boldsymbol{\tau}_\mu$ is arbitrary modulo a lattice vector $\mathbf{R}$. Combining Eqs. (4.56) and (4.58) gives

$$\mathbf{P}_{\text{tot}} = \frac{e}{V_{\text{cell}}} \left[\sum_\mu Z_\mu \, \boldsymbol{\tau}_\mu - \sum_j \frac{\bar{\phi}_j}{2\pi} \mathbf{a}_j \right] . \tag{4.59}$$

The quantity in brackets is well defined modulo $\mathbf{R}$, so that $\mathbf{P}_{\text{tot}}$ is well defined modulo $e\mathbf{R}/V_{\text{cell}}$. The corresponding expression in terms of the reduced polarization variables introduced on p. 161 is

$$\mathbf{P}_{\text{tot}} = \frac{1}{V_{\text{cell}}} \sum_j p_j \mathbf{a}_j \tag{4.60}$$

where

$$p_j = p_{\text{ion},j} + p_{\text{elec},j}$$
$$= \sum_\mu e Z_\mu s_{\mu j} + \sum_n^{\text{occ}} p_{nj} \tag{4.61}$$

with $\boldsymbol{\tau}_\mu = \sum_j s_{\mu j} \mathbf{a}_j$ and $p_{nj} = -e\bar{\phi}_j/2\pi$ (all p_j variables are well defined only modulo e).

If Eq. (4.59) is a valid expression for the polarization, it should be invariant against a translation of the contents of the unit cell by some common vector $\mathbf{t}$ or, equivalently, a translation of the origin by $-\mathbf{t}$. The change of $\mathbf{P}_{\text{ion}}$ under such a shift is $\Delta \mathbf{P}_{\text{ion}} = (Q_{\text{cell}}^+/V_{\text{cell}})\mathbf{t}$, where $Q_{\text{cell}}^+ = e\sum_\mu Z_\mu$ is the total nuclear or ionic charge in the unit cell, so this term by itself is clearly not invariant. To understand what happens to the electronic contribution, it is easiest to return to Eq. (4.44). After the translation, the new Bloch functions are related to the old ones by $\tilde{\psi}_{n\mathbf{k}}(\mathbf{r}) = \psi_{n\mathbf{k}}(\mathbf{r} - \mathbf{t})$, from which it follows that $\tilde{u}_{n\mathbf{k}}(\mathbf{r}) = e^{-i\mathbf{k}\cdot\mathbf{t}} u_{n\mathbf{k}}(\mathbf{r} - \boldsymbol{\tau})$ and thus $\tilde{\mathbf{A}}_{n\mathbf{k}} = \mathbf{A}_{n\mathbf{k}} + \mathbf{t}$ (see Ex. 3.11 and 3.18 for details). The change in the electronic contribution of Eq. (4.44), which we now denote as $\mathbf{P}_{\text{elec}}$, is therefore $\Delta \mathbf{P}_{\text{elec}} = -eN_{\text{occ}}\mathbf{t}V_{\text{BZ}}/(2\pi)^3$. Using $V_{\text{BZ}} = (2\pi)^3/V_{\text{cell}}$ and letting $Q_{\text{cell}}^- = -eN_{\text{occ}}$

be the total electronic charge in the unit cell, this becomes $\Delta \mathbf{P}_{\text{elec}} = (Q_{\text{cell}}^{-}/V_{\text{cell}})\mathbf{t}$. Finally, overall cell neutrality implies that $Q_{\text{cell}}^{-} = -Q_{\text{cell}}^{+}$, so that $\mathbf{P}_{\text{tot}} = \mathbf{P}_{\text{ion}} + \mathbf{P}_{\text{elec}}$ is indeed invariant under global translations of the crystal.

4.3.3 Relation to Wannier Charge Centers

Recall that in Section 3.5.2 we derived Eq. (3.96), which expressed the center of charge of a Wannier function constructed for band n as

$$\bar{\mathbf{r}}_n = \frac{V_{\text{cell}}}{(2\pi)^3} \int_{\text{BZ}} \mathbf{A}_n(\mathbf{k}) \, d^3k. \tag{4.62}$$

You may have noticed the remarkable similarity of this equation to Eq. (4.44), our Berry-phase formula for the electronic polarization. In fact, combining these two equations, one finds that

$$\mathbf{P}_{\text{elec}} = \frac{-e}{V_{\text{cell}}} \sum_{n}^{\text{occ}} \bar{\mathbf{r}}_n, \tag{4.63}$$

so that the electronic contribution to the polarization is determined entirely by the Wannier centers of the occupied bands! In fact, the total polarization of Eq. (4.59) can be written as

$$\mathbf{P}_{\text{tot}} = \frac{e}{V_{\text{cell}}} \left[\sum_{\mu} Z_{\mu} \, \boldsymbol{\tau}_{\mu} - \sum_{n} \bar{\mathbf{r}}_n \right]. \tag{4.64}$$

These expressions assume that all bands n are isolated. For the multiband case discussed in Section 3.6.1, one constructs J Wannier functions $|w_{n\mathbf{R}}\rangle$ that exactly span the J occupied bands of the insulating crystal, although these Wannier functions no longer have a one-to-one correspondence with the individual bands of Bloch eigenstates $|\psi_{n\mathbf{k}}\rangle$, having been constructed as linear combinations of different bands.[8] Nevertheless, Eq. (4.62) can be replaced by

$$\sum_{n} \bar{\mathbf{r}}_n = \frac{V_{\text{cell}}}{(2\pi)^3} \int_{\text{BZ}} \mathbf{A}^{\text{tr}}(\mathbf{k}) \, d^3k, \tag{4.65}$$

as will be shown in Ex. 4.7. Comparing this relation with Eq. (4.45), it follows that Eq. (4.63) still holds.

Equation (4.64) has a remarkably simple interpretation. Recall that in Section 1.1.4 we considered a fictitious physics in which both the positive and negative charges making up a crystal come packaged as point charges carrying an integer multiple of the charge quantum e. For such a system, we argued that we could

---

[8] That is, the index n in $\bar{\mathbf{r}}_n$ does not have the same meaning as in $|\psi\rangle_{n\mathbf{k}}$, although both run from 1 to J.

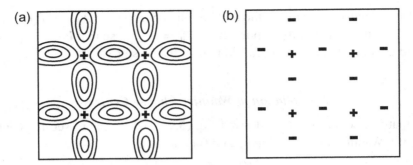

Figure 4.9 Mapping of the true crystalline charge density of an insulator, sketched in (a), onto a point-charge model, as in (b). Contours indicate electronic charge clouds of the real system; '+' symbols denote nuclei carrying charge $+2e$; and '−' symbols indicate integer charges $-e$ located at the Wannier center positions.

define the polarization via Eq. (1.18); namely $\mathbf{P} = (e/V_{\text{cell}}) \sum_j Z_j \boldsymbol{\tau}_j$, where j runs over the charges eZ_j located at $\boldsymbol{\tau}_j$ in the unit cell. Since each $\boldsymbol{\tau}_j$ is only well defined modulo a lattice vector, this definition of $\mathbf{P}$ is only well defined modulo $e\mathbf{R}/V_{\text{cell}}$, as it should be. But Eq. (4.64) takes exactly this form, with the nuclei (or pseudo-ions) appearing as positive charges $Z_j > 0$, and the electrons appearing as point charges with $Z_j = -1$ located at the Wannier centers.

This provides an insightful way to think about the Berry-phase theory of polarization. The true physical charge distribution in an insulating crystal is sketched in Fig. 4.9(a); the nuclei can be treated as integer point charges,[9] but the electrons form smeared-out charge distributions whose probability distribution is given by the Schrödinger equation. It is this latter fact that causes all the trouble, but now we see a way out! We set up a mapping from the true quantum system to one in which the nuclei continue to appear as point charges, but the distributed electron charge clouds are replaced by $-e$ point charges located at the Wannier centers. We then define the polarization of our true system to be that of the mapped system as given by Eq. (1.18). Equation (4.64) tells us that this definition is entirely equivalent to the Berry-phase formulas like Eq. (4.44) or (4.56) given earlier.

Before concluding this section, it is worth pointing out that the theory also takes a natural form when expressed in terms of the reduced coordinates. For discrete bands it is easy to see that if the Wannier centers are expressed in reduced coordinates as $\bar{\mathbf{r}}_n = \sum_j \bar{s}_{nj} \mathbf{a}_j$, then the reduced polarization components p_j defined on p. 161 are just

$$p_{nj} = -e\,\bar{s}_{nj}. \tag{4.66}$$

---

[9] The size scale of nuclei ($\sim$1 fm) is negligible here. In pseudopotential calculations, the nuclear point charges are replaced by spherically symmetric core charges residing inside nonoverlapping core spheres; by Gauss' law, they too can be treated as though they were point charges.

That is, the reduced p_{nj} is really nothing other than $-e$ times the reduced Wannier center position. Thus the quantum of indeterminacy $e\mathbf{R}/V_{\text{cell}}$ of $\mathbf{P}$ corresponds to a quantum $\mathbf{R}$ in $\bar{\mathbf{r}}_n$, a quantum of an integer in $\bar{s}_{nj}$, and a quantum of e in p_{nj}. The same equations apply for the case of the multiband Wannier functions of Section 3.6.1, with the understanding that n now labels Wannier functions and not individual bands.

Moreover, if we return to Eq. (4.54), which defines a 1D Berry phase $\phi_n^{(\kappa_1)}(\kappa_2, \kappa_3)$ in direction κ_1 as a function of (κ_2, κ_3), and use the relation of Eq. (3.97) to replace these by 1D Wannier centers via $\bar{s}_{n1}(\kappa_2, \kappa_3) = \phi_n^{(\kappa_1)}(\kappa_2, \kappa_3)/2\pi$, then the $\bar{s}_{n1}$ can be used in place of the $\phi_n^{(\kappa_1)}$ in Eq. (4.53) (after inserting a factor of 2π). This framework, in which we Wannierize in 1D while keeping wavevector labels in the other dimensions, is known as the *hybrid Wannier representation*, and is developed further in Section 4.5.4.

4.3.4 Practicalities

The comments in this subsection will be of most interest to users of first-principles DFT electronic structure codes developed for the study of crystalline solids. Today, almost all such codes have a capability for computing the Berry-phase polarization. We first review how the calculation is typically implemented in practice, and then discuss some considerations that should be kept in mind in the process of doing the calculations.

Typically most of the computer time in DFT and related methods, as introduced on p. 40, is devoted to the self-consistent field iteration to the ground state for a given set of nuclear coordinates, which is often repeated in an outer iterative loop to allow for structural relaxation. The polarization typically needs to be computed only at the end of this process, and is usually much less expensive because a one-shot calculation using the previously converged Kohn-Sham potential can be used. In this sense, it is similar to the calculation of band structures and densities of states. In particular, the **k**-point sampling need not be the same as the one used in the self-consistent field calculation.

The calculation basically proceeds in the manner of Eq. (4.56), as illustrated in Fig. 4.7. The user chooses a primitive real-space lattice vector $\mathbf{a}_j$ as the direction for which the component of $\mathbf{P}$ (i.e., the average phase $\bar{\phi}_j$) will be calculated. In some cases the choice of $\mathbf{a}_j$ is obvious from the start, as for a tetragonal or rhombohedral ferroelectric where symmetry constrains the polarization to lie along the symmetry axis, or for a superlattice calculation where one is interested only in the polarization normal to the layers. In a more general case, one just starts with the first primitive vector $\mathbf{a}_1$ and then repeats the entire procedure independently for $\mathbf{a}_2$ and $\mathbf{a}_3$, and later assembles the full polarization using Eq. (4.56).

For a given a_j, then, the user is asked to choose a mesh sampling scheme in k-space. Typically this is a regular mesh in the two k-space directions orthogonal to a_j, with a density similar to that used for the self-consistent field (SCF) part of the calculation (shown as 2×2 in Fig. 4.7, but typically somewhat denser). Then, to ensure good convergence for the k-point strings extending along the b_j direction, shown as dashed lines in Fig. 4.7, one usually requests a finer grid, perhaps with 10 or even 20 equally spaced k-points along the path. Empirically it is found that the convergence of the Berry phase with respect to the density of the 1D k-point mesh is slower than that of conventional observables that are sampled in the manner of Eq. (2.46), so some testing of the convergence of the results with respect to this mesh density is recommended.[10]

The code then computes a Berry phase along each k-point string, using the multi-band formulation of Section 3.6.2 to treat the set of all occupied bands as a group. For each neighboring pair of points k_i and k_{i+1} along the string, the code computes the inner product matrix $M_{mn}^{(i,i+1)} = \langle u_{mk_i} | u_{nk_{i+1}} \rangle$ for each $i = \{0, \ldots, N-1\}$. This step is rather trivial in a plane-wave basis, but requires more care in the context of ultrasoft pseudopotential, augmented-wave, muffin-tin–orbital, or localized-orbital approaches. Note that $M_{mn}^{N-1,N}$ is computed as $\langle u_{mk_{N-1}} | e^{-i b_j \cdot r} | u_{nk_0} \rangle$ to account for the fact that the string winds around the BZ as in Eq. (3.74). The Berry phase of the string is usually then just computed from Eq. (3.122) – that is, from the product of the M matrices[11]

$$\phi = -\mathrm{Im}\ln\det \prod_{i=0}^{N-1} M^{(i,i+1)} \tag{4.67}$$

or an equivalent formula such as $-\mathrm{Im}\ln \prod \det M$ or $-\mathrm{Im} \sum \ln\det M$. Note that this formula has the nice feature of being exactly gauge-invariant; it is, insensitive to any unitary rotation among the occupied states at a given k-point, as shown in Ex. 4.6. Thus, there is no need to construct a smooth gauge in advance of this calculation; any set of occupied Bloch states on the mesh, even if they are chosen with random individual phases and random unitary rotations among degenerate states, yields exactly the same result.

As a reminder, in 3D one computes $\phi_1(\kappa_2, \kappa_3)$ on a 2D mesh in (κ_2, κ_3) as illustrated in Fig. 4.7, and then takes the average to get $\bar{\phi}_1$ (and similarly for $\bar{\phi}_2$ and $\bar{\phi}_3$). In calculating this average, the individual values have to be clustered such

---

[10]   The convergence is typically exponential in the case of conventional observables, but only power-law for the Berry phase. Algorithms have been proposed to improve the convergence; see, for example, Stengel and Spaldin (2006).

[11]   As discussed in Section 3.6.2, one can alternatively evaluate the expression in Eq. (3.121), in which the M matrices have been replaced by their unitary approximants $\mathcal{M}$ using the procedure leading to Eq. (B.8). This approach is somewhat closer to the spirit of parallel transport, but involves additional operations. Both formulas tend to the same continuum limit as the mesh density becomes finer.

that the branch choice treats all of them as a group, as illustrated in Fig. 4.8. Most code packages have implemented an algorithm for doing this grouping automatically. Similar considerations apply in 2D.

Incidentally, if one has access to a set of Wannier charge centers $\bar{r}_n$ for all the occupied states – as, for example, if one is using WANNIER90 to compute maximally localized Wannier functions (MLWFs) for some other purpose – then another option is to compute **P** directly from the Wannier-center expression, Eq. (4.63). The result is usually not quite as well converged with respect to **k**-point sampling, however, unless a relatively dense **k** mesh is used in all three dimensions.

In practical calculations, the user should always be on the lookout for artifacts associated with the quantum of polarization. Each code typically uses some algorithm to make a branch choice for each $\bar{\phi}_j$, typically $-\pi < \bar{\phi}_j \leq \pi$, but it may or may not apply a similar reduction to the ionic term of Eq. (4.59), and if so, either before or after combining with the electronic term. Thus, it could be that different code packages might report different **P** values (differing only modulo $e\mathbf{R}/V_{cell}$) using the same physical approximations for the same crystal. Also, when carrying out a series of calculations in which some structural parameter is modified, or as a function of strain or some other external field, the calculated **P** might "jump by a quantum." An additional source of potential confusion is that when performing spin-degenerate calculations, an extra factor of 2 is inserted into the second term of Eq. (4.59) to reflect the spin degeneracy. If at least one of the constituents of the crystal is from an odd-numbered column of the Periodic Table, the ionic term in Eq. (4.59) is only well defined modulo $e\mathbf{R}/V_{cell}$, while the electronic one is modulo $2e\mathbf{R}/V_{cell}$. If all atoms are from even columns, then a quantum of $2e\mathbf{R}/V_{cell}$ can be used consistently for all terms.

Luckily, the changes in polarization that we are interested in are usually small compared to the quantum. Suppose, for example, that you have performed a series of calculations for some ferroelectric Ti compound in which both the tetragonal axis and the polarization are along $\hat{z}$, and you have displaced the Ti sublattice along z in increments of 0.2 Å from one calculation to the next. If you find values of 0.39, 0.43, 0.47, −0.45, and −0.39 μC/cm^2, respectively, you should immediately suspect that there has been a jump by a quantum. Calculating the quantum as ec/a^2, where a and c are the in-plane and out-of-plane lattice constants, respectively, suppose you find this to be 0.97 μC/cm^2. Then you should adjust the last two values by hand to be 0.52 and 0.58 μC/cm^2, respectively, and the polarization difference ΔP_z from the first to the last configuration should be reported as 0.19 μC/cm^2.

In case some doubt remains about the correct branch choice for $\Delta\mathbf{P}$, as when the structural change is substantial and computed **P** values for intermediate configurations are not available, another useful heuristic is to estimate $\Delta\mathbf{P}$ using a rough model based on nominal ionic charges, while treating the crystal as composed of rigid ions carrying these charges. That is, we estimate the change of polarization as

$$\Delta \mathbf{P}_{\text{nom}} = \frac{e}{V_{\text{cell}}} \sum_{\mu} Z_{\text{nom},\mu} \, \Delta \boldsymbol{\tau}_{\mu} \,, \tag{4.68}$$

where the nominal charge $Z_{\text{nom},\mu}$ is identified with the formal ionic charge of atom μ. If we are considering a structural change in Fe_2O_3, for example, we treat Fe as trivalent and assign charges $Z_{\text{nom}}(\text{Fe}) = +3$ and $Z_{\text{nom}}(\text{O}) = -2$. Since we presumably know the path taken by the ions during the structural change, there is no ambiguity about the atomic displacements $\Delta \boldsymbol{\tau}_{\mu}$ and, therefore, no uncertainty about the branch choice of $\Delta \mathbf{P}_{\text{nom}}$. This estimate may be quite rough, but this shouldn't matter, as it is used only as a guide for making the correct branch choice. That is, $\Delta \mathbf{P}$ is actually computed from the Berry-phase theory, and then the uncertainty modulo $e\mathbf{R}/V_{\text{cell}}$ is resolved by choosing the value closest to $\Delta \mathbf{P}_{\text{nom}}$.[12]

Experience has shown that the electric polarization can be computed using standard LDA or GGA versions of DFT with an accuracy comparable to that of other quantities of interest such as energy differences, elastic constants, and phonon frequencies. Moreover, experimental uncertainties in measurements of $\mathbf{P}$ are often often sizable, so questions about the accuracy of standard functionals do not appear to be very pressing in practice.

4.3.5 Polarization in the Presence of Interactions

The preceding discussion has been framed entirely in the context of a single-particle crystalline Hamiltonian of the kind that arises in Hartree–Fock or DFT. What can we say about the case when the real two-particle Coulomb interaction is restored and correlations are treated properly?

We saw in Chapter 1 that statements about quantization and invariance are expected to remain robust in the presence of interactions, at least when these are weak enough that the many-body Hamiltonian can be adiabatically connected to a single-particle Hamiltonian without gap closure. So, for example, we still expect that $\mathbf{P}$ is well defined modulo $e\mathbf{R}/V_{\text{cell}}$, and that adiabatic charge transport is quantized as before. However, the electric polarization itself is not an invariant, and we would like to know how to calculate it in correlated systems. How should we do so? Three kinds of answers to this question have been provided in the literature, as discussed next.

---

[12] If the appropriate branch choice is still not clear when using the nominal ionic charge model, an improved estimate can usually be obtained by replacing the formal charges with the Born effective charges, defined in Eq. (4.19) as the first-order change in $\mathbf{P}$ resulting from a change in the coordinates $\boldsymbol{\tau}_{\mu}$ of sublattice μ. The Born-charge values can either be taken from the literature or computed in a relevant reference configuration.

The first approach goes back to the early work of Niu and Thouless (1984) and considers twisted periodic boundary conditions on a supercell. It was developed by Ortiz and Martin (1994), Resta (1998), and Souza et al. (2000), and is summarized by Resta (2002). Letting $\mathcal{R}_j$ label the supercell lattice vectors and $\mathbf{q}$ be the wavevector in the corresponding small BZ, we assume we have access to the many-body ground state $\Psi_0(\mathbf{q})$ of the finite N-electron system living in the supercell with twisted periodic boundary conditions

$$\Psi(\mathbf{r}_1, \ldots, \mathbf{r}_j + \mathcal{R}_j, \ldots, \mathbf{r}_N) = e^{i\mathbf{q}\cdot\mathbf{r}_j}\, \Psi(\mathbf{r}_1, \ldots, \mathbf{r}_j, \ldots, \mathbf{r}_N)\,. \tag{4.69}$$

Equivalently, we can consider many-body eigenstates obeying ordinary periodic boundary conditions for a Hamiltonian $H_{\mathbf{q}} = e^{-i\mathbf{q}\cdot\hat{\mathbf{r}}}He^{i\mathbf{q}\cdot\hat{\mathbf{r}}}$, which plays a role analogous to $H_{\mathbf{k}}$ of Eq. (2.43) in the single-particle theory. Then a Berry connection can be defined in the small BZ as

$$\mathbf{A}(\mathbf{q}) = i\langle\Psi_0(\mathbf{q})|\nabla_{\mathbf{q}}\Psi_0(\mathbf{q})\rangle \tag{4.70}$$

with corresponding Berry phases and curvatures, and the electric polarization can be expressed as before in terms of the average of $\mathbf{A}(\mathbf{q})$ over the small BZ.

The solution is understood to become exact in the limit that the supercell size becomes large. Since the BZ becomes small in this limit, even sampling a single $\mathbf{q}$-point in the small BZ is equivalent to a dense $\mathbf{k}$-point sampling in the usual BZ, so it is sufficient to compute the polarization with a single-point version of the discrete Berry-phase formula (see Ex. 4.8). This was pointed out by Resta (1998), who also reformulated the theory in terms of replacing the many-body position operator $\hat{x}$ by an exponentiated analogue $\exp(2\pi i\hat{x}/L)$ for a supercell of size L (here written in 1D). Subsequently, Resta and Sorella (1999) argued that this framework provides a fundamental basis for characterizing electron localization and distinguishing between conducting and insulating states.

Unfortunately, currently used methods for computing properties of correlated systems rarely give access to the many-body wave functions. Thus, a second approach is to extend one of the currently used methods for treating correlated systems, such as Green's function or other diagrammatic perturbation schemes, in such a way that the electric polarization can be computed. A recent contribution of this kind is found in the work of Nourafkan and Kotliar (2013), where the dynamical mean-field theory was extended in this way.

The third approach is to invoke exact Kohn–Sham theory. This is not a very practical approach, since we don't know the exact Kohn–Sham functional, but it raises some interesting questions of principle. We know that the exact Kohn–Sham version of DFT should give the exact ground-state charge density and total energy, but it is *not* exact for the ground-state wave function, band gap, and many other observables. What about the Berry-phase polarization? It is closely related to the charge density, and for a finite system the electric dipole clearly does have the sanction of the exact Kohn–Sham theory. But we have emphasized that the Berry-phase polarization depends instead on *currents* that flow during an adiabatic evolution, and these do not have the exact Kohn–Sham sanction. In fact, the question is quite subtle; it seems the answer depends on boundary conditions in a rather nonlocal manner. The interested reader is referred to Vanderbilt (1997) for further discussion.

Exercises

4.6 Demonstrate the multiband gauge invariance claimed just below Eq. (4.67). In particular, suppose we choose one state i' on the chain of $\mathbf{k}$ points and remix the occupied states there by applying a unitary rotation matrix U so that the new states at $\mathbf{k}_{i'}$ become $|\tilde{u}_{n\mathbf{k}_{i'}}\rangle = \sum_m U_{mn} |u_{m\mathbf{k}_{i'}}\rangle$. Show that the new ϕ computed from Eq. (4.67) using the new states is exactly the same as the old one.

4.7 Here we demonstrate Eq. (4.65), assuming we already have a set of multiband Wannier functions in hand. (If you did not previously do Ex. 3.23, do it now, as it will be needed.)

(a) We first evaluate Eq. (4.65) using the particular gauge corresponding to the given set of Wannier functions via Eq. (3.105), whose reverse transform is

$$|\tilde{\psi}_{n\mathbf{k}}\rangle = \sum_{\mathbf{R}} e^{i\mathbf{k}\cdot\mathbf{R}} |w_{n\mathbf{R}}\rangle .$$

Show that this results in Eq. (4.65).

(b) Show that any other multiband gauge choice would give the same result modulo a lattice vector. *Hint:* Use the result of Ex. 3.23 and argue about the boundary conditions on $\beta(\mathbf{k})$.

4.8 When treating systems with defects or disorder, a common practice is to perform the calculation on a large, periodically repeated unit cell ("supercell") containing the structure of interest. Since the corresponding BZ is small, a frequently used approximation is to sample this BZ with only a single $\mathbf{k}$-vector (typically $\mathbf{k} = \mathbf{0}$) when computing properties of the system. The purpose of this problem is to show how even the Berry-phase polarization can be formulated in this single-point context. We work in the multiband context, so the starting point is

$$p_j = \frac{1}{2\pi} \bar{\phi}^{(\kappa_j)},$$

which is Eq. (4.55) except that occupied bands have been summed over and a multiband formula like that in Eq. (4.67) is used to evaluate $\bar{\phi}^{(\kappa_j)}$. Show that if one assumes that only a single **k**-point string needs to be sampled for sufficient accuracy, and that this string passes through the origin, and if it is sufficient to sample only two points along the string (an initial one at **0** and a final one at $\mathbf{b}_j$), then the polarization component p_j is approximated as $p_j = -\mathrm{Im}\ln\det M$, where

$$M_{mn} = \langle \psi_{m0} | e^{-i\mathbf{b}_j \cdot \mathbf{r}} | \psi_{n0} \rangle.$$

See Yaschenko et al. (1998) for further justification.

4.4 Questions of Interpretation

As long as one is interested in computing the change in polarization from one reference configuration to another, the resolution of the quantum in the definition $\Delta\mathbf{P}$ can be carried out following the earlier discussion. However, in practice one would often like to discuss the polarization **P** for a given configuration. In this case, the issues of interpretation of the quantum can become even more confusing.

As an example, suppose we are interested in the spontaneous polarization P_s of KNbO$_3$ in its observed tetragonal structure at 300°C. Experimental literature values of P_s are around $40\,\mu\mathrm{C/cm}^2$, but we would like to check this using a DFT calculation.

The structure of an up-polarized domain of tetragonal KNbO$_3$ is sketched in Fig. 4.10(a). It is a distorted version of a simple cubic perovskite structure in which the K, O, and Nb atoms lie at the face corners, face centers, and body centers, respectively; the ferroelectric distortion corresponds to a displacement of the central

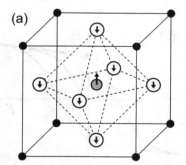 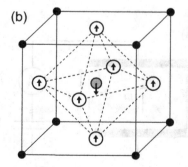

Figure 4.10 Unit cell of KNbO$_3$ in the perovskite structure; filled, gray, and open circles denote K, Nb, and O atoms, respectively. Arrows indicate direction of atomic displacements for (a) up-polarized and (b) down-polarized ferroelectric domains.

Nb along $+\hat{z}$ relative to the surrounding oxygen atoms. The opposite polarization state is illustrated in Fig. 4.10(b).

Taking the distorted atomic coordinates from an experimental X-ray structural refinement, we carry out a DFT calculation of the Berry-phase polarization and find a result of $86\,\mu C/cm^2$. This is nowhere close to the observed value! The quantum $eR_{[001]}/V_{cell} = e/a^2$ is reported as $102\,\mu C/cm^2$, so adding or subtracting this value from $86\,\mu C/cm^2$ clearly will not help the agreement with experiment.

What went wrong?

4.4.1 How Is Polarization Measured?

Let's first be sure we understand the meaning of the experimental value of the spontaneous polarization P_s. There is no way to measure P_s directly; in practice it is always measured by observing the *switching current* that flows during a reversal of the polarization by an external bias. That is, it is determined by hysteresis-loop measurements.

An example is shown in Fig. 4.11. The experimental setup is sketched in Fig. 4.11(a); a time-varying bias $V(t)$ is applied to a ferroelectric film of thickness d and the resulting current $I(t)$ flowing through the ammeter is measured. The measured hysteresis curve is illustrated in Fig. 4.11(b); the vertical axis is really just a measure of the integrated current $Q(t) = \int I(t)\,dt$ that passed from one capacitor plate to the other. Starting from an unpoled sample (equal populations of up and down domains) at the origin of Fig. 4.11(b), a positive voltage is applied ($\mathcal{E}$ along $\hat{z}$) until the polarization P rises and saturates, at which point the entire sample is uniformly up-polarized. Reversing the bias carries the system along the top-left path; reversing it again results in the bottom-right path. E_c marks the coercive field, which is the value of the reversing electric field needed to transform half the sample

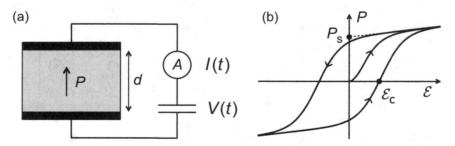

Figure 4.11 (a) Ferroelectric of thickness d in a capacitor configuration with voltage supply and ammeter indicated. (b) Typical hysteresis loop showing change of polarization P with applied electric field $\mathcal{E} = V/d$. Coercive field $\mathcal{E}_c$ and spontaneous polarization P_s are indicated.

into the opposite-domain state. The spontaneous polarization P_s is marked on the diagram; it is an extrapolated estimate of the polarization that a single uniform up domain would have at $\mathcal{E} = 0$.

In short, what is *actually measured* is the current that flows during the polarization reversal; since this is essentially $2P_s$, one divides by 2 (and by the area of the plates) to get the spontaneous polarization. So, we are back to polarization as a difference. Theoretically, we should interpret P_s as being half of the ΔP in going from the structure of Fig. 4.10(b) to that of Fig. 4.10(a); or equivalently, the ΔP to go from a centrosymmetric cubic reference structure to the structure of Fig. 4.10(a).

Let's return to the discussion of the hypothetical DFT results on p. 174, where the experiments suggested $P_s \simeq 40\,\mu C/cm^2$ but the calculated P of an up domain was $86\,\mu C/cm^2$. Now we carry out a corresponding calculation of P in the down-domain state of Fig. 4.10(b) and find that the code reports it as $16\,\mu C/cm^2$. From one perspective this is good news: Combining this result with our previous result of $86\,\mu C/cm^2$ for the structure of Fig. 4.10(a), we get $P_s = (86 - 16)/2 = 35\,\mu C/cm^2$, in quite reasonable agreement with experiment.

But we may also be more confused then ever. Figures 4.10(a) and (b) are inversion images of one another, so the polarization ought to be reversed in going from one to the other. How can the computed values of 86 and $16\,\mu C/cm^2$ make sense in this context?

As a last-ditch effort, we try setting the coordinates of our structure halfway between those in Fig. 4.10, in the centrosymmetric structure, thinking that by symmetry we must get $P = 0$. Instead, the code reports $P = 51\,\mu C/cm^2$. Is something terribly wrong?

4.4.2 *"Formal" versus "Effective" Polarization*

Let's see what we get if we treat centrosymmetric $KNbO_3$ using the nominal ionic model discussed on p. 170. (In fact, this was precisely the object of Ex. 1.3 in Chapter 1.) In the spirit of Eq. (1.18) we are now obtaining the polarization itself via

$$\mathbf{P}_{nom} = \frac{e}{V_{cell}} \sum_{\mu} Z_{nom,\mu}\, \boldsymbol{\tau}_{\mu}, \tag{4.71}$$

rather than just a change in polarization as in Eq. (4.68). We adopt nominal ionic charges of 1, 5 and -2 for K, Nb, and O, respectively, and set $\boldsymbol{\tau}_\mu$ to the ideal coordinates. Taking the origin on the K ion, we get $P_z = (1/a^2 c)[(e)(0) + (5e)(c/2) + (-2e)(c/2 + c/2 + 0)] = e/2a^2$. This is half the quantum, or $102/2 = 51\,\mu C/cm^2$, exactly what we obtained from the DFT calculation!

The resolution of all these apparently confusing results comes from remembering that the Berry-phase polarization is inherently a multivalued quantity, as emphasized on p. 160. Let's restate the results reported earlier in this language. For the up-domain state, the "value" of P_z is the lattice $\{\ldots, -118, -16, 86, \ldots\}$ $\mu C/cm^2$, while for the down-domain state it is $\{\ldots, -86, 16, 118, \ldots\}\,\mu C/cm^2$. Sure enough, these "values" are interchanged by a sign reversal, as they must be under inversion.

Regarding the centrosymmetric configuration, note that there are only two possible sets of P_z values that are mapped onto themselves by inversion symmetry. If the quantum is $102\,\mu C/cm^2$ as assumed previously, then one is $\{\ldots, -102, 0, 102, \ldots\}\,\mu C/cm^2$, while the other is $\{\ldots, -51, 51, \ldots\}\,\mu C/cm^2$. Evidently centrosymmetric KNbO$_3$ belongs to the latter case, as does our classical calculation within the extreme ionic limit. These two calculations were destined either to agree perfectly, or else differ by exactly half a quantum (as it happens, the former case applies).

This is a remarkable result. We are saying that the polarization of centrosymmetric KNbO$_3$ is not zero! More precisely, we are saying that the lattice-valued polarization of KNbO$_3$ does not contain zero as one of its values, but instead straddles zero in a symmetric fashion. The situation is shown graphically as projected onto the x–z plane in Fig. 4.12. Panels (a) and (b) of that figure show two possibilities for **P** that are both consistent with inversion symmetry; KNbO$_3$ belongs to case (b). When the crystal becomes spontaneously polarized along the $\hat{z}$ direction, the polarization lattice shifts upward to the configuration shown in Fig. 4.12(c), and the change in polarization is indicated by the arrow at left in Fig. 4.12(c).

The formal conclusion, that the polarization of centrosymmetric material such as KNbO$_3$ need not be zero, is likely to run up against strong resistance from

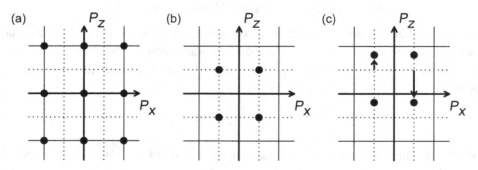

Figure 4.12 Polarization of a tetragonal insulator viewed as a lattice-valued object (lattice of solid dots). Solid (dashed) lines are at integer (half-integer) multiples of the quantum $e\mathbf{R}/V_{cell}$. Two possible "values" of **P** that respect inversion symmetry are shown in (a) and (b). Centrosymmetric KNbO$_3$ has polarization as in (b), while ferroelectrically polarized KNbO$_3$ corresponds to (c) (displacements exaggerated). Arrows in (c) indicate two possible assignments of "effective polarization," with the shorter upward and longer downward ones corresponding to $+35$ and $-67\,\mu C/cm^2$, respectively (see text).

practically minded experimental physicists and materials scientists. With some justification, they are likely to insist that polarization has to be defined such that it vanishes for a centrosymmetric insulator. With this point in mind and to provide a context-sensitive framework for discussing electric polarization, Resta and Vanderbilt (2007) introduced a distinction between two definitions of polarization:

- The *formal polarization* is defined by the Berry-phase theory, or in the classical context, by a formula like Eq. (1.18) or Eq. (4.71). It is intrinsically multivalued modulo a quantum $e\mathbf{R}/V_{cell}$.
- The *effective polarization* is instead defined to be the *change* in polarization relative to a nearby centrosymmetric[13] state.

The practicing computational theorist should be able to read and interpret the results of a calculation in the formal context, and then translate the results into the language of effective polarization for comparison with experiment and discussion with experimental colleagues.

In the case outlined previously, the results for the formal polarizations of the centrosymmetric and up-domain states are $\{\ldots, -51, 51, \ldots\}\,\mu C/cm^2$ and $\{\ldots, -118, -16, 86, \ldots\}\,\mu C/cm^2$, respectively, so that ΔP should be chosen from the values $\{\ldots, -67, 35, 137, \ldots\}\,\mu C/cm^2$. Since $35\,\mu C/cm^2$ is the smallest in magnitude, and the change is also in the expected direction, we can be reasonably confident that $35\,\mu C/cm^2$ is the correct branch choice for ΔP_z. To provide more certainty, we try computing P_{nom} for the up-domain state using the nominal ionic model of Eq. (4.71). Suppose we find it to be $73\,\mu C/cm^2$. Subtracting the previous value of $51\,\mu C/cm^2$ we get $\Delta P_{nom} = 22\,\mu C/cm^2$, pointing to $35\,\mu C/cm^2$ as the closest value. This is the value of effective polarization that we should focus on and report to our more practically minded colleagues.[14]

Of course, there is a potential problem with the definition of effective polarization: It has to be defined relative to a "nearby" nonpolar reference structure. The choice of this reference structure is usually obvious, but there can be exceptions. For example, $BiFeO_3$ is a perovskite ferroelectric that has been of considerable interest in recent years (in part because it is also magnetic and has interesting couplings

---

[13] More precisely, the reference should be nonpolar. This does not necessarily require inversion symmetry; a counterexample is GaAs, in which the tetrahedral symmetry is also sufficient.

[14] Three comments are in order here. (1) Because of spin pairing, the electronic quantum is really $204\,\mu C/cm^2$ in this example. If all calculations are done with a consistent choice of ionic cores, the lattice of possible ΔP values is really spaced by intervals of $204\,\mu C/cm^2$, making the selection of $35\,\mu C/cm^2$ even more obvious. (2) The underestimate made by the nominal ionic model is related to the fact that the Born dynamical charges tend to be larger in magnitude than their nominal values in this class of transition-metal oxides. (3) It may be tempting to inspect the individual changes of ionic and electronic polarization [Eqs. (4.58) and (4.56)] reported by the code, but this can be confusing because there are often large and opposite contributions in the two terms coming from included semicore shells, and as a result the branch choices made for the individual contributions are often rather arbitrary. The strategies described here are usually more productive.

of electric, magnetic, and elastic properties). By averaging the atomic coordinates of $BiFeO_3$ over its two oppositely polarized configurations, we can arrive at a centrosymmetric reference. Unfortunately, this reference structure is metallic (Neaton et al., 2005), so it cannot be used to define a reference polarization. In such cases the safest approach is to consider carefully what is being measured experimentally. If the physical system exhibits a well-defined switching polarization, then the system must be following some kind of insulating path, and the theory should concentrate on identifying this path and computing the change of polarization along it.

4.4.3 Symmetry Considerations

It is worth expounding briefly on the symmetry properties of lattice-valued quantities. For 3D systems in general, and 3D crystals in particular, one usually classifies physical properties as follows:

Scalar: The single element is invariant under spatial rotations
Vector: Triplet of three elements that transforms according to the standard 3×3 rotation matrix $\mathcal{R}$
Second-rank tensor: 3×3 matrix with both rows and columns transforming according to $\mathcal{R}$
Symmetric second-rank tensor: Same as above but with $A_{\mu\nu} = A_{\nu\mu}$
Antisymmetric second-rank tensor: Similar but with $A_{\mu\nu} = -A_{\nu\mu}$

and so on for higher-rank tensors. Implicitly these are all "single-valued" in the sense that there is only one scalar, vector, or matrix assigned to a physical quantity. In the crystalline context, for example, mass density is a scalar, magnetization is a vector,[15] and the dielectric permittivity is a symmetric second-rank tensor. We now expand the list by including

Lattice-valued scalar: 1D lattice of scalars
Lattice-valued vector: 3D lattice of vectors

and so on. The formal polarization **P** is an example of the latter.

Let's return briefly to the case of asynchronicity introduced on p. 160, which is an example of a lattice-valued scalar. To briefly recap the scenario, two clocks each tick once per second, one loudly and one softly, and the asynchronicity A is the time delay from a soft tick to a loud one. Suppose we make an audio recording and play it backward; if we cannot tell the difference, then the physical

---

[15] Quantities such as magnetization that undergo a sign reversal under time reversal or improper spatial rotations are often labeled as "pseudoscalars," "pseudovectors," and so on. We are not interested in this distinction at the moment.

situation respects time-reversal (TR) symmetry. Which values of asynchronicity are consistent with TR symmetry? Clearly there are *two* allowed values: the ticks can either be synchronized, $A = \{\ldots, -1, 0, 1, \ldots\}$ s, or perfectly staggered, $A = \{\ldots, -1/2, 1/2, \ldots\}$ s. Unlike the ordinary time difference $t_2 - t_1$ between two discrete events, there are now two "values" of A that are consistent with TR symmetry.

In the context of crystal physics, the basic symmetry principle governing the allowed values of any physical quantity is that the value of this quantity must transform into itself under any symmetry operator in the crystal point group. This principle applies equally to ordinary and lattice-valued quantities, but as illustrated by the asynchronicity example, the symmetry-allowed possibilities are typically more numerous for lattice-valued quantities.

Let's return to the case of a centrosymmetric tetragonal crystal (cell dimensions $a \times a \times c$) with axis along $\hat{z}$. Let's also assume the point group includes inversion, the fourfold rotation C_4^z about the tetragonal axis, and other operations generated by these; this is the C_{4h} point group. If this crystal is ferromagnetic, its magnetization **M** must be of the form $M\hat{z}$, since any component in the x–y plane would not transform into itself under C_4^z. (Because magnetization is a pseudovector, its z component is not forced to vanish by inversion symmetry.) Its dielectric permittivity has to take the form

$$\epsilon = \begin{pmatrix} u & 0 & 0 \\ 0 & u & 0 \\ 0 & 0 & v \end{pmatrix},$$

and so on. Although an ordinary vector quantity would be forced to vanish, in this case there are four (and only four) possibilities for the formal polarization **P** that are consistent with the C_{4h} point group. Since the quantum of polarization is $e\mathbf{R}/V_{\text{cell}}$, it is convenient to report these in the form $(V_{\text{cell}}/e)\mathbf{P}$. Then the four possibilities consistent with the point-group symmetry are $(V_{\text{cell}}/e)\mathbf{P} = \mathcal{L}_{(0,0,0)}$, $\mathcal{L}_{(a/2,a/2,0)}$, $\mathcal{L}_{(0,0,c/2)}$, and $\mathcal{L}_{(a/2,a/2,c/2)}$, where $\mathcal{L}_\tau$ refers to the set of vectors $\mathbf{R} + \tau$ where **R** runs over all lattice vectors. The first and last of these four possibilities are illustrated in Fig. 4.12(a–b), respectively.

In terms of the reduced polarization variables p_j of Eq. (4.60), the four possibilities are given by setting $p_1 = p_2 = 0$ or $e/2$ and $p_3 = 0$ or $e/2$. In fact, quite generally, each p_j must either be 0 or $e/2$ (mod e) in any centrosymmetric crystal.

The analysis becomes a little trickier to carry out for some other crystal symmetries, but the basic principles are the same. For example, consider the zincblende structure of GaAs. The lattice is face-centered cubic (fcc) with lattice vectors $\mathbf{a}_1 = (0, a/2, a/2)$, $\mathbf{a}_2 = (a/2, 0, a/2)$, and $\mathbf{a}_3 = (a/2, a/2, 0)$. We take the Ga atom to be at $\tau_1 = (0, 0, 0)$ and the As atom at $\tau_2 = (a/4, a/4, a/4)$. The point group is

T_d, which is the full symmetry group of the tetrahedron including diagonal mirrors. The rescaled polarization lattice $(V_{\text{cell}}/e)\mathbf{P}$ is fcc, and the symmetry requires that it should either include the origin or be symmetrically disposed about the origin. Here some investigation reveals that there are again four possibilities: $(V_{\text{cell}}/e)\mathbf{P} = \mathcal{L}_{(0,0,0)}$, $\mathcal{L}_{(a/2,a/2,a/2)}$ (which can also be written as $\mathcal{L}_{(a/2,0,0)}$), $\mathcal{L}_{(a/4,a/4,a/4)}$, and $\mathcal{L}_{(-a/4,-a/4,-a/4)}$.[16] A first-principles calculation confirms what we might guess: namely, that the formal polarization is the same as that of an ideal ionic model with ionic charges $+3e$ and $-3e$ on the Ga and As atoms, respectively.[17] This implies that $(V_{\text{cell}}/e)\mathbf{P} = \sum_\mu Z_\mu \boldsymbol{\tau}_\mu = (-3a/4, -3a/4, -3a/4)$. Using our freedom to translate by $\mathbf{R}$, we find that the correct choice is $(V_{\text{cell}}/e)\mathbf{P} = \mathcal{L}_{(a/4,a/4,a/4)}$. This is valid for the particular orientation of the GaAs crystal assumed here; applying a C_4^z or inversion operator to the GaAs crystal would change the result to $\mathcal{L}_{(-a/4,-a/4,-a/4)}$. However, no rotation or inversion of GaAs can access the other two symmetry-consistent values; these are simply not applicable to GaAs, though they might apply to other insulating crystals having the same point group.

Note that we can convert wurtzite GaAs into diamond-structure Ge by moving a proton from the As nucleus at $(a/4, a/4, a/4)$ to the Ga nucleus at the origin. This changes the result to $\mathcal{L}_{(0,0,0)}$, as might be expected in view of the fact that the diamond structure is intrinsically nonpolar.

The situation becomes even more complicated if we are motivated by spin degeneracy to try to keep track of the polarization modulo $2e\mathbf{R}/V_{\text{cell}}$, instead of just modulo $e\mathbf{R}/V_{\text{cell}}$. For a case like GaAs, where there are nuclei with odd-integer charges, a different choice of atomic basis (i.e., the set of chosen representative atoms $\boldsymbol{\tau}_\mu$ associated with the home unit cell) can shift $\mathbf{P}$ by half of this quantum. In this context, the correct symmetry principle is that a point-group operation acting on the polarization lattice must return it to itself, or to an alternative resulting from another valid choice of atomic basis. When only nuclei of even charge are present, we don't have to worry about that subtlety, but the results can still be surprising. For Si or Ge, for example, the polarization is given by Eq. (4.64) with ionic charges $+4e$ at the atomic sites and $-2e$ at each of the four midbond positions, so that $\sum_\mu Z_\mu \boldsymbol{\tau}_\mu - \sum_n \bar{\mathbf{r}}_n = (-a, -a, -a)$. The result is that $(V_{\text{cell}}/2e)\mathbf{P} = \mathcal{L}_{(a/2,a/2,a/2)}$. So, with spin degeneracy taken into account, the formal polarization does not vanish even for Si or Ge!

---

[16] Intuitively, these are connected with the high-symmetry sites in the unit cell of a simple fcc crystal such as Cu. These are the Cu site itself at $(0, 0, 0)$; the octahedral interstitial site at $(a/2, a/2, a/2)$; and the tetrahedral interstitial sites at $\pm(a/4, a/4, a/4)$. However, $(V_{\text{cell}}/e)\mathbf{P}$ should not be thought of as "located in" any physical unit cell. For example, $(V_{\text{cell}}/e)\mathbf{P}$ is independent of choice of origin, but the atomic coordinates are not. The quantity $(V_{\text{cell}}/e)\mathbf{P}$ just happens to have the same units and mathematical structure as a position in a unit cell.

[17] This can be rationalized in terms of the Wannier-center positions. A maximal localization procedure applied to GaAs gives four Wannier function, each of which lies on one of the four Ga–As bonds, but closer to the As than to the Ga. If we choose the Wannier centers to be the four lying close to the As atom at $(a/4, a/4, a/4)$, and recall that each carries charge $-2e$, we can group these together with the As core charge of $+5e$ to make a unit of net charge $-3e$ centered at the As position, leaving an ionic charge of $+3e$ on the Ga site.

The interested reader can find further discussion of these symmetry considerations in Section III.E of Vanderbilt and King-Smith (1993). From the perspective of a practitioner of first-principles calculations, however, the main thing to keep in mind is that a nonzero formal polarization in a high-symmetry structure is not necessarily a sign that anything is amiss. If you focus instead on computing changes in formal polarization with respect to a high-symmetry reference structure – that is, the effective polarization – then it should not be necessary to delve deeply into these issues.

Exercises

4.9 In Ex. 3.20 we considered a nonsymmorphic 2D insulating crystal with a glide mirror symmetry $\{M_y | \frac{1}{2} a \hat{x}\}$ and an $a \times b$ rectangular unit cell.

(a) Argue that if this insulator has N electrons per unit cell, its polarization changes by $(-Ne/2b)\hat{x}$ under the action of the glide mirror.

(b) The polarization lattice must return to itself under the action of any operation that commutes with H, including this glide mirror. Deduce from this that the number of occupied bands N must be an even integer. *Note:* In Ex. 3.20 similar conclusions were reached by reasoning that Wannier functions should come in pairs related by the nonsymmorphic operation.

4.10 Consider an hexagonal insulator with lattice vectors $\mathbf{a}_1 = (a/2, -a\sqrt{3}/2, 0)$, $\mathbf{a}_2 = (a/2, a\sqrt{3}/2, 0)$, and $\mathbf{a}_3 = (0, 0, c)$.

(a) Assume that the system has C_{3h} point-group symmetry, such that the symmetry group includes the three-fold rotation C_3^z about the z axis, the M_z mirror ($z \leftrightarrow -z$), and other operations generated by these. Show that there are six possible locations for the polarization lattice. What are they?

(b) Now enlarge the point group to C_{6h}, by including the sixfold C_6^z rotation in the list of generators. What are the allowed polarization lattice locations now?

(c) If we keep C_6^z but drop all the operations involving the M_z mirror, we get the C_6 point group. What is the locus of possible polarization lattices now?

4.5 Surface Charge Theorem

In Section 1.1 we introduced the expectation that the electric polarization $\mathbf{P}$ should be connected with the macroscopic surface charge σ_{surf} according to an equation with a form similar to $\sigma_{surf} = \mathbf{P} \cdot \hat{n}$, which is a standard equation for surface bound charge in elementary electrostatics. This would be a good time for the reader to

review Section 1.1, where physical arguments were given in arriving at Eq. (1.6) as
a more precise statement of this relationship. Now that we have carefully introduced
the theory of polarization in the earlier sections of this chapter, it is time to return to
its connection with surface charge, in the form of the *surface charge theorem* first
introduced by Vanderbilt and King-Smith (1993). This theorem applies to a surface
of a crystalline insulator under the following conditions:

1. The surface is also insulating, such that the Fermi level falls in a gap common
   to both the bulk and the surface.
2. The surface is free of defects and disorder, and is periodic with the smallest
   possible primitive cell area A_{surf} consistent with bulk periodicity and surface
   unit normal $\hat{\mathbf{n}}$.
3. The macroscopic electric field in the interior of the bulk crystal vanishes, so
   that $\mathbf{P}$ represents the spontaneous bulk polarization. (In general this requires the
   presence of a nonzero electric field normal to the surface in the vacuum region.)
4. The system is noninteracting (possibly in a DFT mean-field context), or is
   interacting but is adiabatically connected to a noninteracting system by a path
   along which the preceding conditions are satisfied.

Under these conditions, the theorem states that

$$\sigma_{surf} := \mathbf{P} \cdot \hat{\mathbf{n}}. \qquad (4.72)$$

This is Eq. (1.6), but now with the understanding that $\mathbf{P}$ represents the *formal
polarization* as defined in Section 4.4.2. As a reminder, the ':=' carries the meaning
that the left-hand side, which has a definite value for a given surface, is equal to one
of the values on the right-hand side, treating the formal polarization $\mathbf{P}$ as a lattice-
valued object with lattice spacings $e\mathbf{R}/V_{cell}$. Equivalently, if one particular $\mathbf{P}$ value
is chosen on the right-hand side, then Eq. (4.72) states that σ_{surf} is equal to $\mathbf{P} \cdot \hat{\mathbf{n}}$
modulo e/A_{surf}.

Concerning Condition 4 in this list, we shall concentrate on demonstrating the
theorem for the case of a noninteracting system, since the generalization to an
adiabatically connected interacting one is straightforward using arguments similar
to those given in Section 1.1.5. We shall also concentrate on proving it for the
case of a 1D "crystal" (e.g., polymer chain). We can do this because Condition 2
supports the introduction of a 2D surface wavevector $\mathbf{k}_{\parallel}$ as a good quantum
number, such that the 3D case is demonstrated by treating the states at each $\mathbf{k}_{\parallel}$
independently as those of an effective 1D system. Condition 1 ensures that there are
no sudden changes of occupation of these effective 1D systems as a function of $\mathbf{k}_{\parallel}$.
Finally, Condition 3 is needed because, up to now, we have implicitly restricted the
definition of polarization to systems in the absence of macroscopic electric fields
(but see Section 4.6 for an indication of how nonzero fields might be handled).

Let us first restate the surface theorem in 2D as

$$\lambda_{\text{surf}} := \mathbf{P} \cdot \hat{\mathbf{n}} \tag{4.73}$$

with quantum e/L_{surf}, where λ_{surf} is the linear charge density at the "surface" and L_{surf} is the "surface" lattice constant; and in 1D as

$$Q_{\text{surf}} := P \tag{4.74}$$

with quantum e, where Q_{surf} is the net excess charge at the "surface." One normally refers to the surfaces of 2D and 1D objects as "edges" and "ends," respectively, so that a usage like λ_{edge} and Q_{end} might appear more natural. We adopt here a broader view in which "edges" and "ends" are regarded as special cases of "surfaces" for systems of reduced dimensionality. In the same spirit, we shall refer to the bulk linear charge density of a 1D system as $\rho(x)$ in the discussion that follows, even if $\lambda(x)$ might otherwise have been chosen.

Before attempting a proof of Eq. (4.74), we first need to discuss how to define and compute macroscopic surface charges.

4.5.1 Definition of Macroscopic Surface Charge

Consider some infinite and periodic 1D system with lattice constant a and charge density $\rho_{\text{per}}(x) = \rho_{\text{per}}(x - a)$, extending over the entire x axis as shown in Fig. 4.13(a). We also assume bulk neutrality, such that $\int_0^a \rho_{\text{per}}(x)\,dx = 0$. (Our model in Fig. 4.13 has unit lattice constant $a = 1$.) We now imagine truncating the system around $x = 0$ by removing the right half of the chain, creating a "surface" near $x = 0$. A possible shape of the resulting charge density $\rho(x)$ is illustrated in Fig. 4.13(b). We shall assume exponential decay of $\rho(x)$ in the vacuum region ($x \gg 0$), and of $\rho(x) - \rho_{\text{per}}(x)$ in the bulk region ($x \ll 0$).[18]

How can we compute the excess surface charge appearing on the dangling end near $x = 0$? We proceed to discuss two methods that may look different at first sight, but that give mathematically identical results in the end.

Broadening approach. A familiar approach is to compute a macroscopically smoothened density by applying some coarse-graining procedure first, and then integrate this smoothened charge density. Typically one defines a broadening function $s(x)$ with unit weight ($\int s(x)\,dx = 1$) such as the Gaussian function

---

[18] We are imagining a strictly 1D model system, but the theory that follows can equally well be applied to a polymer running along the $\hat{\mathbf{x}}$-direction, or even a 3D crystal with lattice planes lying perpendicular to $\hat{\mathbf{x}}$. In such cases the 1D function $\rho(x)$ represents a spatial integral or average of the physical $\rho(x, y, z)$ over the y–z plane at each x.

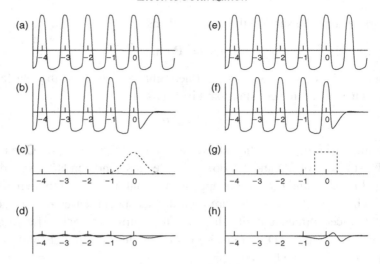

Figure 4.13 Illustration of the broadening-based definition of surface charge at the end of a 1D chain. (a) Periodic bulk charge density $\rho_{\text{per}}(x)$ of the infinite chain. (b) Charge density $\rho(x)$ of semi-infinite chain with surface near $x = 0$. (c) Gaussian broadening function $s(x)$ (dashed). (d) Smoothened surface charge density $\bar{\rho}(x)$ obtained by convoluting $\rho(x)$ with $s(x)$. (e–h) Same, but for rectangular broadening function shown in (g). Surface charge Q is given by integrating the curve in (d) or (h).

shown in Fig. 4.13(c), and convolutes this with $\rho(x)$ to obtain the macroscopically smoothened density

$$\bar{\rho}(x) = \int \rho(x + x')\, s(x')\, dx', \tag{4.75}$$

yielding a result like that shown in Fig. 4.13(d). The surface charge is then obtained by integrating $\bar{\rho}(x)$ over all space.

There is a problem with using a Gaussian function, however, in that $\bar{\rho}(x)$ does not vanish completely in the bulk region ($x < 0$) for any given width σ of the broadening function. In practice, then, the surface charge has to be obtained by a limiting procedure along the lines of

$$Q_{\text{surf}} = \lim_{\sigma \to \infty} \int_{-b}^{b} \bar{\rho}^{(\sigma)}(x)\, dx \tag{4.76}$$

where $\bar{\rho}^{(\sigma)}$ is the result of smoothening with a Gaussian function of width σ, and where the region of integration specified by b also has to grow with σ. This approach to computing the surface charge is therefore not quite as straightforward as one would like.

A clever modification of this approach, which takes advantage of the periodicity in the bulk region, is the *sliding window average* method corresponding to the adoption of the rectangular broadening function shown in Fig. 4.13(g), defined as $s(x) = 1/a$ for $|x| < a/2$ and zero otherwise.[19] Because $\rho_{per}(x)$ is a periodic and neutral charge density, convoluting it with $s(x)$ as in Eq. (4.75) yields zero. As a result, $\bar{\rho}(x)$ now vanishes exponentially for both $x \ll 0$ and $x \gg 0$, as shown in Fig. 4.13(h). Then it is straightforward to obtain

$$Q_{surf} = \int_{-b}^{b} \bar{\rho}(x) \, dx, \tag{4.77}$$

where b is chosen to be large enough to cover the surface region.

There is some flexibility in how we implement this sliding-window algorithm. For example, we may choose a point x_0 that is sufficiently deep in the bulk that the density is bulk-like to some desired accuracy, and let the window function $s(x)$ be $1/a$ in the interval $[-a, 0]$ and zero otherwise. Then if we obtain $\bar{\rho}(x)$ from Eq. (4.75) and evaluate Q_{surf} by integrating Eq. (4.77) from x_0 to infinity, it follows that[20]

$$Q_{surf} = \int_{-\infty}^{\infty} f_{ramp}(x - x_0) \, \rho(x) \, dx \tag{4.78}$$

where $f_{ramp}(x)$ is the ramp-up function

$$f_{ramp}(x) = \begin{cases} 0 & \text{if } x < -a \\ \dfrac{x+a}{a} & \text{if } x \in [-a, 0] \\ 1 & \text{if } x > 0. \end{cases} \tag{4.79}$$

In practice, of course, the limits on Eq. (4.78) just have to be taken from $x_0 - a$ to some point well outside the surface. This is an easy algorithm to implement in a practical calculation.

Tiling approach.   An apparently different approach, but one that will prove more useful for our purposes, is illustrated in Fig. 4.14. Panel (a) shows the same surface charge density $\rho(x)$ as in Fig. 4.13(b) and (f). Now imagine that we can somehow find a localized charge distribution $\rho_{tile}(x)$ (a "tile") with the property that the bulk charge density $\rho_{per}(x)$ is exactly given by a superposition of $\rho_{tile}(x)$ and all of its periodic images:

$$\rho_{per}(x) = \sum_{l=-\infty}^{\infty} \rho_{tile}(x - la). \tag{4.80}$$

---

[19] For a discussion of this window averaging method and higher-order generalizations thereof, see Resta (2010).
[20] See Ex. 4.11.

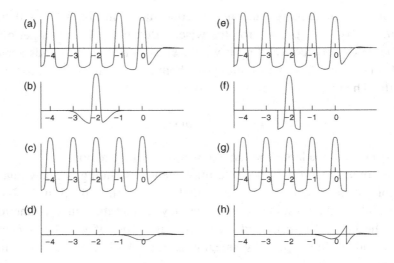

Figure 4.14 Illustration of the tiling-based definition of surface charge at the end
of a 1D chain. (a) Charge density $\rho(x)$ of semi-infinite chain. (b) Tiling charge
density $\rho_{\text{tile}}(x)$ that reproduces the bulk density when superposed with its periodic
images. (c) Model surface density $\rho_0(x)$ obtained by superposing only those tiles
centered at $x = 0$ and below. (d) Difference $\Delta\rho(x)$ between $\rho(x)$ and $\rho_0(x)$. (e–h)
Same, but with $\rho_{\text{tile}}(x)$ chosen as a single-unit-cell slice of the bulk density $\rho(x)$
as in (f). A contribution ΔQ to the surface charge is given by integrating the curve
in (d) or (h).

An example of such a charge distribution, which evidently must be net neutral, is
illustrated in Fig. 4.14(b). Then we can imagine dividing the charge density of the
surface system into two parts, $\rho(x) = \rho_0(x) + \Delta\rho(x)$, where

$$\rho_0(x) = \sum_{l=-\infty}^{0} \rho_{\text{tile}}(x - la) \tag{4.81}$$

and

$$\Delta\rho(x) = \rho(x) - \rho_0(x) \tag{4.82}$$

by definition. Then it is clear that $\rho_0(x)$ will exactly match $\rho_{\text{per}}(x)$ deep in the bulk
region, so that $\Delta\rho(x)$ is guaranteed to decay rapidly in both directions away from
the surface. The $\rho_0(x)$ and $\Delta\rho(x)$ contributions are illustrated in Fig. 4.14(c) and
(d), respectively.

    How does this help us compute the surface charge? Well, let's compute the *dipole
moment* of $\rho_{\text{tile}}(x)$ and the *monopole moment* of $\Delta\rho(x)$:

$$D_{\text{tile}} = \int \rho_{\text{tile}}(x) \, x \, dx, \tag{4.83}$$

$$\Delta Q = \int \Delta \rho(x) \, dx. \tag{4.84}$$

Now the system described by $\rho_0(x)$ is just a simple superposition of localized entities with zero net charge and dipole moments D_{tile} spaced a distance a apart. Classically this gives rise to a macroscopic bulk polarization $P_{\text{tile}} = D_{\text{tile}}/a$ and, therefore, to a surface bound charge of the same magnitude. Taking into account the leftover charge density $\Delta \rho(x)$, we obtain the important result

$$Q_{\text{surf}} = \frac{1}{a} D_{\text{tile}} + \Delta Q. \tag{4.85}$$

One especially simple choice is to let $\rho_{\text{tile}}(x)$ be just the truncation of $\rho_{\text{per}}(x)$ to one unit cell: $\rho_{\text{tile}}(x) = \rho_{\text{per}}(x)$ for $x \in [-a/2, a/2]$ and zero otherwise. This is illustrated in Fig. 4.14(f), leading to the $\rho_0(x)$ and $\Delta \rho(x)$ plotted in Fig. 4.14(g) and (h), respectively. As a variation, we can stop the tiling at some depth x_0 chosen such that $\rho(x)$ is bulk-like for $x < x_0$. That is, we choose a tile corresponding to a slice of the bulk $\rho_{\text{per}}(x)$ over the interval $[x_0 - a, x_0]$ and define $\rho_0(x)$ to be the superposition of these tiles up to and including the interval $[x_0 - a, x_0]$ but not beyond. It follows that $\Delta \rho(x)$ of Eq. (4.82) vanishes for $x < x_0$ and is equal to $\rho(x)$ in the region $x > x_0$ beyond the last tile. It is easy to show that this procedure exactly reproduces the formulation in Eqs. (4.78) and (4.79). After all, the integration over $x > x_0$ in Eq. (4.78) is nothing other than ΔQ, and the integration over $x_0 - a < x < x_0$ is just $(1/a)D_{\text{tile}}$ (see Ex. 4.11).

This result illustrates the mathematical equivalence of the broadening-based and tiling-based definitions of surface charge. At the end of the day, the integration of any one of the bottom panels of Fig. 4.13, or those of Fig. 4.14 (after combining with D_{tile}), is guaranteed to yield the same result for Q_{surf}.

4.5.2 Surface Theorem

The preceding discussion was essentially classical. More precisely, the methods discussed there provide a means to define and calculate the macroscopic surface charge of any surface, classical or quantum, metallic or insulating, provided that the charge distribution is known. We now specialize to the case of the surface of an insulating crystal described by quantum mechanics, and return our attention to proving the surface theorem of Section 4.5 for such a system.

We do this using the tiling approach with a judicious choice of tile. Specifically, we construct a set of Wannier functions from the bulk Bloch bands, and let the bulk tile be composed of all the nuclear (or core) charges and the charge densities of all the Wannier functions belonging to the home unit cell. That is,

$$\rho_{\text{tile}}(x) = e \sum_{\mu} Z_{\mu} \delta(x - \tau_{\mu}) - e \sum_{n} |w_n(x)|^2 , \tag{4.86}$$

where the notation is similar to that of Eq. (4.64) except that we are working in 1D. The dipole of this distribution is

$$D_{\text{tile}} = e \left(\sum_{\mu} Z_{\mu} \tau_{\mu} - \sum_{n} \bar{x}_n \right) . \tag{4.87}$$

Comparing this relation to Eq. (4.64), we see that $P_{\text{tile}} = D_{\text{tile}}/a$ is nothing other than the bulk formal polarization P as given by the full quantum theory. To summarize, Eq. (4.85) can now be written as

$$Q_{\text{surf}} = P + \Delta Q \tag{4.88}$$

where P is the formal Berry-phase polarization computed with this specific choice of bulk tile. We are trying to prove Eq. (4.74), so we want to show that ΔQ is necessarily an integer multiple of the charge quantum.

We will prove this by considering a long but finite chain of lattice constant a and length L, with one end near $x = -L$ and the other end near $x = 0$. (This should be sufficient since the ends are not expected to affect one another for large L.) Starting with the N occupied eigenstates $|\psi_j\rangle$, we follow the discussion on p. 133 in constructing the finite-chain Wannier functions for this system by diagonalizing the position operator in the occupied Hamiltonian subspace. That is, we diagonalize the $N \times N$ position matrix $X_{ij} = \langle \psi_i | x | \psi_j \rangle$, obtaining N real-space Wannier centers $\bar{x}'_i$ and corresponding Wannier functions $|w'_i\rangle$ from the eigenvalues and eigenvectors, respectively. The prime symbols have been attached here to emphasize that the $|w'_i\rangle$ are the finite-system Wannier functions, as opposed to the bulk Wannier functions $|w_n\rangle$ in Eq. (4.86). However, as discussed in Section 3.6.3 and illustrated there via the PYTHTB code of Appendix D.8, these two sets of Wannier functions rapidly become identical as one goes deep into the chain.

We now use the tiling approach by extending the left end of the chain to infinity and writing the charge density of the semi-infinite chain as

$$\rho(x) = \sum_{l=-\infty}^{0} \rho^{(l)}(x) \tag{4.89}$$

where

$$\rho^{(l)}(x) = e \sum_{\mu \in l} Z_{\mu} \delta(x - \tau_{l\mu}) - e \sum_{n \in l} |w'_n(x)|^2 \tag{4.90}$$

accounts for the ionic point charges and electronic Wannier charges associated with cell l. The number of core and Wannier charges in each cell l is kept the same as in the bulk, so that each is neutral, except for the last cell $l = 0$, which accommodates any leftover charges. We then choose an integer number l_0 of unit cells near the right end of the chain and decompose the charge distribution as $\rho(x) = \rho_0(x) + \Delta\rho(x)$, where

$$\rho_0(x) = \sum_{l=-\infty}^{-l_0} \rho^{(l)}(x) \qquad (4.91)$$

comes from the bulk-like region extending to the left and

$$\rho_{\text{surf}}(x) = \sum_{l=-l_0+1}^{0} \rho^{(l)}(x) \qquad (4.92)$$

describes the remainder at the right end of the chain. This is exact, but since both the ionic locations and the Wannier charge distributions converge rapidly to the bulk ones deep in the sample, we can choose l_0 large enough that $\rho_0(x)$ is equal to a sum of bulk tiles $\rho_{\text{tile}}(x)$ given by Eq. (4.86) to any desired accuracy. We already established that D_{tile}/a for this distribution is nothing other than the formal bulk polarization P, so that $\rho_0(x)$ contributes a net charge P to the right end of the chain. The total surface charge is then given by $Q_{\text{surf}} = P + \Delta Q$ as in Eq. (4.88), with ΔQ being the total charge contained in $\rho_{\text{surf}}(x)$ or, because all but the last cell is neutral, the total charge in $\rho^{(l=0)}(x)$. This is an integer multiple of e, thus completing the proof of the surface theorem of Eq. (4.74) for 1D systems.

Note that this analysis provides the macroscopic surface charge completely, not just modulo a quantum e, provided that the numbers of leftover ionic charges and Wannier functions at the surface are known. Also note that a different choice of bulk tiling, in which one or more of the ions or Wannier functions considered to belong to the bulk unit cell is shifted by a lattice constant, will shift P and ΔQ by equal and opposite amounts. Thus, this "choice of basis," as it is referred to in Vanderbilt and King-Smith (1993), has no effect on the surface charge; all choices will result in the same Q_{surf}. The situation is illustrated in Fig. 4.15, where panel (a) indicates the positions of the nuclei (boxes) and finite-chain Wannier centers (minus signs) for some simple model system that terminates at the right end of the figure. The bulk tile (rounded box) has been chosen in different ways in panels (b) and (c). In going from (b) to (c), D_{tile} clearly decreases by ea, P decreases by e, and ΔQ increases by e, such that the Q_{surf} predicted by Eq. (4.88) remains unchanged.

To summarize, the surface charge theorem says that if we want to know Q_{surf} only modulo the quantum e, it is sufficient to use bulk-only methods to compute P from bulk Berry phases or, equivalently, bulk Wannier centers. If we want to

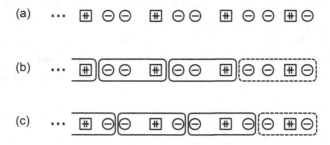

Figure 4.15 (a) Sketch of Wannier analysis of a semi-infinite insulating chain ('...' at left indicates extension to $-\infty$) with two occupied bands (two electrons per unit cell). Squares with '++' symbols indicate nuclei carrying charge $+2e$; minus signs denote finite-chain Wannier centers $\bar{x}'$ as determined by the diagonalization procedure described in the text. (b) One possible assignment of nuclei and Wannier centers to unit cells (i.e., one possible choice of basis) as indicated by the rounded boxes. The dashed box at right indicates the last cell $l = 0$ (see text). (c) Same, but with a different choice of basis.

know the correct branch choice as well, then an analysis in terms of the finite-chain Wannier functions near the surface of a long but finite chain will give us the needed additional information.

4.5.3 A PythTB Example

Some of the previously described concepts are illustrated by the chain_alt_surf.py program provided in Appendix D.10, which returns to the alternating site model of Eq. (2.83) and Fig. 2.6 (see also Appendices D.4 and D.7), but now focuses on the surface properties of a finite chain cut from the bulk. As a reminder, alternating hoppings have strength $t \pm \delta$ and alternating sites have energies $E_p \pm \Delta$. We now reparametrized δ and Δ as

$$\delta = \delta_0 \cos \lambda ,$$
$$\Delta = \Delta_0 \sin \lambda , \tag{4.93}$$

so that δ_0 and Δ_0 set the energy scales for the differences between neighboring hoppings and site energies, while λ modulates their relative importance. In the following discussion, we have adopted a unit lattice constant and somewhat arbitrarily chosen $t = -1$, $\delta = -0.4$, and $\Delta_0 = -0.3$. We study the model as λ is varied.

At half-filling, this bulk model exhibits quantized adiabatic charge transport, since the Wannier center of the bottom band shifts one lattice constant to the right as λ cycles from 0 to 2π. This is not hard to understand: At $\lambda = 0$, the hoppings are symmetric and site 0 is deeper in energy, so the Wannier function begins by being centered on that site. By the time $\lambda = \pi/2$, the site asymmetry has disappeared and the Wannier center has shifted to sit midway between sites 0 and 1 (at $\bar{x} = 0.25$ in the program), so that it is centered on the stronger bond. As the cycle progresses, the Wannier center shifts to site 1, then to the next midbond position, and so on, picking up an additional shift of 0.25 in $\bar{x}$ for every $\pi/2$ in λ. (This kind of charge pump was also explored in Ex. 3.15.)

Program `chain_alt_surf.py` constructs a finite chain composed of 20 cells (40 orbitals) cut from this bulk model. It also introduces a second parameter ΔE_{surf} (`en_shift` in the program) describing an additional site energy shift applied to the last (rightmost) site of the chain. Figure 4.16 shows the computed behavior of various bulk and surface properties as one of the parameters ($\Delta E_{\text{surf}}, \lambda$) is varied

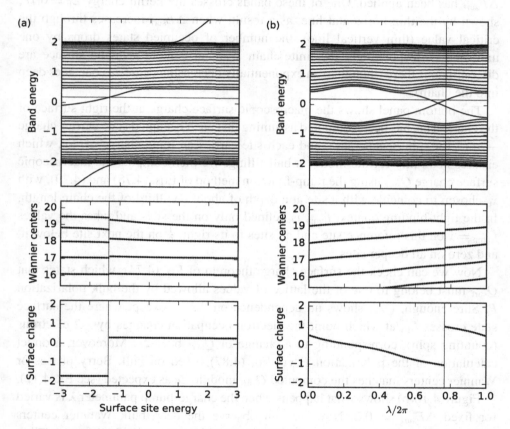

Figure 4.16 (a) Finite-chain band energy, Wannier center locations, and surface (end) charge computed for the alternating site chain model as a function of a shift of the orbital energy on the surface site while keeping the bulk Hamiltonian fixed. (b) Same but for a cyclic variation of the parameter λ in Eq. (4.93) that drives a quantized charge transport in the bulk.

while the other is held fixed. We arbitrarily set the Fermi energy to be $E_F = 0.18$, slightly above the middle of the bulk gap.

Figure 4.16(a) demonstrates the behavior of the system as ΔE_{surf} is varied for fixed $\lambda = 0.15\pi$. On the one hand, since the bulk Hamiltonian is not changing, the bulk band edges (borders of "shaded" regions near ± 2.0 and ± 0.5 in the top panel) do not change with ΔE_{surf}, and the centers of the finite-chain Wannier functions, computed in the same manner as in Section 3.6.3, all tend to the same ΔE_{surf}-independent bulk values as one goes deeper into the chain (middle panels). On the other hand, for some ranges of ΔE_{surf} the applied energy shift at the surface causes the appearance of one or more surface states, corresponding to the curves that depart from the bulk-band regions in the top panel. A further inspection (not shown) indicates that these surface states are strongly confined near the right surface, where ΔE_{surf} has been applied. One of these bands crosses the Fermi energy $E_F = 0.18$, shown by the thin horizontal line; as a result, when ΔE_{surf} increases through the critical value (thin vertical line), the number of occupied states drops by one unit, and correspondingly the finite-chain Wannier centers near the surface are discontinuous there (though to an exponentially decreasing degree as one goes deep into the chain).

The bottom panel shows the macroscopic surface charge at the right surface of the chain. This has been computed assuming that each occupied state carries charge $-2e$ (2 for spin degeneracy), and each site carries an ionic charge of $+e$, which ensures neutrality for the model at half-filling. We then compute the macroscopic surface charge Q_{surf} using the ramp-function method of Eqs. (4.78) and (4.79), with x_0 chosen to coincide with a site at a depth of about one third of the chain length. In the tight-binding context f_{ramp} is defined only on the sites and takes the values $f_{ramp} = 1$ on this reference site and all sites to its right, $\frac{1}{2}$ on the next site below it, and zero on all deeper sites.

Now we can check the surface charge theorem of Eq. (4.74), which states that Q_{surf} must belong to one of the lattice of values allowed by the bulk polarization P. Sure enough, Q_{surf} shows no dependence on ΔE_{surf} except where the surface state crosses E_F, at which point the electron occupation changes by -2 electrons (counting spin), corresponding to a change of Q_{surf} by $+2e$. Moreover, a direct calculation of the polarization P via Eq. (4.87) based on bulk Berry phases or Wannier centers matches the computed Q_{surf} modulo $2e$ as expected (see Ex. 4.12).

Figure 4.16(b) shows what happens when the charge-pump parameter λ is varied for fixed $\Delta E_{surf} = 0.2$. Now we can observe the deep bulk Wannier centers shifting in the positive direction as λ increases (middle panel), with a corresponding decrease in Q_{surf} (bottom panel). The situation corresponds to a Chern number of $+1$ for the charge pump ($+2$ counting spin), with ϕ increasing by 2π, $\bar{x}$ increasing by a, and P decreasing by $2e$. But there is a potential paradox here. If the cycle pumps one set of spin-paired electrons from the left end of the chain to the right, it seems the surface charge must have changed by $-2e$ at the right surface and by $+2e$ at the left after the cycle has completed. However, the chain Hamiltonian is identical at $\lambda = 0$ and $\lambda = 2\pi$, so the surface charge cannot change! Where did the excess charge go?

The answer is provided by the behavior of the surface states that are visible crossing the gap in the top panel. One of these, the up-crossing one (positive $dE/d\lambda$), is localized at the right end of the chain, as is evident from the fact that the nearby Wannier centers are disturbed. The other is localized on the left end of the chain, where it has a negligible effect on the right-end properties shown. The presence of these surface states, and their gap-crossing behavior, is actually *required* for charge conservation, since the depopulation of the right-end surface state is needed to dispose of the charge pumped to that end, and the population of the left-end surface state is needed to inject an electron pair into the bulk, so that the charge distribution can return to its initial state after the cycle.

Here we have learned an important lesson. The Chern number associated with the adiabatic charge transport *in the bulk* manifests itself *at the surface* in the form of a surface state that crosses from the occupied to the unoccupied manifold. More generally, the same kind of charge-conservation argument demonstrates that the excess number of up-crossing versus down-crossing surface states at the right end of the chain must equal the bulk Chern number, and similarly for the left end (but with up-crossing and down-crossing interchanged). We have seen hints of this kind of *bulk-boundary correspondence* in Section 1.3.2, and as will become evident in the next chapter, it turns out to be central to the theory of topological insulators.

4.5.4 Hybrid Wannier Functions and Surfaces of 3D Insulators

The discussion presented to this point has been in 1D. How can we generalize the proof of the surface theorem to 3D, as it was given in Eq. (4.72)? A derivation directly in terms of the 3D Wannier functions in the cells near the surface was provided in Vanderbilt and King-Smith (1993). Here, however, we take the opportunity to develop a very useful theoretical construct that we hinted at earlier on p. 167, namely the *hybrid Wannier representation*, and use it to demonstrate the surface theorem in 3D. The hybrid Wannier representation will also play an important role in the discussion of topological insulators in Chapter 5.

For a general crystal, such as one having rhombohedral or monoclinic lattice symmetry, the lattice vectors may not be orthogonal, and as a result it is inconvenient to work in Cartesian coordinates. Instead, we shall work here in lattice (or "reduced") coordinates. For the surface of interest, let $\mathbf{b}_3$ be the primitive reciprocal lattice vector normal to the surface, and let $\mathbf{a}_1$ and $\mathbf{a}_2$ be two primitive real-space vectors parallel to the surface, obeying the usual relation $\mathbf{a}_i \cdot \mathbf{b}_j = 2\pi \delta_{ij}$. To model the surface we will consider a slab consisting of some number M_3 of unit cells in the $\mathbf{b}_3$ direction, but repeated indefinitely in the $\mathbf{a}_1$ and $\mathbf{a}_2$ directions. As a reminder, in lattice coordinates the wavevectors $\mathbf{k}$ are rewritten via Eq. (2.31) in terms of $\kappa = (\kappa_1, \kappa_2, \kappa_3)$, and the real-space lattice vectors $\mathbf{R}$ are written via Eq. (2.26) in terms of integer-valued vectors $\ell = (\ell_1, \ell_2, \ell_3)$. We also now introduce $\mathbf{s} = (s_1, s_2, s_3)$ as the lattice-vector coordinates of a real-space position $\mathbf{r} = s_1 \mathbf{a}_1 + s_2 \mathbf{a}_2 + s_3 \mathbf{a}_3$. This framework allow us to imagine the crystal to have been transformed onto a simple cubic lattice with an (001) surface.

Now the idea of the hybrid Wannier representation is to carry out the Wannier transformation in one direction only, here the $\mathbf{b}_3$ or surface-normal direction. Before considering the surface, however, let's see how this works in the bulk. The transformations between Bloch functions and Wannier functions were given in Eq. (3.80), and after translating into lattice coordinate this becomes

$$|w_{n\ell}\rangle = \frac{1}{(2\pi)^3} \int_{\mathrm{BZ}} e^{-i\kappa \cdot \ell} |\psi_{n\kappa}\rangle \, d^3\kappa \,, \tag{4.94a}$$

$$\Updownarrow \text{ FT}$$

$$|\psi_{n\kappa}\rangle = \sum_{\ell} e^{i\kappa \cdot \ell} |w_{n\ell}\rangle \,. \tag{4.94b}$$

Each Wannier function $|w_{n\ell}\rangle$ is localized in all three directions and is labeled by the indices (ℓ_1, ℓ_2, ℓ_3) of the unit cell associated to it, in addition to a band-like index n.

We can instead obtain *hybrid Wannier functions* by carrying out the Fourier transform in only one direction, say the $\mathbf{b}_3$ direction:

$$|h_{n,\kappa_1\kappa_2\ell_3}\rangle = \frac{1}{2\pi} \int_0^{2\pi} e^{-i\kappa_3\ell_3} |\psi_{n,\kappa_1\kappa_2\kappa_3}\rangle \, d\kappa_3 \,, \tag{4.95a}$$

$$\Updownarrow \text{ FT}$$

$$|\psi_{n,\kappa_1\kappa_2\kappa_3}\rangle = \sum_{\ell_3} e^{i\kappa_3\ell_3} |h_{n,\kappa_1\kappa_2\ell_3}\rangle \,. \tag{4.95b}$$

The resulting hybrid Wannier functions (HWFs) $|h_{n,\kappa_1\kappa_2\ell_3}\rangle$ are localized in the vicinity of the ℓ_3'th layer of unit cells (i.e., decaying exponentially with $|s_3 - \ell_3|$), but remain extended and Bloch-like in the other directions.

This is very similar in spirit to what was done in the paragraph leading up to Eq. (4.56) (where the special direction was κ_1 instead of κ_3). Recalling that Berry phases and Wannier centers are essentially the same thing, we can reinterpret the discussion of Berry phases presented there in terms of hybrid Wannier centers

$$\bar{s}_{n3}(\kappa_1, \kappa_2) = \langle h_{n,\kappa_1\kappa_20} | s_3 | h_{n,\kappa_1\kappa_20} \rangle \tag{4.96}$$

in the present context (here written for "home" HWFs at $\ell_3 = 0$). If we also express the polarization in reduced coordinates via Eq. (4.49), $\mathbf{P} = (-e/V_{\text{cell}}) \sum_j p_j \mathbf{a}_j$, then the contribution to p_3 from band n is just

$$p_{n3} = \frac{1}{(2\pi)^2} \int \bar{s}_{n3} \, d\kappa_1 d\kappa_2 . \tag{4.97}$$

This is the average of $\bar{s}_{n3}$ over the 2D projected BZ, so we can also write this as $p_{n3} = \bar{\bar{s}}_{n3}$, where one bar is for the quantum expectation value in Eq. (4.96) and the other is for the BZ average in Eq. (4.97). This equation is just Eq. (4.55), reexpressed in terms of Wannier centers instead of Berry phases.

So far, this discussion has been presented as though for the case of isolated energy bands labeled by index n, but in fact the entire theory becomes more natural in the multiband formulation of Section 3.6.1. That is, for a given (κ_1, κ_2), we use the methods of Section 3.6.3 to calculate the multiband Berry phases $\phi_n^{(\kappa_3)}$ or, equivalently (after dividing by 2π), the maximally localized 1D Wannier centers $\bar{s}_{3,n}$, and insert these into Eq. (4.97) for the polarization. Note that n now labels Wannier centers, not energy bands. From now on we shall imagine that this is how the bulk 1D Wannier centers have been obtained.

Now let's return to the case of a slab built by stacking M_3 layers of cells along the $\mathbf{b}_3$ direction. The system retains periodicity along lattice vectors $\mathbf{a}_1$ and $\mathbf{a}_2$ so that κ_1 and κ_2 remain good quantum numbers, but we can no longer work with κ_3 since the periodicity is broken in the surface-normal direction. In Section 3.6.3, we learned how to construct finite-chain Wannier-like functions by diagonalizing the position operator, and we can do the same here. That is, for each (κ_1, κ_2) we treat the system as a finite chain in the third direction and diagonalize the s_3 operator in the space of occupied states to obtain the Wannier centers $\bar{s}_{j3}(\kappa_1, \kappa_2)$, where j runs up to the number N of occupied bands of the slab ($N \simeq JM_3$ where J is the number of occupied 3D bulk bands).

This is very similar to what was done in the middle panels of Fig. 4.16, where the chain Wannier centers were plotted versus some parameter ΔE_{surf} or λ appearing in the chain Hamiltonian. Here there are two important differences. First, the dependence is not upon a Hamiltonian parameter, but rather upon the other wavevector coordinates (κ_1, κ_2). Second, we are trying to prove the surface theorem in the case of an insulating surface, which means that there are no crossings of surface energy bands through E_F, and therefore no sudden discontinuities in the Wannier centers of the kind that appeared in Fig. 4.16. We thus expect the "Wannier sheets" $\bar{s}_{j3}(\kappa_1, \kappa_2)$ of the slab to be smooth and periodic functions of (κ_1, κ_2), and to approach exponentially to the bulk sheet positions given by Eq. (4.96) as one goes deep into the slab.

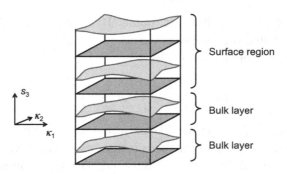

Figure 4.17 Sketch of the surface region of a crystalline insulator from the viewpoint of the hybrid Wannier representation. Lightly shaded warped sheets show the locations of the hybrid Wannier centers (real space plotted vertically) as a function of the in-plane reduced wavevector (reciprocal space plotted horizontally). Heavily shaded rectangles indicate positions of positive ionic charge. The system is periodic in the (κ_1, κ_2) directions and extends periodically into the bulk of the slab at the bottom of the figure.

This situation is loosely sketched for a bulk insulator with a single occupied band in Fig. 4.17, which shows the top region of the slab. The darkly shaded flat surfaces denote the positions of the positively charged nuclei or ions, while the lighter gray sheets are plots of the electronic Wannier centers $\bar{s}_{j3}(\kappa_1, \kappa_2)$ near the surface. There is vacuum above the top Wannier sheet, and the pattern at the bottom of the figure is understood to repeat periodically into the bulk of the slab. The Wannier sheets in the lower layers of the figure are shown as having converged to a bulk-like behavior, but the ones at the surface may look quite different depending on details of the surface Hamiltonian.

Now the point is that, just as the bulk polarization can be decomposed into an average over (κ_1, κ_2) as in Eq. (4.97), so can the surface charge. We have already demonstrated the surface theorem for a 1D system in Section 4.5.2, so we just apply the same reasoning to connect the polarization contribution at (κ_1, κ_2) to the corresponding surface charge contribution from (κ_1, κ_2), and then average over the 2D BZ to demonstrate the desired 3D surface charge theorem. In particular, we now rewrite Eq. (4.88) for the case of a surface charge density as

$$\sigma_{\text{surf}} = \frac{1}{A_{\text{surf}}} p_3 + \Delta\sigma \tag{4.98}$$

where $\Delta\sigma$ accounts for all of the ionic and electronic sheets included in the "surface region" in Fig. 4.17, and $p_j = p_{\text{ion},j} + p_{\text{elec},j}$ as in Eq. (4.61), with

$$p_{\text{elec},3} = \frac{1}{(2\pi)^2} \sum_n \iint p_{n3}(\kappa_1, \kappa_2) \, d\kappa_1 d\kappa_2 , \tag{4.99}$$

accounts for the bound charge coming from all of the deeper sheets (A_{surf} is the surface cell area). Recalling that the p_j are well defined modulo e and noting that $\Delta\sigma$ is an integer multiple of e/A_{surf}, we can write this as

$$\sigma_{surf} := \frac{1}{A_{surf}} p_3 \qquad (4.100)$$

where ':=' indicates equality modulo e/A_{surf}.

We can reexpress this in terms of **P** by noting that $\hat{\mathbf{b}}_3$ is in the surface normal $\hat{\mathbf{n}}$ direction and the unit cell spacing along $\hat{\mathbf{n}}$ is $2\pi/b_3 = V_{cell}/A_{surf}$. Then from Eq. (4.48) we have that $p_3 = (V_{cell}/2\pi)(b_3\hat{\mathbf{n}}) \cdot \mathbf{P} = A_{surf} \mathbf{P} \cdot \hat{\mathbf{n}}$, and Eq. (4.100) turns into the desired result:

$$\sigma_{surf} := \mathbf{P} \cdot \hat{\mathbf{n}}. \qquad (4.101)$$

This equation is sufficient if one wants to know only the surface charge modulo the quantum e/A_{surf}. Instead, if the surface Wannier sheets have been computed, one can go back to Eq. (4.98) to resolve the correct choice of branch.

As a final comment, the reader is reminded that this expression assumes an insulating surface. If the surface is metallic (while the bulk remains insulating), Eq. (4.101) has to be generalized to

$$\sigma_{surf} := \mathbf{P} \cdot \hat{\mathbf{n}} - \frac{e}{(2\pi)^2} \int_{BZ} f(\mathbf{k}) \, d^2k \qquad (4.102)$$

where $f(\mathbf{k})$ is the occupation function (0 or 1) for the surface band at wavevector $\mathbf{k}$. The last term just measures the occupied area fraction of the 2D surface BZ. Referring back to Fig. 1.3, if the Fermi energy starts in the gap under an isolated surface band and is continuously raised until the surface band becomes filled, the second term in Eq. (4.102) tracks the change in σ_{surf} in the intermediate metallic state, resulting in an overall change by the quantum $-e/A_{surf}$.

4.5.5 Interface Theorem

The surface charge theorem given earlier has an obvious extension to the case of an insulating interface between two crystalline bulk materials, in which Eq. (4.72) generalizes to

$$\sigma_{int} := (\mathbf{P}_1 - \mathbf{P}_2) \cdot \hat{\mathbf{n}} \qquad (4.103)$$

where $\hat{\mathbf{n}}$ points into material 2 and away from material 1. For this to make sense, the interface has to be free of defects and *coherent*, such that the entire structure (interface and two bulk materials) must all share the same 2D periodicity described by an interface unit cell of area A_{int}. The "equality modulo a quantum" expressed by the ':=' symbol in Eq. (4.103) then refers to equality modulo e/A_{int}. Of course, the system also has to be globally insulating, with a Fermi level in a gap common to the interface as well as both bulk materials.

There is also an implicit assumption here, as well as in Eq. (4.72), connected with the fact that the polarization $\mathbf{P}$ as defined so far has to be that of a material in zero macroscopic electric field. However, if a surface charge σ_{surf} or σ_{int} is present, it generally does give rise to macroscopic fields. In the case of the surface theorem of Eq. (4.72), we can always insist on describing a case in which the macroscopic field vanishes inside the material and resides only in the vacuum region outside, but clearly such a strategy is not available in the case of an interface. In the next section we briefly discuss how to define $\mathbf{P}$ in the presence of an $\mathcal{E}$-field. In practice, however, it is usually enough to use Eq. (4.103) to compute the bound charge that would be present at the interface if the interface were somehow neutralized by some compensating charges, and then use the theory of linear dielectric media to estimate the effect of removing the compensating charges.

4.5.6 Is the Formal Polarization Experimentally Measurable?

The question sometimes arises whether the formal polarization, as given by the Berry-phase theory of Eq. (4.59), is experimentally measurable. Our derivation of the theory was based on computing changes in $\mathbf{P}$ as a Hamiltonian is varied, and experimentally what is measured in practice is almost always the change in polarization as indicated by a measurable current flow. Nevertheless, we can argue that the formal polarization is experimentally measurable *in principle*, albeit perhaps not in practice. The surface and interface theorems provide support for this claim by suggesting the following recipe.

To measure the formal polarization in lattice direction j, start with a sample of the material and prepare a perfectly ordered and defect-free surface for the surface oriented normal to $\mathbf{b}_j$. By "perfectly ordered," we mean that it should have the maximum possible translational periodicity parallel to the surface (an "unreconstructed" or "1×1" surface). If possible, prepare the surface so as to be insulating, and then measure the surface charge, say by observing the discontinuity in the electric field at the surface. This allows you to deduce one component p_j of the reduced polarization via Eq. (4.100). Alternatively, prepare the surface to be neutral, and measure the fraction of the 2D BZ that is occupied by metallic electron or hole pockets (e.g., via angle-resolved photoemission), and deduce p_j from Eq. (4.102). If preparation of a perfectly ordered vacuum-exposed surface proves too difficult, prepare a perfectly ordered interface with another crystal whose formal polarization is already known, and use Eq. (4.103) to obtain the needed polarization component. After repeating this procedure for two other surface orientations, you will have determined the formal polarization $\mathbf{P}$.

There are so many practical objections to these procedures that it is not worth elaborating them. Nevertheless, this line of argument does support the claim that the formal polarization can be said to be experimentally measurable *in principle*.

Exercises

4.11 Finish the demonstration of the equivalence between the ramp-function approach, Eqs. (4.78–4.79), and the tiling approach, Eq. (4.85), for defining the macroscopic surface charge. That is, fill in the steps needed to complete the argument given at the very end of Section 4.5.1.

4.12 Modify the `chain_alt_surf.py` program so that it calculates the bulk polarization P from the k-space Berry-phase expression [i.e., from the bulk $\bar{x}$ using Eq. (4.87), inserting a factor of 2 for spin in the appropriate place], and compare with the results of the existing implementation to check the truth of the surface theorem, Eq. (4.74). Also count the excess number of Wannier centers and ionic charges remaining after a tiling that terminates a few cells before the end of the chain, and check that the value of Q_{surf} is correctly given (absolutely, not just modulo a quantum) by Eq. (4.88).

4.13 (a) Show that an ideal surface of an insulator with unit normal $\hat{n}$ can be nonpolar only if the 1D lattice of values $\mathbf{P} \cdot \hat{n}$ includes zero when running over the 3D lattice of $\mathbf{P}$ values. By "ideal" we mean the surface is defect-free and has no broken in-plane periodicity, and by "nonpolar" we mean simultaneously insulating and charge-neutral.

(b) Review the discussion of the polarization lattice of GaAs on p. 180, and consider ideal surfaces normal to (001), (011), and (111). Which of these can be nonpolar?

4.6 Uniform Electric Fields

Up to this point, our treatment has been limited to the case of insulators in a vanishing macroscopic electric field. How can we treat the case when a uniform macroscopic field $\mathcal{E}$ is present? The answer to this question is far from obvious, since the presentation until now has been built on the basis of Bloch's theorem, which requires that the crystal potential be periodic. This rules out the presence of a macroscopic electric field $\mathcal{E}$, since this would imply a change by $e\mathcal{E} \cdot \mathbf{R}$ of the electron potential under a translation by a lattice vector $\mathbf{R}$. Moreover, by following the electric field "downhill" the electron potential energy $V(\mathbf{r})$ decreases without bound, and under these conditions we cannot even properly speak of a well-defined ground state!

Nevertheless, if we start with an insulating crystal in its ground state and adiabatically apply a modest electric field, there should be a reasonably well-defined "state" that we can solve for. Indeed, perturbative treatments of the application of an electric field have long been known, as reviewed briefly on p. 149. What we seek is an approximate state of the system, corresponding not to a true ground state but rather to a very long-lived resonance, with the property that the charge density (and other single-particle expectation values) remain periodic despite the fact that $V(\mathbf{r})$ is not strictly periodic. The decay of this state occurs by *Zener tunneling*, a process in which an electron in the valence band tunnels into a lower-energy state in the conduction band some distance away. For small to moderate electric fields, the characteristic tunneling time can be exponentially long, so that the state is well defined for all practical purposes.

These physical arguments do not tell us how to find such a state in practice. A solution to this problem was proposed by Souza et al. (2002) and Umari and Pasquarello (2002). In the density-functional context, they suggested searching for a state that minimizes the electric enthalpy functional

$$F = E_{\mathrm{KS}}(\{\psi_{n\mathbf{k}}\}) - \boldsymbol{\mathcal{E}} \cdot \mathbf{P}(\{\psi_{n\mathbf{k}}\})\,. \tag{4.104}$$

Here E_{KS} is the ordinary zero-field Kohn–Sham energy functional while $\mathbf{P}$ is the Berry-phase polarization, both expressed in terms of a set of wave functions $\{\psi_{n\mathbf{k}}\}$ of Bloch form on a $\mathbf{k}$-point grid in the same way as for a zero-field calculation. In this context, however, the $\{\psi_{n\mathbf{k}}\}$ are *no longer eigenstates* of the Hamiltonian. Instead, they can be interpreted as providing a convenient representation of the one-particle density matrix, which is expected to remain periodic in the presence of the perturbing electric field (Souza et al., 2002, 2004).

The implementation of this scheme has some unconventional aspects, since Bloch functions at neighboring $\mathbf{k}$-points, which are normally only indirectly coupled via charge self-consistency, become directly coupled through the Berry-phase expression for $\mathbf{P}$. We shall not pursue this theory further here, except to mention that features for treating finite electric fields are now available in several widely used electronic structure codes including ABINIT and QUANTUM ESPRESSO, and that further details of the theory can be found in Souza et al. (2002, 2004) and Stengel et al. (2009).

5

Topological Insulators and Semimetals

Beginning in about 2005, the investigation of topological phases of matter emerged as an exciting theme, and has since become established as an important new subfield in the condensed matter physics research community. The seeds for this development were planted by Thouless, Haldane, and other investigators[1] in the years following the discovery of the quantum Hall effect with the theoretical prediction of a possible insulating 2D state now known as a *quantum anomalous Hall insulator*. However, the field didn't really take off until the appearance of the papers of Kane and Mele (2005a, 2005b) and Fu and Kane (2006), who identified a class of 2D nonmagnetic *spin Hall insulator*s whose Bloch states are topologically twisted in a certain sense. The term *topological insulator* (TI) came to be attached to both types of states. The concept of TIs was then rapidly extended to describe 3D nonmagnetic insulators, where "weak" and "strong" TIs were identified (see, for example, Fu et al., 2007; Moore and Balents, 2007), and eventually to topological crystalline insulators, topological superconductors, topological semimetals, and other topological phases.

The impact of these developments has been dramatic in part because they reflect a paradigm shift in the understanding of the classification of phases of matter. Until recently, the dominant framework was that of phase transitions governed by symmetry breaking and order parameters. A canonical example is that of a ferromagnet such as bcc iron as it is cooled through its Curie temperature, below which there is a reduction of symmetry and the appearance of an order parameter, the magnetization, which was absent in the high-temperature paramagnetic state. Identified with the school of Landau, this paradigm has been used with outstanding success to describe an enormous variety of phase transitions in condensed matter physics. In the case of TIs such as the quantum spin Hall state, however, there is

---

[1] The 2016 Nobel Prize in Physics was awarded to Thouless, Haldane, and Kosterlitz for early developments in the theory of topological phases of matter.

no symmetry that distinguishes the topological from the trivial state, and no order parameter that becomes nonzero on the topological side.

Instead, the basic paradigm for defining TIs is quite different. Two insulators are said to belong to the same topological class if and only if their Hamiltonians can be continuously connected, at least in principle, in such a way that the gap never closes at any point along the connecting path. Here there is a close analogy to the classification of closed geometric surfaces by genus, where two surfaces are said to be topologically equivalent if they can be deformed into one another without any "violent events" such as poking new holes or healing old ones. In this sense, a coffee cup and a doughnut belong to the same classification (genus 1) but are distinct from a sphere (genus zero). In the classification of TIs, the "violent events" are gap closures at which the system becomes metallic at some points or in some intervals along the connecting path. An insulating crystal is said to be *trivial* if it can be adiabatically connected without gap closure to an atomic limit (crystal of well-separated atoms with discrete sets of occupied and unoccupied atomic orbitals), and is *topological* otherwise.

A systematic review of the developments of the theory of TIs and related topological states will not be attempted here. The reader is referred to excellent reviews by Kane (2008), Hasan and Kane (2010), Moore (2010), Qi and Zhang (2011), and Bansil et al. (2016), and the book by Bernevig (2013). The goal in the present chapter is to introduce the concept of TIs by building, in a natural way, on the theory of Berry phases and curvatures developed in the previous chapters, but now applied to crystalline insulators in 2D and 3D. We shall begin with the quantum anomalous Hall (QAH) state, which was introduced phenomenologically in Section 1.3.2, and which can be regarded as the parent state from which other topological states are derived. We shall then build from there to understand the topological classifications of nonmagnetic and magnetic systems in 2D and 3D more generally.

5.1 Quantum Anomalous Hall Insulators

The physics of the QAH effect is closely related to the phenomenon of 1D quantum adiabatic transport discussed in Section 4.2.3. There we saw that if there is a nonzero Chern number C associated with the Bloch bands on the 2-torus (λ, k), where λ labels progress around an adiabatic loop and k is the Bloch wavevector, then this corresponds to a charge pump of C electrons by a unit cell per cycle, as illustrated, for example, by the winding of the Wannier centers in Fig. 4.5. We now focus instead on the 2D parameter space (k_x, k_y) describing the Bloch wavevector of a 2D crystalline insulator, and see what physics can be attached to the Chern index defined on the 2-torus in this case.

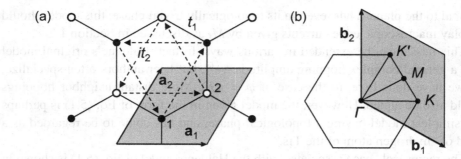

Figure 5.1 (a) Haldane model in real space. Lattice vectors $\mathbf{a}_1$ and $\mathbf{a}_2$ span the shaded parallelogram unit cell containing orbitals 1 (filled circles) and 2 (open circles). Real hoppings t_1 connect nearest neighbors, and imaginary hoppings it_2 connect second-neighbor sites in the directions indicated by arrows (or $-it_2$ in the reverse directions). (b) Parallelogram Brillouin zone spanned by reciprocal vectors $\mathbf{b}_1$ and $\mathbf{b}_1$; high-symmetry special points are labeled.

5.1.1 2D Chern Insulators

In a landmark paper, Haldane (1988) introduced a simple tight-binding model demonstrating the possibility of such a nonzero Chern number in the 2D Brillouin zone (BZ). This model describes spinless electrons hopping between sites as sketched in Fig. 5.1(a). There are two sites (labeled '1' and '2' and shown as filled and open circles, respectively) on a honeycomb lattice, with an s-like orbital on each site. The real-space lattice vectors $\mathbf{a}_1$ and $\mathbf{a}_2$ form the shaded parallelogram unit cell. The Hamiltonian written in second quantized notation [c.f. Eq. (2.60)] is

$$H = \Delta \sum_i (-)^{\tau_i} c_i^\dagger c_i + t_1 \sum_{\langle ij \rangle} (c_i^\dagger c_j + \text{h.c.}) + t_2 \sum_{\langle\langle ij \rangle\rangle} (i c_i^\dagger c_j + \text{h.c.}) \qquad (5.1)$$

where i and j run over all sites, $\tau_i = \{1, 2\}$ is the sublattice index of the site, and t_1 is the nearest-neighbor hopping strength for first-neighbor pairs indicated by $\langle ij \rangle$ (each pair counted once). The model is assumed to be half filled. In the absence of the t_2 term, this Hamiltonian describes a nonmagnetic system like the p_z orbitals of single-layer graphene ($\Delta = 0$) or boron nitride ($\Delta \neq 0$).

Crucially, Haldane added the last term, which ascribes a complex hopping amplitude it_2 to each second-neighbor pair $\langle\langle ij \rangle\rangle$, where the sites are arranged such that the dashed arrows in Fig. 5.1(a) point from site j to site i; the Hermitian conjugate term indicated by 'h.c.' describes corresponding hopping amplitudes $-it_2$ for hops in the reverse direction. This term breaks time-reversal (TR) symmetry, since complex conjugation reverses its sign. We have essentially broken TR "by hand" by adding this term, so we now have to think of this as a model of a ferromagnetic system with some orbital magnetization M_z, where z is the direction normal to the plane. Thus, even in its topologically trivial phase, this model should display macroscopic edge currents given by M_z as discussed in Section 1.2.

This model can be extended in various ways. In fact, Haldane's original model had a general complex hopping amplitude $t_2 e^{i\varphi}$, but later authors often specialize, as we have done here, to the case of $\varphi = \pi/2$. More distant-neighbor hoppings could also be added. However, the model given in the form of Eq. (5.1) is perhaps the simplest model having a topological phase, and has come to be regarded as a kind of "hydrogen atom of the TIs."

The reciprocal space associated with the Haldane model of Eq. (5.1) is shown in Fig. 5.1(b). The Wigner-Seitz BZ is shown by the hexagonal outline, but we prefer to focus on the parallelogram BZ indicated by the shaded region, and to imagine it as wrapped into a 2-torus by the primitive reciprocal lattice vectors $\mathbf{b}_1$ and $\mathbf{b}_2$.

The model is solved by the PYTHTB program `haldane_bsr.py` appearing in Appendix D.11. Typical results are shown in Fig. 5.2, where $\Delta = 0.7$ and $t_1 = -1$ were held fixed while t_2 was set to 0, -0.06, -0.1347, and -0.24, respectively, in panels (a–d). The interesting action is at the K and K' points, where it happens

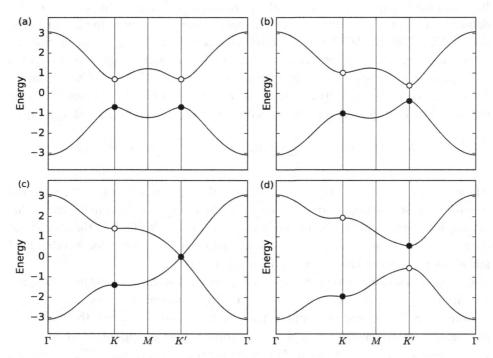

Figure 5.2 Band structures for the Haldane model of Eq. (5.1), with $\Delta = 0.7$, $t_1 = -1.0$, and (a) $t_2 = 0$; (b) $t_2 = -0.06$; (c) $t_2 = -0.1347$; (d) $t_2 = -0.24$. Filled (open) circles mark states of pure site-1 (site-2) character; band inversion at K' is evident in (d).

that the contribution of t_1 to $H_{\mathbf{k}}$ drops out due to a cancellation of phases, and the energies are given by the analytic solutions

$$E_{\mathbf{k}} = \pm \begin{cases} \Delta - 3\sqrt{3}t_2 & \text{at } K, \\ \Delta + 3\sqrt{3}t_2 & \text{at } K'. \end{cases} \tag{5.2}$$

When $t_2 = 0$, the states at K and K' are related by TR symmetry and so necessarily have the same energies, but as t_2 is turned on the gaps at K and K' develop an asymmetry. We take $t_2 < 0$, in which case the gap grows at K and shrinks at K' as t_2 gets more negative. Eventually the gap at K' closes at the critical value $t_{2c} = -\Delta/3\sqrt{3} \simeq -0.1347$, and then reopens for $t_2 < t_{2c}$.

There is also an important change of character of the band-edge states at t_{2c}. Because t_1 has no influence at K or K', the states at these locations are either of pure atom-1 character (filled circles), or atom-2 character (open circles), as shown in Fig. 5.2. Significantly, a reversal between the character of the valence-band maximum and the conduction-band minimum is visible at K' for $t_2 < t_{2c}$.

It turns out that such a *band inversion* is typical of many kinds of topological transition. Indeed, for $t_2 > t_{2c}$, we can adiabatically send t_1 and t_2 to zero at fixed Δ and convert the system into a trivial atomic state (completely flat bands of simple atom-1 or atom-2 character), corresponding to our notion of an "atomic limit" in this system. For $t_2 < t_{2c}$, however, this cannot be done without a gap closure at K', suggesting that the system is in a nontrivial topological class in this region of parameter space.

Let us now see what the connection is to the theory of Berry curvatures and Chern numbers developed in the earlier chapters. Since the system is at half-filling, we concentrate on the properties of the occupied lower band. At $t_2 = 0$ the system has TR symmetry, and as discussed on p. 105, this enforces that $\Omega(\mathbf{k}) = -\Omega(-\mathbf{k})$; that is, the Berry curvature is an odd function of wavevector.[2] As a result, the integrated Berry flux vanishes due to cancellations between contributions at $\mathbf{k}$ and $-\mathbf{k}$, and the Chern number of the band (total Berry flux divided by 2π) is zero. This is confirmed in Fig. 5.3(a), which shows a contour plot of the Berry curvature in the BZ for this case. The plot was generated by program `haldane_bcurv.py` in Appendix D.12, and shows that there are positive and negative concentrations of Berry curvature near K and K', which will obviously cancel when integrated over the BZ. Program `haldane_bcurv.py` also integrates the Berry curvature over the 2D BZ so as to report the Chern number, which is zero as expected for this case.

Because it is constrained to be an integer, the Chern number must remain zero as long as t_2 is gradually decreasing without gap closure. Fig. 5.3(b) shows how this

---

[2] At $\Delta = 0$ the system also has inversion symmetry, so that Ω vanishes identically at all $\mathbf{k}$ according to the rules given on p. 105. The plots shown in Fig. 5.3 are at nonzero Δ, so that inversion symmetry is broken.

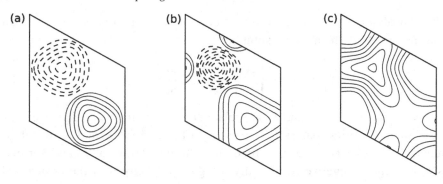

Figure 5.3 Contour plots of Berry curvature $\Omega(\mathbf{k})$ for the Haldane model with $\Delta = 0.7$, $t_1 = -1.0$, and (a) $t_2 = 0$; (b) $t_2 = -0.06$; (c) $t_2 = -0.24$. Parallelograms indicate the same region of reciprocal space as in Fig. 5.1(b). Full and dashed lines denote positive and negative contour levels, respectively; contours begin at $|\Omega| = 0.05$ and increase in geometric progression in powers of 2. Panels (a–c) here correspond to panels (a), (b) and (d) of Fig. 5.2. Panel (c) is the basis for the plot appearing on the front cover of this book.

comes about: The concentration of Berry curvature becomes more diffuse near K where the gap is increasing, while it becomes more strongly peaked near K' where it is decreasing, in such a way that the positive and negative concentrations continue to cancel. As t_2 approaches t_{2c} where the gap closure occurs, the concentration of Ω near K', which amounts to a total Berry flux of about $-\pi$, approaches the form of a delta function. When the gap opens again for t_2 just below t_{2c}, the $-\pi$ Berry flux is replaced by a $+\pi$ Berry flux, which becomes more diffuse as t_2 decreases further as shown in Fig. 5.3(c). The program also reports the Chern number for each case, revealing a jump from $C = 0$ to $C = 1$ as t_2 passes through t_{2c}. A 2D insulator having a nonzero Chern index, as in Fig. 5.3(c), is often termed a *Chern insulator*.

As discussed in Sections 3.2.3 and 3.2.4, the Chern index on a torus in (μ, ν) parameter space can also be obtained by observing the winding of the Berry phase $\phi^{(\nu)}$ as μ is carried around its cycle. In the case of a simple 2D insulator with Cartesian lattice vectors, the corresponding parameter space would be that of the wavevector components (k_x, k_y), and we would seek to track the Berry phase in the k_y direction, or equivalently the hybrid Wannier center $\bar{y}(k_x)$ (see Section 4.5.4), as k_x is carried around the loop from 0 to $2\pi/a_1$. As we shall see, this viewpoint helps clarify what is "topological" about the Chern phase where $C \neq 0$.

In the case of the Haldane model, the lattice vectors are not orthogonal, so it is more convenient to work with the reduced wavevector components (κ_1, κ_2) related to $\mathbf{k}$ by Eq. (2.31). Thus, we shall compute $\phi^{(2)}(\kappa_1)$, or equivalently the hybrid Wannier center $\bar{s}_2(\kappa_1)$, as κ_1 is carried adiabatically around the loop from 0 to 2π. Here we are using the notation of Eq. (4.96) in Section 4.5.4, except that we now have a 2D bulk with a 1D "surface" (edge). The PYTHTB program

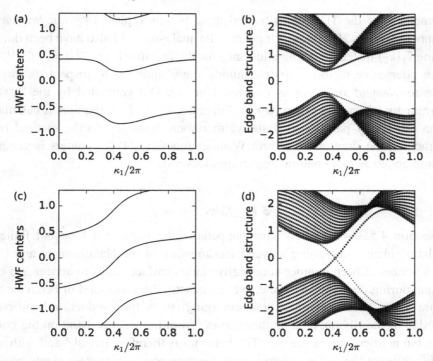

Figure 5.4 Flow of (a) hybrid Wannier centers, and (b) edge state energies on a ribbon cut from the Haldane model with $\Delta = 0.7$ and $t_1 = -1.0$ in the trivial phase, $t_2 = -0.06$. (c–d) Same, but in the topological phase, $t_2 = -0.24$. Surface states on the top and bottom edges of the ribbon are indicated by full and reduced intensity, respectively.

`haldane_topo.py` in Appendix D.13 performs the needed calculation for the cases of $t_2 = -0.06$ and $t_2 = -0.24$, which we know are in the trivial and topologically phases, respectively, from Fig. 5.3(b–c).

In the trivial case, the hybrid Wannier centers plotted in Fig. 5.4(a) are simply periodic in κ_1; the lack of winding again confirms that $C = 0$. For the topological state, Fig. 5.4(c) shows the corresponding behavior. Now the hybrid Wannier centers $\bar{s}_2(\kappa_1)$ shift up by one unit as κ_1 cycles through 2π, confirming that the Chern number is $C = 1$. We also see that the flow of the hybrid Wannier centers becomes most rapid at about $\kappa_1 \simeq 2\pi/3$, at the projection of the point K' where the gap closure occurred, and where there is still a significant concentration of Berry curvature visible in Fig. 5.3(c).

Within the hybrid Wannier viewpoint, then, the meaning of the topological index C is just that it records the number of times the Wannier center $\bar{s}_2$ wraps by a lattice vector as κ_1 cycles around the BZ. It is very much like the story of the quantized adiabatic charge transport discussed in Section 4.2.3, except that the adiabatic parameter λ of the 1D system treated there is now replaced by one wavevector component of our 2D system. Of course, the analysis could also have been done by tracing $\bar{s}_1(\kappa_2)$ instead, making allowance for the fact that $C = C^{(12)} = -C^{(21)}$.

An alternative method for determining the $\mathbb{Z}$ index is to inspect the flow of the *entanglement spectrum* as illustrated in Fig. D.3 generated by the PYTHTB program `haldane_entang.py` in Appendix D.14. This requires a calculation on an artificially partitioned finite-width ribbon, however, so the method based on tracking the flow of the hybrid Wannier function (HWF) centers is generally recommended as more natural and straightforward.

5.1.2 Edge States

In Section 4.5.3 we studied what happens at the ends of a long but finite 1D insulating chain undergoing a cyclic modification of the Hamiltonian, and found that a nonzero Chern number necessarily causes surface states to appear and cross the gap during the adiabatic cycle. We see exactly the same kind of behavior here. Figure 5.4(b) shows a surface state emerging out of the conduction band on the top edge of the ribbon, and another coming from the valence band on the bottom edge, but neither crosses the gap. The behavior is therefore trivial,[3] and with some modification of the surface Hamiltonian the appearance of surface states could be eliminated entirely. For the QAH system with $C = 1$ in Fig. 5.4(d), by contrast, we see that there is an up-crossing of one surface state on the top edge of the ribbon and a down-crossing of another surface state on the bottom edge, signaling a topological behavior.

The qualitative similarity between the flow of the Wannier centers and the surface states that is visible in Fig. 5.4 is not an accident; it is a consequence of the bulk–boundary correspondence discussed at the end of Section 4.5.3. As κ_1 evolves through 2π, the effective 1D system along direction s_2 defined by that κ_1 acts like a quantized adiabatic charge pump, as illustrated in Fig. 4.16(b). The cyclic adiabatic parameter λ appearing there is the analogue of the cyclic wavevector κ_1 in the present case. In the last few paragraphs of Section 4.5.3 we pointed out that the flow of the Wannier centers *requires* a corresponding flow of the surface states based on charge conservation. To repeat the argument in the present context, we know that a cycle of κ_1 pumps one electron from the bottom to the top edge of the ribbon ($C = 1$), but the contribution of this κ_1 to the edge charge is the same at the beginning and end of the cycle; this is possible only if there is one up-crossing surface state at the top edge that can serve to remove an electron when the energy of

---

[3] In this case we can have an insulating edge, and the surface charge λ_{surf} could be computed using the methods of Section 4.5.4 but in one lower dimension.

this state rises past the Fermi energy, and a down-crossing one at the bottom surface to inject an electron there.

The parameters chosen for Fig. 5.4(c–d) produce a state with Chern number $C = +1$, but it is trivial to check that reversing the sign of t_2 produces a state with $C = -1$. Further modifications to the model (Fruchart and Carpentier, 2013) can easily generate states with $C = \pm 2$, and so on. By making plots like those in Fig. 5.4, one can easily check that the flow of hybrid Wannier centers and edge states is consistent in every case, with the Chern number corresponding to the net number of up-crossings on these diagrams.

5.1.3 Connection to Quantum Anomalous Hall Physics

This is an excellent time to review the discussion in Section 1.3, since the up-crossing surface state in Fig. 5.4(b) is exactly like the one we anticipated in Fig. 1.10(b). From the discussion there, we argued that if an edge has a single up-crossing surface state (or in general an excess of one up-crossing surface state over down-crossing ones) in the clockwise direction, then this must be true on all edges of the sample, reflecting a bulk integer index C that we tentatively called "chirality" there. We now finally see that C is just the Chern number!

We have been illustrating the issues involved using the Haldane model, for which the relevant Chern number is just that of a single occupied band, but the generalization to multiband insulators is straightforward. If the bands are isolated, then the "chirality" determining the count of crossing edge states is just the total Chern number $C = \sum_n^{\text{occ}} C_n$; more generally it is given in terms of the BZ integral of the band-traced Berry curvature in Eq. (3.136).

Importantly, in Section 1.3 we argued on physical grounds that the "chirality," which we now know to be the Chern number, is directly related to the anomalous Hall conductivity (AHC) by

$$\sigma_{\text{AHC}} = \sigma_{yx} = C \frac{e^2}{h}. \tag{5.3}$$

This finally provides an answer to the question of what physical response corresponds to the Chern number. The arguments given there were based on charge conservation at the edge of the sample when subjected to a weak electric field, but we can also give two other demonstrations of this relation. The first is heuristic, and the second is based on the linear response theory developed in Section 4.2.

The heuristic derivation proceeds as follows. We consider a single-band Chern insulator (we assume $C = 1$) with an $a \times b$ rectangular unit cell, although relaxing these restrictions is not difficult (see, for example, Ex. 5.2). According to the semi-classical Boltzmann transport theory, which will be developed later in Section 5.1.7, a wave packet centered at wavevector $\mathbf{k}$ evolves in response to a weak electric field by translating in reciprocal space at a rate[4]

$$\dot{\mathbf{k}} = -\frac{e}{\hbar}\mathcal{E}. \qquad (5.4)$$

Letting the field point to the left, $\mathcal{E} = \mathcal{E}_x\hat{x} = -\mathcal{E}_0\hat{x}$, this means that k_x shifts to the right by one reciprocal lattice vector $2\pi/a$ in a "Bloch oscillation time" $T_B = (2\pi/a)/(e\mathcal{E}_0/\hbar) = h/ea\mathcal{E}_0$. In this period of time, every occupied electron is shifted in the real-space $\hat{y}$ direction by a distance b according to the Wannier flow picture of Fig. 5.4(c), generating a sheet current $K_y = -eb/T_B A_{\text{cell}} = (e^2/h)\mathcal{E}_x$. In other words, $\sigma_{yx} = e^2/h$ as claimed!

A second, and more formal, derivation of this result is obtained using the linear response formalism of Section 4.2.1. There we arrived at Eq. (4.22), which in 2D takes the form

$$\frac{\partial_\nu\langle\mathcal{O}\rangle}{A_{\text{cell}}} = \frac{-e\hbar}{(2\pi)^2}\sum_n^{\text{occ}}\int_{\text{BZ}} 2\text{Im}\,\langle u_{nk}|\mathcal{O}T_{nk}^2 v_{k,\nu}|u_{nk}\rangle\,d^2k. \qquad (5.5)$$

This describes the linear response of some operator $\mathcal{O}$ (or more precisely, its expectation value density) to a first-order electric field applied in Cartesian direction ν. From this it is straightforward to compute the induced current in direction μ:

$$\begin{aligned}
\sigma_{\mu\nu} &= \partial_\nu K_\mu \\
&= \frac{\partial_\nu\langle -ev_\mu\rangle}{A_{\text{cell}}} \\
&= \frac{e^2\hbar}{(2\pi)^2}\sum_n^{\text{occ}}\int_{\text{BZ}} 2\text{Im}\,\langle u_{nk}|v_{k,\mu}T_{nk}^2 v_{k,\nu}|u_{nk}\rangle\,d^2k.
\end{aligned} \qquad (5.6)$$

For $\mu = \nu$ the expectation value on the right-hand side is clearly pure real, so that the longitudinal conductivities σ_{xx} and $\sigma_{yy} = 0$ vanish when the imaginary part is taken, as they must. In contrast, the off-diagonal elements do not. Using Eq. (4.29) twice to remove velocity operators in favor of wavevector derivatives, we find

$$\sigma_{yx} = \frac{e^2}{(2\pi)^2\hbar}\sum_n^{\text{occ}}\int_{\text{BZ}} 2\text{Im}\,\langle\partial_y u_{nk}|Q_{nk}|\partial_x u_{nk}\rangle\,d^2k. \qquad (5.7)$$

But $Q_{nk} = 1 - |u_{nk}\rangle\langle u_{nk}|$ can be replaced here by the identity operator because the additional term entering Eq. (5.7) is of the form $\langle\partial_y u|u\rangle\langle u|\partial_x u\rangle = (iA_y)(-iA_x)$, which is pure real. With the Q_{nk} factor gone, we use the Berry curvature formula of Eq. (3.67) to write this as

---

[4] This is a standard result covered in elementary solid state physics texts, but we can also understand it by comparing Eqs. (2.23) and (2.45), where we see that $(e/c)\mathbf{A}$ and $\hbar\mathbf{k}$ enter $H_\mathbf{k}$ in the same way. Thus, perturbations $\mathbf{k}$ and $(e/\hbar c)\mathbf{A}$ act in the same way, and using $\mathcal{E} = (-1/c)\dot{\mathbf{A}}$ leads to the claimed result.

$$\sigma_{yx} = \frac{e^2}{(2\pi)^2 \hbar} \sum_n^{\text{occ}} \int_{\text{BZ}} \Omega_{n,xy} \, d^2k. \tag{5.8}$$

For a Chern insulator, we can use Eq. (3.68) and $\hbar = h/2\pi$ to write this as

$$\sigma_{yx} = \frac{e^2}{h} \sum_n^{\text{occ}} C_n, \tag{5.9}$$

which is again just Eq. (5.3) with C being the total Chern number of the occupied band space. Eq. (5.9) was originally derived by D. J. Thouless et al. (1982) in the context of the integer quantum Hall effect in the presence of a periodic potential. The presence of a nonzero transverse induced current $\partial \langle v_y \rangle / \partial \mathcal{E}_x$ in Eq. (5.6) came to be referred to as an *anomalous velocity* effect.

To summarize, we saw in Chapter 1 that the presence of an AHC of $\sigma_{\text{AHC}} = Ce^2/h$ in a 2D insulator is not ruled out on physical grounds as long as the "chirality" C is an integer, in which case C must correspond to the count of chiral edge states at the boundaries. In Section 5.1.1 we found that any 2D insulator is characterized by a "Chern number" C describing the total Berry flux of the occupied bands integrated over the 2D BZ or, equivalently, the flow of Berry phases (or hybrid Wannier centers) in one lattice direction as the wavevector component in the other lattice direction is varied. We have now also shown, using two independent arguments, that these two constants C are one and the same. We thus have in hand our first kind of TI, namely the "Chern insulator" or "QAH insulator" in 2D. Systems with $C = 0$ are said to be "normal" or "trivial," while those with $C \neq 0$ are "topological." Because the topological index runs over the set $\mathbb{Z}$ of all integers, this is referred to as a topological "$\mathbb{Z}$ classification."

5.1.4 Experimental Realizations

While the QAH or Chern-insulating state is arguably the simplest paradigmatic example of a TI, attempts at an experimental realization of this state were slow to bear fruit. While the theoretical prediction of the existence of 3D nonmagnetic TIs (Fu and Kane, 2007; Fu et al., 2007; Moore and Balents, 2007) was rapidly followed by experimental verification (Hsieh et al., 2008; Xia et al., 2009), and despite the fact that the Haldane model dates to 1988, it was not until 2013 that the first 2D QAH state was experimentally demonstrated.

Theorists were not idle in the interim. An excellent review of theoretical efforts to propose materials systems that could realize the QAH state can be found in Weng et al. (2015). The essential ingredients are as follows:

- The system has to be two-dimensional. It may occur naturally at the surface of some crystal or at the interface between two crystals, or in the form of a naturally 2D material such as graphene. It may also be engineered by some chemical addition or substitution to such 2D systems, or by the intentional growth of ultrathin films with a chosen composition.
- It must be magnetic, with the TR symmetry spontaneously broken in such a way as to allow the appearance of a magnetization component normal to the plane. This constraint arises because the conditions allowing the integrated Berry flux $\int_{\text{BZ}} \Omega_{xy} \, d^2k$ to be nonzero are the same as those allowing a nonzero M_z. The paradigmatic case is a ferromagnet with magnetization normal to the plane, but a ferrimagnetic system (having two or more distinct magnetic sublattices such that the magnetic moments do not cancel) and certain noncollinear magnetic states are also possibilities. In any case, long-range magnetic order must be established.
- It must be insulating. Unfortunately, antiferromagnetic insulators and magnetic metals seem to be more common in nature than ferromagnetic insulators.
- The spin-orbit coupling (SOC) must be strong enough to induce the desired band inversion (discussed later in this section). Systems containing heavy atoms from the lower region of the Periodic Table are therefore favored.
- The system must be sufficiently stable that it can form, or be synthesized, with good structural and chemical order.

It is not easy to satisfy all of these constraints simultaneously.

The need for strong SOC arises as follows, at least in the conventional framework in which a collinear magnetic state develops in response to ferromagnetic exchange couplings in the spin sector.[5] In the limit that the SOC strength $\lambda_{\text{SOC}} = 0$ vanishes, such a system is well described by the local spin-density approximation (LSDA) of Eq. (2.18), in which the up-spin and down-spin systems are independent and each is governed by a real Hamiltonian (i.e., obeying $KH = HK$, where K is complex conjugation). From this it follows that the Berry curvature obeys $\Omega_{xy}(\mathbf{k}) = -\Omega_{xy}(-\mathbf{k})$, so that its integral vanishes ($C = 0$) and a Chern-insulating state is impossible.[6] Essentially, the spontaneously broken TR symmetry in the spin channel is not communicated to the orbital channel, which is the one that governs the AHC.

If we now imagine that λ_{SOC} is gradually tuned from zero to its physical value, and if the final state is to be a QAH state, then it must be the case that the gap closes and reopens, establishing a band inversion, in the manner of Figs. 5.2 and 5.3. Since the maximum shift of the band-edge states that can occur during this process is set by the SOC strength, the size of the final inverted gap is bounded in the same way. The heaviest commonly occurring elements in the Periodic Table

---

[5] In the Haldane model, by contrast, TR is broken by hand by introducing complex hoppings in a model of scalar electrons.

[6] A similar argument was given on p. 205.

(e.g., Bi, Pb, and $5d$ transition metals) have SOC strengths on the order of 0.5 eV or less, providing a kind of practical upper limit on the gap that can be expected in a Chern-insulating state.

Unconventional mechanisms that avoid the limitations associated with the intrinsic SOC strength have also been discussed. One direction is to consider the possible effects of strong correlation, which could lead to an effective magnification of the SOC strength, or even to a spontaneous generation of an effective SOC-like field in connection with a breaking of TR symmetry directly in the orbital sector (Rüegg and Fiete, 2011; Yang et al., 2011). Hints of such a state appeared in a recent first-principles calculation (Liu et al., 2016). A different strategy is to search for a noncollinear spin system that can take advantage of the "spin chirality" mechanism proposed by Taguchi et al. (2001), in which an electron acquires a complex phase factor as it hops among sites for which the underlying magnetization directions are noncoplanar (see, for example, the triangular-loop Berry phase discussed on p. 85). This is sometimes referred to as a "topological Hall effect," and is the basis for some recent theoretical proposals for a QAH state (Hamamoto et al., 2015; Zhou et al., 2016). However, at the time of this writing, none of these unconventional avenues have yet led to an experimentally realized QAH state.

The one route that has proved successful to date involves the growth of thin films of a 3D TR-invariant TI of the Bi_2Se_3 class doped with magnetic impurities such as Cr or V. In a remarkably farsighted theoretical paper, Yu et al. (2010) proposed a mechanism whereby the magnetic impurities, embedded in a host displaying a large Van Vleck paramagnetic susceptibility, develop a long-range ferromagnetic state, which in turn gaps the topological surface states and gives rise to a Chern-insulating 2D film. In an experimental tour de force by Chang et al. (2013) just a few years later, this scenario was implemented experimentally in Cr-doped $(Bi,Sb)_2Te_3$ thin films, and the QAH behavior was demonstrated. Subsequent work has refined this approach, with some of the best results to date having been demonstrated for V-doped films by Chang et al. (2015).

Figure 5.5(a) shows the longitudinal and transverse resistivities of a $(Bi_{0.29}Sb_{0.71})_{1.89}V_{0.11}Te_3$ thin film measured at 25 mK as a function of an applied magnetic field. The film is ferromagnetic with a coercive field of approximately 1.0 T, with the magnetization (not shown) following a hysteretic behavior similar to that of the ρ_{yx} curve. The longitudinal resistivity shoots up during domain reversal, but at zero field $\rho_{xx}e^2/h$ and $\rho_{yx}e^2/h$ differ from 0 and 1, respectively, by only about one and two parts per 10,000, showing good quantization and dissipationless transport. Figure 5.5(b) shows the same quantities as a function of temperature, showing that the quantization is already poor at 1 K.

The use of magnetically doped TIs to realize the QAH state thus seems to be limited to rather low temperature. The reason is that the gap E_g is not on the energy scale of the SOC, but rather on the energy scale of the magnetically

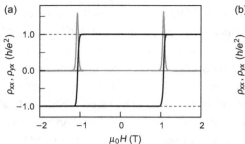
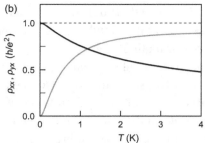

Figure 5.5 Demonstration of QAH in a V-doped topological-insulator thin film. (a) Longitudinal (ρ_{xx}, light curves) and transverse (ρ_{yx}, dark curves) resistivities obtained while sweeping H at fixed $T = 25$ mK, in units of h/e^2. (b) Same, but as a function of temperature at zero field. Data courtesy of Chang et al. (2015) and replotted with permission of the authors.

induced splitting of the surface states, with E_g/k_B falling in the sub-Kelvin range. For this reason, many researchers are continuing to search for higher-temperature realizations of the QAH effect as of this writing.

Several possible applications of QAH insulators have been been suggested. One is high-precision metrology, connected with the quantized nature of σ_{AHC}. Perhaps the most discussed application arises from the fact that the current carried by the edge states is chiral and dissipationless, making these edge states possible candidate "nanowires" for dissipationless interconnects in integrated circuits. The fact that the chiral direction is potentially reversible by an applied magnetic field suggests the possibility of giant magnetotransport effects. The proximity of the QAH state to superconductors or other exotic states could lead to new effects. The QAH effect could also enable the fabrication of strong magnetoelectric couplers, as will be discussed in Section 6.4.2. These and other potential applications are drivers for the community to identify new QAH systems that are physically and chemically robust, and that exhibit larger gaps so as to achieve highly quantized performance at higher temperatures. There does not appear to be any reason in principle why the effect could not occur at room temperature, but the research community still has a long way to go to make that goal a reality.

5.1.5 Wannier Representability

Before moving on to 3D systems, it is worth pointing out an interesting feature of the Chern-insulating state – namely, that there is a topological obstruction to the construction of Wannier functions for the valence bands. (Here we refer to true Wannier functions that are well localized in both lattice directions, not the hybrid Wannier functions discussed earlier.) The Wannier functions are constructed from a 2D version of Eq. (3.80a) for an ordinary insulator, but the construction is guaranteed to fail in the topological case.

The issue is that the localization of the Wannier function $w_{n\mathbf{R}}(\mathbf{r})$ in real space depends on a choice of a gauge that is *smooth* and free of singularities in $\mathbf{k}$-space, and also *periodic*, $|\psi_{n\mathbf{k}}\rangle = |\psi_{n,\mathbf{k}+\mathbf{G}}\rangle$. In other words, the gauge must be singularity-free on the 2D BZ regarded as a 2-torus. However, the presence of a nonzero Chern number presents an obstruction to the construction of such a gauge, as we saw at the very end of Section 3.2.2 on p. 93; there our manifold was the unit sphere, but the logic applies in the same way to the 2-torus. Indeed, if we have a smooth and periodic gauge on the 2D BZ, then the Berry phase around the boundary is zero (due to cancellations at the top and bottom, and on the left and right). Stokes' theorem then implies that the integrated Berry curvature, and thus the Chern number, must vanish.

If we try to construct Wannier functions using Eq. (3.80a) for a gauge that is periodic but not smooth, the singularity in $\mathbf{k}$ translates into poor localization in $\mathbf{r}$ (power-law tails such that position matrix elements like $\langle w_{n\mathbf{R}}|x|w_{n\mathbf{R}}\rangle$ are ill defined). Conversely, if one assumes the existence of well-localized Wannier functions, then the Bloch functions constructed from them via Eq. (3.80b) are necessarily smooth and periodic, and cannot belong to a nontrivial insulator.

In short, it is topologically impossible to construct Wannier functions for a Chern insulator. In a multiband case it may be possible to do a Wannier construction for a subset of valence bands whose total Chern number vanishes, but an obstruction will always be encountered when trying to finish the construction for the entire occupied manifold.

5.1.6 3D Chern Insulators and Their Surface States

Up to now we have been discussing the case of Chern insulators only in 2D, but 3D realizations are also possible. In general we have a lattice described by three primitive lattice vectors $\mathbf{b}_1$, $\mathbf{b}_2$ and $\mathbf{b}_3$, so that the 3D BZ is a 3-torus parametrized by $(\kappa_1, \kappa_2, \kappa_3)$, each running from 0 to 2π. At fixed κ_3 we can compute the Chern number $C_{12}(\kappa_3)$ obtained by integrating the Berry curvature Ω_{12} over the 2D (κ_1, κ_2) BZ at that κ_3. But if we then gradually change κ_3, there can be no gap closure anywhere on the corresponding 2D cross section of the BZ (since, by assumption, there is no gap closure anywhere in the BZ). Thus there can be no change in the Chern number of the slice as κ_3 is varied, and we conclude that C_{12}, which we relabel as C_3, is a topological invariant of the crystal. By the same argument, C_1 and C_2 are also topological invariants, corresponding to the integral of Ω_{23} or Ω_{31} over (κ_2, κ_3) or (κ_3, κ_1) planes, respectively.

This structure was discussed by Kohmoto et al. (1993) and Haldane (2004), who concluded that any 3D insulator is characterized by a triplet $\mathbf{C} = (C_1, C_2, C_3)$ of Chern integers. Once again, the system has to have broken TR symmetry to achieve a nontrivial state, one in which at least one of these indices is nonzero.

The simplest case is that of a layered system composed of weakly interacting stacked QAH layers. Let the periodicity of each layer be described by lattice vectors $\mathbf{a}_1$ and $\mathbf{a}_2$, and let $\mathbf{a}_3$ be a third lattice vector indicating how the layers are stacked. By "weakly," we mean that it is possible to separate the layers without gap closure by increasing the layer spacing until the layers are isolated. If the Chern number of each such layer is denoted as C_3, then the system has index triplet $\mathbf{C} = (0, 0, C_3)$. The 3D AHC of this system is, therefore, just the 2D one multiplied by the density of layers. But the interlayer distance is just $2\pi/b_3$ and we find that $\sigma_{\mathrm{AHC}} = (e^2/h) C_3 b_3 / 2\pi$ in the plane normal to lattice vector $\mathbf{b}_3$. Note that this response is no longer quantized to quite the degree as it was in 2D, because the result depends on a dimensionful material parameter (the interlayer spacing) as well as constants of nature.

Conversely, if we find a bulk system having Chern triplet $(0, 0, C_3)$, we can think of this as topologically connected to a set of QAH layers weakly stacked along the $(0, 0, 1)$ direction. A similar statement holds for any Chern triplet (C_1, C_2, C_3): Constructing the reciprocal lattice vector $\mathbf{G_C} = C_1 \mathbf{b}_1 + C_2 \mathbf{b}_2 + C_3 \mathbf{b}_3$, we can treat the system as composed of layers with Miller index (C_1, C_2, C_3) stacked along the direction of $\mathbf{G_C}$. To see this, let $\mathbf{a}'_1$ and $\mathbf{a}'_2$ be primitive vectors spanning the 2D subspace of all real-space lattice vectors orthogonal to $\mathbf{G_C}$, and let $\mathbf{a}'_3$ be a lattice vector connecting the origin to one of the lattice sites in the neighboring $(\mathbf{a}'_1, \mathbf{a}'_2)$ plane. Then in this primed frame we will find that $\mathbf{C}' = (0, 0, C'_3)$, similar to the case just considered. Thus, any Chern insulator in 3D can be thought of as topologically connected to a weakly stacked set of QAH layers. In this sense, the 3D QAH state is a sort of trivial extension of the 2D one.

What about the surface states? For definiteness, consider QAH layers with $C=1$, and first stack them along the z direction to make a 3D Chern insulator with $\mathbf{C} = (0, 0, 1)$. In this case, the (001) surface is not likely to show any special surface states, since each layer is fully gapped in its interior. However, the side surfaces are required to have surface states. To describe this situation, let's rotate the layers and stack them along the y direction, so that now $\mathbf{C} = (0, 1, 0)$, and consider the (001) surface at $z=0$. Each QAH plane is now cut by the surface so as to expose its gap-crossing edge channel, giving rise to a metallic surface. But it is a metallic surface of a rather peculiar kind!

Figures 5.6(a) and (b) show possible Fermi loop structures in the 2D BZ of a system, such as an atomically thin layer or thin slab, that is extended in x and y but finite in z. We use the term "Fermi loop" in analogy with the term "Fermi

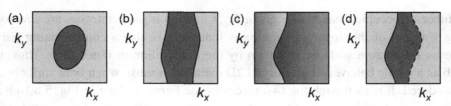

Figure 5.6 (a–b) Sketch of possible Fermi loop structures in the 2D BZ for a metallic 2D layer or a metallic 2D surface of a normal 3D insulator. The heavily shaded side of a Fermi loop is the side with one more surface band occupied. (c) Same, but for the (001) surface of a 3D Chern insulator with indices $(C_x, C_y, C_z) = (0, 1, 0)$, showing a band-filling anomaly. (d) Same as (c), but for an (001) slab of finite thickness in the z direction; solid and dashed lines indicate Fermi loops on the top and bottom surfaces, respectively.

surface" for 3D metals. The boundary of the electron pocket in Fig. 5.6(a) is clearly a loop, but we apply the term as well to each of the zone-crossing boundaries in Fig. 5.6(b). This is quite natural when we regard the BZ as a 2-torus, in which case these really are closed loops. Since the system is finite in z, the differently shaded regions refer to different numbers of occupied bands – for example, 17 and 18 occupied bands in the lightly and darkly shaded regions, respectively. These figures could also refer to a metallic surface of an ordinary, topologically trivial insulator. In this case we can no longer associate an absolute integer filling to each region, since the crystal extends indefinitely in the $-z$ direction. Nevertheless, the *relative* filling is globally consistent. For example, as k_x runs across the BZ in Fig. 5.6(b), the surface band filling increases by one unit, then decreases, returning to a value at the right boundary that is consistent with the value at the left boundary.

This does not happen, however, for the surface of the 3D Chern insulator described previously, which had Chern indices $\mathbf{C} = (0, 1, 0)$. Figure 5.6(c) shows a possible Fermi loop structure for such a surface. At any fixed k_y, the Hamiltonian $H_{\mathbf{k}}$ is that of a 2D Chern insulator with $C_{zx} = 1$, implying the existence of an up-crossing edge channel in the $-k_x$ direction on the top surface. This must cross the Fermi energy once, no matter where we place the Fermi energy in the gap, and for any value of k_y. As a result, we get a Fermi loop structure like that shown in Fig. 5.6(c), in which there is one and only one zone-crossing Fermi loop, and in which the band occupations cannot be consistently labeled in the BZ as a whole. The shading in Fig. 5.6(c) still indicates the side of the Fermi line on which the occupation is higher (Fermi velocity pointing to the lighter-shaded side), but the shading has to be gradually modified as one loops around the BZ in the k_x direction to get a consistent picture.

This peculiar state of affairs becomes less puzzling if we consider a slab of large but finite thickness along z. In this case it must again be possible to count the total number of occupied bands at a given (k_x, k_y), even if large integers are involved. The way out of the apparent paradox is that there is also a compensating Fermi line on the *bottom* surface, as shown by the dashed line in Fig. 5.6(d). Thus, the slab as a whole behaves like a normal 2D metal, as it must, when both surfaces are considered. It is as though the two pieces of the Fermi surface in Fig. 5.6(b) have been divorced, with one residing on the top surface and the other transported to the bottom surface.

The presence of a Fermi line structure like that in Fig. 5.6(c) is an example of an *anomaly*. This term is used for a physical feature that cannot occur for a d-dimensional system in isolation, but can occur at the d-dimensional surface of a $(d + 1)$-dimensional system, provided that this higher-dimensional system is nontrivial in some sense. The Fermi-loop anomaly of Fig. 5.6(c) is an example for $d = 2$, and the existence of a chiral channel at the edge of a 2D QAH state is an example in $d = 1$. For that matter, we can interpret the existence of a fractional surface charge at the end of a finite 1D chain with nontrivial electric polarization (see Sections 4.5.2–4.5.3) as an example of a zero-d anomaly, since fractional charges cannot occur in isolation. The concept of anomaly is closely linked to the concept of topology, and will recur in other contexts before we are done.

5.1.7 Anomalous Hall Conductivity in Metals

Historically, the phenomenon of AHC arose in the context of ferromagnetic metals, and has only recently been applied to insulators. The AHC was discovered in metallic ferromagnetic films by Hall (1881), just two years after his discovery of the ordinary Hall effect (Hall, 1879). The term "anomalous" refers to the fact that the phenomenon occurs in the absence of an external magnetic field, with the magnetization of the film playing the role of breaking TR symmetry. A good review of the physics of the AHC and its somewhat long and controversial history is provided by Nagaosa et al. (2010). Very briefly, Karplus and Luttinger (1954) introduced a theory of the AHC based on a Kubo-Greenwood formulation of the SOC-mediated response of the perfect crystalline material to an applied electric field. This Karplus-Luttinger component is often referred to as the "intrinsic" contribution because it depends only on the electronic band structure and is not connected with carrier scattering processes. We arrived at essentially the same formula in Eq. (5.6), which in 3D takes the form

$$\sigma_{\mu\nu} = \frac{e^3 \hbar}{(2\pi)^3} 2\text{Im} \sum_n^{\text{occ}} \int_{\text{BZ}} f_{n\mathbf{k}} \langle u_{n\mathbf{k}} | v_{\mathbf{k},\mu} T_{n\mathbf{k}}^2 v_{\mathbf{k},\nu} | u_{n\mathbf{k}} \rangle \, d^3k, \qquad (5.10)$$

where the integral is now over the 3D BZ and $f_{n\mathbf{k}}$ is the Fermi occupation function.[7]

---

[7] In 3D, $\sigma_{\mu\nu} = \partial J_\mu / \mathcal{E}_\nu$ has units of e^2/h divided by length.

The complete absence of scattering in this theory made it controversial, and competing theories based on defect scattering mechanisms were advanced. In particular, Smit (1955, 1958) argued for a "skew scattering" mechanism in which the scattering develops an asymmetry with respect to left-handed and right-handed scattering components in the plane normal to the magnetization. Subsequently, Berger (1964, 1970) focused on a "side-jump" mechanism in which the scattered electron acquires an extra transverse translation after the scattering event. These mechanisms also rely on SOC, and apply to the scattering of spin-polarized electrons at the Fermi energy even for a symmetric scattering potential. Similar mechanisms can be considered in connection with electron-phonon scattering. Luttinger (1958) revisited the subject using the recently developed transport theory of Kohn and Luttinger (1957), providing a uniform framework for the treatment of both intrinsic and defect mechanisms.

According to the Boltzmann transport theory, an applied electric field induces a drift of the Bloch wavevectors according to $\dot{\mathbf{k}} = -e\mathcal{E}/\hbar$ [c.f. Eq. (5.4)] that is balanced by scattering mechanisms operating at a rate $1/\tau$ governed by the relaxation time τ. In the steady state this results in a net shift of the occupied Fermi sea by an amount proportional to $\mathcal{E}$, as sketched in Fig. 5.7(a–b), and a longitudinal conductivity ρ_{xx} that scales as τ. In a ferromagnet the skew-scattering mechanisms also generate a transverse average shift of the Fermi sea, as shown in Fig. 5.7(c), giving rise to a transverse component σ_{yx} of the conductivity tensor that also increases with τ. By contrast, the intrinsic AHC of Karplus and Luttinger arises from the electric field perturbation of the occupied-state wave functions in the *undeformed* Fermi sea of Fig. 5.7(a), independent of τ. Side-jump scattering behaves in a similar way, allowing it to be regarded as a kind of impurity contribution that comes along with the intrinsic Karplus–Luttinger AHC.

A peculiar result of the foregoing is that the skew-scattering mechanism dominates for clean samples, in which the defect density is low and the scattering

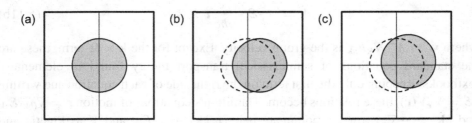

Figure 5.7 Sketch of the shift of the Fermi sea in response to an in-plane electric field $\mathcal{E} = -\mathcal{E}_0 \hat{x}$. (a) Unperturbed Fermi sea. (b) Fermi sea shifted along $\hat{k}_x$ (shaded region) by an amount proportional to $\mathcal{E}_0/\tau$, where τ is the scattering time, drawn for a nonmagnetic system. (c) Same for a ferromagnet with magnetization normal to the plane, showing a sideways shift associated with skew scattering. Shading indicates regions in which the Fermi occupation function $f_{n\mathbf{k}}$ exceeds $\frac{1}{2}$.

time τ is long; conversely, it becomes negligible in the dirty limit. After inverting the conductivity tensor, the intrinsic mechanism implies a transverse resistivity ρ_{yx} scaling as the square of the longitudinal resistivity ρ_{xx}, in contrast to the linear dependence expected for skew scattering. However, the side-jump contribution scales similarly to the intrinsic one, making an experimental identification of the dominant mechanisms difficult. Uncertainties associated with the treatment of phonon and spin-wave contributions add to the confusion, so that the matter remained unsettled for many years.

In the absence of an intuitive physical picture of the predicted anomalous velocity and the lack of reliable predictions from band structure calculations, the Karplus–Luttinger theory was initially not well accepted. The first of these obstacles began to dissipate as a series of authors (Chang and Niu, 1996; Haldane, 2004; Jungwirth, 2002; Onoda and Nagaosa, 2002; Sundaram and Niu, 1999) reexamined the Karplus–Luttinger theory in the modern language of Berry phases and curvatures, by making the transition from Eq. (5.6) to Eq. (5.8) such that the AHC is expressed in terms of the Berry curvature integrated over the occupied Fermi sea. Furthermore, using Stokes' theorem σ_{yx} can be reexpressed in terms of the Berry phase computed around the boundary of the occupied Fermi sea in 2D, or the k_z average of Berry phases on fixed-k_z slices in 3D (Haldane, 2004; Jungwirth, 2002; Wang et al., 2007).

The meaning of this Berry-curvature formulation was further clarified in the context of a semiclassical formulation of electron dynamics by Chang and Niu (1996) and Sundaram and Niu (1999), as nicely reviewed by Chang and Niu (2008) and Xiao et al. (2010). Denoting the real-space and reciprocal-space centroids of an electron wave packet in a crystal as $\mathbf{r}$ and $\mathbf{k}$, respectively, these authors showed that these variables evolve under weak electric and magnetic fields according to the equations of motion

$$\dot{\mathbf{r}} = \mathbf{v}_g - \dot{\mathbf{k}} \times \boldsymbol{\Omega}, \tag{5.11a}$$

$$\dot{\mathbf{k}} = -\frac{e}{\hbar}\boldsymbol{\mathcal{E}} - \frac{e}{\hbar c}\dot{\mathbf{r}} \times \mathbf{B}, \tag{5.11b}$$

where $\mathbf{v}_g = \hbar^{-1}\nabla_{\mathbf{k}}E_{n\mathbf{k}}$ is the group velocity. Except for the $\dot{\mathbf{k}} \times \boldsymbol{\Omega}$ term, these are the standard equations of semiclassical transport theory found in elementary textbooks. Keeping only the first term on the right side of each equation and writing $e\boldsymbol{\mathcal{E}} = \nabla_{\mathbf{r}}V(\mathbf{r})$, these relations become Hamilton's equations of motion $\dot{\mathbf{r}} = \nabla_{(\hbar\mathbf{k})}E_{n\mathbf{k}}$ and $\hbar\dot{\mathbf{k}} = -\nabla_{\mathbf{r}}V$ for the conjugate pair $(\mathbf{r}, \hbar\mathbf{k})$ with $E_{n\mathbf{k}}$ and V as kinetic and potential energies. The second term in Eq. (5.11b) accounts for the Lorentz force $(q/c)\mathbf{v} \times \mathbf{B}$ (written in Gaussian units).

Importantly, there is now also a second term in Eq. (5.11a), known as the *anomalous velocity* term, involving the Berry curvature $\boldsymbol{\Omega}$. Note that

$\mathbf{\Omega} = \nabla_{\mathbf{k}} \times \mathbf{A}_{n\mathbf{k}}$, where $\mathbf{A}_{n\mathbf{k}}$ is the Berry connection, while $\mathbf{B} = \nabla_{\mathbf{r}} \times \mathbf{A}$, where $\mathbf{A}$ is the electromagnetic vector potential. The presence of the anomalous velocity term in Eq. (5.11) thus restores an elegant symmetry to the semiclassical equations of motion of electrons in a crystal. In ordinary metals such as Cu and Al having both inversion and TR symmetry, the neglect of this term is entirely justified because $\mathbf{\Omega}$ vanishes identically in that case (see p. 105). If inversion alone is broken, TR symmetry continues to enforce that $\mathbf{\Omega}(\mathbf{k})$ and $\mathbf{\Omega}(-\mathbf{k})$ are equal and opposite; the effects of this term then cancel for many purposes, although not necessarily for higher-order responses (see, for example, Sodemann and Fu, 2015). However, neglect of this term is rarely justified in the case of magnetic crystals, especially when SOC is important.

The presence of the anomalous velocity term provides an immediate explanation for the intrinsic AHC. After all, an applied $\mathcal{E}_x$ induces a $\dot{k}_x$ via the first term of Eq. (5.11b), which in turn generates a $\dot{y}$ and thus a K_y through the second term of Eq. (5.11a). Spelling this out for the case of a 2D metal with a single partially filled band with occupation function $f(\mathbf{k})$, we get

$$
\begin{aligned}
K_y &= \frac{-e}{(2\pi)^2} \int_{\text{BZ}} f(\mathbf{k}) \, \dot{y}(\mathbf{k}) \, d^2 k \\
&= \frac{-e}{(2\pi)^2} \int_{\text{BZ}} f(\mathbf{k}) \, \dot{k}_x \, \Omega(\mathbf{k}) \, d^2 k \\
&= \frac{-e}{(2\pi)^2} \left(\frac{-e}{\hbar} \mathcal{E}_x \right) \int_{\text{BZ}} f(\mathbf{k}) \, \Omega(\mathbf{k}) \, d^2 k,
\end{aligned} \tag{5.12}
$$

so that

$$
\sigma_{yx} = \frac{e^2}{(2\pi)^2 \hbar} \int_{\text{BZ}} f(\mathbf{k}) \, \Omega(\mathbf{k}) \, d^2 k, \tag{5.13}
$$

consistent with Eq. (5.8). Working at zero temperature, Stokes' theorem can be used to write this relation as

$$
\sigma_{yx} = \frac{e^2}{h} \frac{\phi_{\text{F}}}{2\pi} \tag{5.14}
$$

where ϕ_{F} is the Berry phase around the Fermi loop bounding the occupied region. In 3D, Eq. (5.13) becomes

$$
\sigma_{\text{AHC}} = \frac{e^2}{(2\pi)^3 \hbar} \int_{\text{BZ}} f(\mathbf{k}) \, \mathbf{\Omega}(\mathbf{k}) \, d^3 k \tag{5.15}
$$

where σ_{AHC} is a pseudovector such that $\mathbf{J} = \sigma_{\text{AHC}} \times \mathcal{E}$.

In an insulator the scattering mechanisms are obviously frozen out, and the anomalous-velocity contribution in Eq. (5.13) or (5.15) is the only one present in the low-temperature limit. Setting $f(\mathbf{k}) = 1$ for an occupied band, the integral in Eq. (5.13) or (5.15) gets extended over the full 2D or 3D BZ. In 2D, the integral of Ω over the BZ is just $2\pi C$, so that Eq. (5.13) becomes

$$\sigma_{yx} = \frac{e^2}{h} C, \qquad (5.16)$$

which clearly describes the QAH or Chern-insulator state discussed in Section 5.1.3. The generalizations to multiband insulators via a summation over occupied bands n, and to 3D via an integral over k_z, are straightforward. Thus, the sideways current occurring as a result of $\dot{\mathbf{k}}$ discussed earlier following Eq. (5.4) is just a consequence of the anomalous velocity of the wave packets making up the filled bands.

Pioneering density-functional theory calculations of σ_{AHC} were carried out for SrRuO$_3$ by Fang et al. (2003) and for Fe by Yao et al. (2004) using a direct evaluation of the Kubo formula of Eq. (5.6). Soon thereafter, Wang et al. (2006), Yates et al. (2007), and Wang et al. (2007) developed methods based on Wannier interpolation to compute and integrate the Berry curvature efficiently even in difficult cases. Realistic calculations of the intrinsic AHC for ferromagnetic materials are challenging in practice because the Berry curvature, which we need to integrate over the BZ, is typically a very sharply varying function of $\mathbf{k}$. It fact, it may often depend strongly on contributions from a few "hot spots" in $\mathbf{k}$-space where spin-orbit–induced avoided crossings between bands are found close to the Fermi energy. Such features often require an extremely dense $\mathbf{k}$-point sampling, and the calculations have to be carried out with care. As an example, Fig. 5.8 shows the band structure in the top panel, and the Berry curvature in the bottom panel, plotted along some high-symmetry lines in the BZ of bcc Fe (Wang et al., 2006). It is clear that special care is needed to integrate such an ill-behaved function as the one shown in the bottom panel.

As a result of these and other calculations, coupled with renewed interest in experimental investigations, there is a growing consensus that while defect scattering typically dominates at low temperature, the intrinsic mechanism may be the dominant contribution at room temperature in at least some materials. For example, in elemental Fe and Co, the AHC computed by Yao et al. (2004) and Wang et al. (2006) was in reasonable agreement with the experimental result in the temperature regime where intrinsic effects are thought to dominate. In Ni, the calculated value was only about 30% of the experimental one. However, a subsequent careful study of the thickness dependence of thin-film samples by Ye et al. (2012) led to a higher experimental value, which proved to be in good agreement with revised theoretical values of Fuh and Guo (2011) using a GGA plus Hubbard U (GGA+U) exchange-correlation potential.

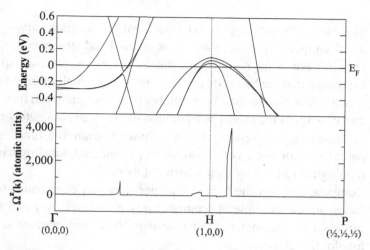

Figure 5.8 Band structure and total Berry curvature (summed over occupied bands) along a series of high-symmetry lines in the BZ of bcc Fe. Reprinted with permission from Wang et al. (2006). Copyright 2006 by the American Physical Society.

Gradhand et al. (2012) have provided a nice review of methods for first-principles calculations of Berry curvature, and of applications to the computation of anomalous Nernst and spin Hall as well as anomalous Hall conductivity.

Exercises

5.1 Suppose a 2D crystal has an isolated group of bands with bulk band gaps separating the group from both the bands below and the bands above, and consider its edge band structure. Following the spirit of the arguments given in Section 5.1.1, relate the total Chern number of the group to the numbers of edge states crossing these two gaps.

5.2 Consider a 2D QAH insulator with $C = 1$ and having nonorthogonal lattice vectors $\mathbf{a}_1$ and $\mathbf{a}_2$. Using Eqs. (5.11a–5.11b):

  (a) Show that $\dot{\kappa}_j = (-e/2\pi\hbar)\,\mathbf{a}_j \cdot \boldsymbol{\mathcal{E}}$.

  (b) Show that the formula $\sigma_{\text{AHC}} = e^2/h$ for the AHC is consistent with a picture in which one electron per cell acquires a real-space shift by lattice vector $\mathbf{a}_2$ when κ_1 is adiabatically increased by one unit ($\kappa_1 \rightarrow \kappa_1 + 1$ at fixed κ_2) for all electrons in the 2D BZ.

  (c) What is the corresponding statement regarding an adiabatic increase of κ_2 by one unit?

5.3 Program `checkerboard.py` in Appendix D.15 sets up and solves a 2D checkerboard model with some complex hoppings along diagonal bonds.

    (a) Make a sketch showing the 2D model that the program solves.

    (b) Set t' temporarily to zero and look at the band-structure plot as you vary Δ. For what range of Δ values do the bands overlap? What happens when the bands do overlap and you gradually turn t' back on?

    (c) Add a calculation of the Chern number to the program, and then explore the behavior of the model as a function of the three parameters t_0, t', and Δ. Based on your results, sketch a phase diagram as a function of the parameters (or some combinations of parameters), labeling trivial and topological regions by their Chern numbers.

    (d) Compute and plot the hybrid Wannier flow and edge band structures following the example of program `haldane_topo.py` and Fig. 5.4, for a set of parameters in the normal phase and another set in the topological phase.

5.4 The sketch below shows a portion of a 2D kagome lattice, which has three sites per unit cell as shown. There is 3-fold rotational symmetry and the unit cell is shaded.

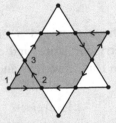

Xu et al. (2015) introduced a tight-binding model for this lattice, with first-neighbor complex hoppings $t_1 + it_2$ in the directions indicated by the arrows. Let t_1 be positive and the site energies be zero.

    (a) Plot the band structure along the path Γ–K–M–Γ [see Fig. 5.1(b)] first for $t_2 = 0$ and then for nonzero t_2, showing that the system is necessarily metallic in the first case and insulating for any integer band filling in the second case. (*Note:* If you got a perfectly flat band for $t_2 = 0$, it is not a mistake; it is a peculiar artifact of this model.)

    (b) Plot the hybrid Wannier flow for each band separately. What are the Chern numbers?

    (c) Plot the two hybrid Wannier centers obtained by treating the two lowest band as a group, for a Fermi level in the upper gap. Is this flow consistent with what you found in part (c)?

5.5 Another simple model of a QAH insulator is that described by Qi et al. (2006), which is a two-band model at half-filling. From the tight-binding perspective, it has one spinor basis orbital on each site of a simple square lattice. The on-site Hamiltonian is $b\sigma_z$, and the nearest-neighbor hoppings in the $+\hat{\mathbf{x}}$ and $+\hat{\mathbf{y}}$ are $t(\sigma_z+i\sigma_x)$ and $t(\sigma_z+i\sigma_y)$, respectively, where σ_j are the Pauli matrices. Using the spinor features of PYTHTB (you may need to look ahead at program `kanemele_bsr.py` in Appendix D.16 for an example of how to use these features), program the model and explore its phase diagram by computing the Chern number for various values of b and t. Plot the Wannier flow for two parameter sets, one in the trivial and one in the QAH phase.

5.6 Clarify the consistency between Eqs. (5.13) and (5.15), making use of the pseudovector form of the Berry curvature introduced on p. 89. How do the units of σ_{yx} and σ_{AHC} compare?

5.2 Quantum Spin Hall Insulators

We have seen that 2D insulators with broken TR symmetry can be topologically classified according to their Chern number C; this is known as a "$\mathbb{Z}$ classification" since C runs over the group of all integers $\mathbb{Z}$. Nonmagnetic insulators, for which TR is a good symmetry, automatically belong to the trivial class with $C=0$. Does this mean that nonmagnetic 2D materials do not admit any topological classification?

Let's recall what this would mean. A topological classification exists if there are two or more classes such that no member of one class can be converted into a member of another by following some adiabatic path without gap closure. However, in setting up the ground rules for the classification, we have to specify the symmetry conditions that are imposed along the connecting path. If we work entirely in the context of TR-invariant systems, meaning that we insist on TR invariance everywhere along the path, then it turns out that a topological classification *does* exist – but *not* a $\mathbb{Z}$ classification as was the case previously.

5.2.1 The $\mathbb{Z}_2$ Classification

In Section 5.1.1 we introduced the Haldane model as a simple paradigmatic system with a nonzero Chern number. This was a model of scalar (spinless) electrons with spontaneously broken TR symmetry. In reality electrons are always spinors, so a more realistic model Hamiltonian should have contributions governing the spin-up and spin-down components separately, as well as the couplings between them. This is what Kane and Mele (2005b) did in a pathbreaking paper; we shall roughly follow in their footsteps in the discussion that follows.

Let's begin at a simple level of approximation in which the spin-up and spin-down systems are completely decoupled, so that spin S_z is a good quantum number. In this case we can define Chern numbers $C_\uparrow$ and $C_\downarrow$ separately for each subsystem, and we would have a topological classification in which each system is labeled by a pair $(C_\uparrow, C_\downarrow)$ of integers (technically, a "$\mathbb{Z} \times \mathbb{Z}$ classification"). If SOC is neglected entirely, then the effective Hamiltonians for spin-up and spin-down electrons take a form similar to that of the LSDA approximation in Eq. (2.18). At this level of approximation the potentials appearing in Eq. (2.18) are all real, so that each spin system acts like a system of scalar electrons with unbroken TR symmetry (where TR is just complex conjugation K), and we expect $C_\uparrow = C_\downarrow = 0$. So far, then, we don't expect anything interesting.

We can take one small step forward and include only the σ_z terms in the SOC coupling of Eq. (2.21), while discarding all σ_x and σ_y terms. This hybrid approximation is sometimes denoted as the "diagonal SOC" approximation. In this case spin-up and spin-down systems are again decoupled; they obey $H_\uparrow |\psi_{i\uparrow}\rangle = E_{i\uparrow}|\psi_{i\uparrow}\rangle$ and $H_\downarrow |\psi_{i\downarrow}\rangle = E_{i\downarrow}|\psi_{i\downarrow}\rangle$ independently, as in Eq. (2.18), except that now $H_\uparrow$ and $H_\downarrow$ contain terms like $(\partial_x V)p_y - (\partial_y V)p_x$ coming from the SOC coupling of Eq. (2.21). Because $\mathbf{p}$ is odd under TR, $H_\uparrow$ and $H_\downarrow$ are no longer TR-invariant, and the above-mentioned $\mathbb{Z} \times \mathbb{Z}$ classification applies; $(C_\uparrow, C_\downarrow)$ can take any integer values for a general magnetic insulator, and $C_{\text{tot}} = C_\uparrow + C_\downarrow$ determines the AHC through $\sigma_{\text{AHC}} = (e^2/h)C_{\text{tot}}$.

This approximation is not very physical, however; once the spin-mixing terms involving σ_x and σ_y in Eq. (2.21) are restored, only the C_{tot} index survives, and we are back to the $\mathbb{Z}$ classification of 2D magnetic systems studied in the last section.[8] Nevertheless, let's stick with the diagonal SOC approximation a little while longer, and now consider nonmagnetic systems in which TR is a symmetry. Since TR reverses $\mathbf{p}$, we now have that $KH_\uparrow K = H_\downarrow$ and $KH_\downarrow K = H_\uparrow$; that is, TR interchanges the spin-up and spin-down Hamiltonians. But TR also interconverts the identities of up and down electrons, so overall TR symmetry still exists! Now the spin-up system can have a nonzero Chern number $C_\uparrow$, but if so, necessarily $C_\downarrow = -C_\uparrow$. As a result, $C_{\text{tot}} = 0$, and the system cannot be a QAH insulator.

Now let's consider a 2D insulator with $(C_\uparrow, C_\downarrow) = (1, -1)$. It is still an interesting system, for if an electric field $\mathcal{E}$ is applied along the x direction, spin-up electrons flow in the $+\hat{y}$ direction contributing a current $+(e^2/h)\mathcal{E}$, while spin-down electrons flow along $-\hat{y}$ and provide an exact cancellation of the charge current. We can describe this situation by saying that an induced spin current is flowing along $+\hat{y}$, corresponding to a *spin Hall effect*. Moreover, at least within the diagonal

---

[8] Some authors have introduced a topological classification according to "spin Chern numbers" even in the presence of spin-mixing terms, as discussed by Prodan (2009). Essentially this involves diagonalizing S_z in the occupied band space. Then, provided the resulting spectrum is gapped at zero, one can decompose the band space into the subspaces with positive and negative eigenvalues and compute the Chern number of each separately. However, this type of classification does not apply to all TR-invariant 2D insulators, since the gap in the spin spectrum may not exist, and it imposes a second and rather more artificial requirement on the behavior along the adiabatic path. The treatment provided here follows a more conservative convention, according to which topological classes are defined only in terms of nonclosure of the fundamental bulk energy gap.

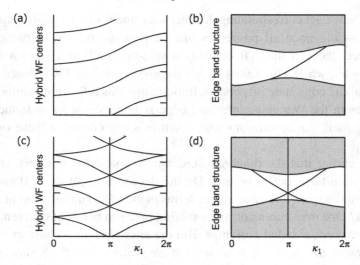

Figure 5.9 (a–b) Illustration of the flow of (a) hybrid Wannier centers and (b) edge-state dispersion for a single-band Chern insulator consisting of only a single species of scalar electrons. (c–d) Same, but for a TR-invariant system composed of spin-up electrons with $C_\uparrow = +1$ and complex-conjugated spin-down electrons with $C_\downarrow = -1$.

SOC approximation, this spin current is quantized, strongly hinting at a topological classification. After all, it must be impossible to convert a system with index pair $(C_\uparrow, C_\downarrow) = (1, -1)$ into one with $(C_\uparrow, C_\downarrow) = (0, 0)$ without a gap closure, for how else could the quantized spin-current response change discontinuously? Diagonal-SOC systems of this kind are then characterized by a single integer, which we may take to be $C_\uparrow$, and we have a $\mathbb{Z}$ classification.

We can realize a system of this kind by letting the spin-up electrons be described by a Haldane-like model with a single occupied band and $C_\uparrow = +1$, while the spin-down electrons are described by the complex-conjugated partner Hamiltonian having $C_\uparrow = -1$. Figure 5.9 shows a sketch of the expected hybrid Wannier flow and edge-state dispersions for such a system. Panels (a) and (b) show the behavior for the spin-up system alone; since this has $C = +1$, we expect the winding of the HWF centers and the existence of a crossing surface state as shown, very much as in Fig. 5.4(c–d). Because of the breaking of "TR symmetry" (complex conjugation) in the spin-up sector, neither of these panels shows left–right mirror symmetry.

Panels (c) and (d) indicate the corresponding behavior when both up and down spins are included. Here we are treating the two occupied bands as a group and plotting the maximally localized HWF centers. The patterns are now mirror-symmetric about $\kappa_1 = \pi$ because of the TR symmetry. In the 2D bulk, TR converts a spin-up Bloch state at $\mathbf{k}$ into a complex-conjugated spin-down one at $-\mathbf{k}$ with $E_{n\mathbf{k}} = E_{n,-\mathbf{k}}$. As a consequence, when considering projections as a function of κ_1

as in Fig. 5.9, we get corresponding degeneracies under $\kappa_1 \to -\kappa_1$ (or equivalently, $\kappa_1 \to 2\pi - \kappa_1$, since all properties are 2π-periodic in κ_1). In particular, Fig. 5.9(c) reflects the fact that TR converts a spin-up HWF at κ_1 into a spin-down one at $2\pi - \kappa_1$, while Fig. 5.9(d) shows the corresponding TR-induced reflection symmetry of the edge state dispersion. Importantly, this reflection symmetry leaves degeneracies in the Wannier-center and edge-state curves at $\kappa_1 = 0$ and $\kappa_1 = \pi$, "gluing together" the up-crossing and down-crossing curves at these two special values of κ_1.

We said earlier that the diagonal SOC approximation is not very physical. A crucial question turns out to be this: Do the degeneracies at $\kappa_1 = 0$ and $\kappa_1 = \pi$ survive when the spin-mixing σ_x and σ_y terms in the SOC Hamiltonian of Eq. (2.21) are restored? One may guess not, since mixing between bands of different character typically generates avoided crossings. But the answer is yes! These crossings at 0 and π represent *Kramers degeneracies* and are protected by the *Kramers theorem*, which applies to spinor electrons[9] in the presence of TR symmetry.

The Kramers theorem was introduced in Section 2.1.6. For a finite system, it says that all energy eigenstates are doubly degenerate. In the case of a general 3D crystal, it states that each Bloch state $|\psi_{n\mathbf{k}}\rangle$ is degenerate with a partner $|\psi_{n,-\mathbf{k}}\rangle$. As we saw on p. 51, this also means that at the eight special time-reversal invariant momenta (TRIM), at which $-\mathbf{k}$ is identical to $\mathbf{k}$ modulo a reciprocal lattice vector, all states must occur in degenerate pairs.

In terms of the reduced wavevectors κ introduced in Eq. (2.27), the TRIM occur when each κ_j is either 0 or π. Thus, for the 1D edge band structure of a 2D TR-invariant insulator, we can be sure that if there are any edge states in the gap at $\kappa_1 = 0$ or π, they must be doubly degenerate. We find such a degeneracy at $\kappa_1 = \pi$ in Fig. 5.9(d). If we imagine making some gradual TR-conserving perturbation to the Hamiltonian at the edge, this point of degeneracy can be raised or lowered in energy, but can never be gapped. In fact, no perturbation that conserves TR symmetry – even one that includes off-diagonal parts of the SOC interaction – is capable of splitting these degeneracies. When edge bands emerge out of the valence and conduction bands and meet at $\kappa_1 = 0$ or π in this way, the Kramers degeneracy keeps them "glued together" at the boundaries of the half BZ.

The Kramers argument also applies to the spectrum of TR-even Hermitian operators other than the Hamiltonian. In particular, when we compute the Wannier centers of a finite 1D chain with TR symmetry as described on p. 133, the eigenvalues $\bar{x}_i$ obtained in Eq. (3.130) are also guaranteed to come in Kramers-degenerate pairs. Recall that the hybrid Wannier centers sketched in Fig. 5.9(c) can

---

[9] Note that the entire discussion here and in Section 5.3 assumes spinor electrons. The present considerations do not apply to the case of spinless scalar particles described by Eq. (2.50).

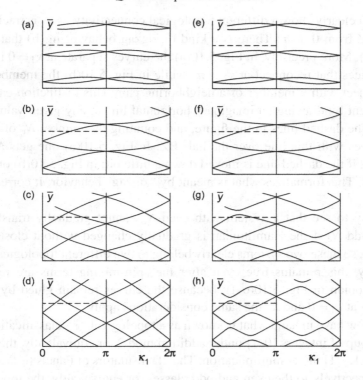

Figure 5.10 Sketch of Hybrid Wannier flow for 2D TR-invariant insulators constructed from single-band spin-up and spin-down systems with Chern indices $(C_\uparrow, C_\downarrow) = (n, -n)$. (a–d) Behavior for uncoupled spin sectors with $n = \{0, 1, 2, 3\}$, respectively. (e–h) Corresponding behavior after spin-mixing terms are included in the Hamiltonian. Horizontal dashed lines are reference lines for counting crossings as discussed in the text.

be regarded as those of an effective 1D system extending along the $\mathbf{a}_2$ direction, plotted as a function of κ_1, and that this effective system is TR-invariant at $\kappa_1 = 0$ and π. This means we must have Kramers degeneracies in the hybrid Wannier centers at the boundaries of the half BZ. The situation parallels that of the edge band structure, except that whereas the edge states may exist only in a restricted range of κ before disappearing into the bulk bands, here we can always follow the hybrid Wannier curves all the way to the BZ boundary.

All this gives us grounds to introduce a topological classification of 2D TR-invariant insulators. To see how, consider Fig. 5.10, which shows sketches of the Wannier flow expected if we build a TR-invariant model from spin-up and spin-down systems having Chern indices $(C_\uparrow, C_\downarrow) = (n, -n)$.

Panels (a–d) show the expected behavior when the spin-mixing terms involving σ_x and σ_y are omitted from the Hamiltonian. The systems with Chern indices $(0, 0)$

and $(1, -1)$ clearly show a different topological connectivity as κ_1 is tracked along the half BZ from 0 to π: There is a kind of zigzag behavior in (b) that does not occur in (a). More precisely, in Fig. 5.10(a) the curves separate at $\kappa_1 = 0$ into a pair of eigenvalues that reconnect at $\kappa_1 = \pi$, while in Fig. 5.10(b) the members of the pair reconnect with a member of a neighboring pair. This distinction can also be made evident by drawing an imaginary horizontal line at any given value of $\bar{y}$, as shown by the dashed lines in the figure, and counting the number N_c of crossings of the curves with this line over the half BZ. In Fig. 5.10(a) one gets $N_c = 0$ as drawn, or 2 if the dashed line is moved down a little bit; in Fig. 5.10(b) one always gets $N_c = 1$. This formalizes what is meant by "zigzag" behavior: It corresponds to odd N_c.

It is easy to see that a system with even N_c cannot gradually transform into one with odd N_c if the Hamiltonian is gradually changed without closure of the energy gap, so these two systems clearly belong to two different topological classes. Importantly, this remains true even after the spin-mixing terms are restored to the Hamiltonian in Fig. 5.10(e–f); because the curves remain glued by Kramers degeneracy at $\kappa_1 = 0$ and π, the same considerations apply.

So we now have in hand what is known as a topological "$\mathbb{Z}_2$ classification." $\mathbb{Z}_2$ is just the group of integers $\{0, 1\}$ under addition mod 2, or equivalently, the group of integers $\{+1, -1\}$ under multiplication. The 2D insulators of Panels (e-f) are said to belong respectively to the even and odd classes, or equivalently, the topologically trivial and nontrivial classes.[10]

For the case of Chern insulators with broken TR, we had a $\mathbb{Z}$ classification in which the topological index ran over all integers, with an infinite number of topological classes. Can't we do the same here? The answer is no! To see why not, let's ask whether the system built from spin-up and spin-down systems with $(C_\uparrow, C_\downarrow) = (2, -2)$ is topologically distinct from that built from $(0, 0)$. The behavior in the $(2, -2)$ system is shown in Fig. 5.10(c) and (g). In (c), where spin-mixing terms are absent, there does appear to be a distinction. In (g), however, where spin-mixing terms introduce avoided crossings between the curves (except at 0 and π), the distinction disappears. Indeed, the HWF flow curves in (g) are topologically equivalent to those in (e), and it is easy to imagin e one being converted into the other as the Hamiltonian is carried along some adiabatic path. Counting of crossings yields that N_c is even, confirming that this system belongs

---

[10] We encountered another $\mathbb{Z}_2$ classification when discussing inversion symmetry in Chapter 4, although we didn't identify it as such at the time. For a centrosymmetric crystal such as that in Fig. 4.12, each reduced polarization component p_j must either vanish or be equal to $e/2$ mod e, as discussed on p. 179. This establishes a topological classification for centrosymmetric insulators, since two systems with different sets of p_j values can never be adiabatically connected without gap closure. Thus, 1D, 2D, and 3D centrosymmetric insulators are characterized by one, two, and three $\mathbb{Z}_2$ indices, respectively.

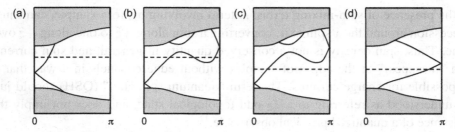

Figure 5.11 Sketch of edge band structures (vertical axis) of a 2D $\mathbb{Z}_2$-odd insulator plotted over the half BZ (horizontal axis). Shaded regions indicate projected bulk states; curves are edge states. Evolution from panel (a) to (b–c) and then to (d) is via a smooth modification of the Hamiltonian; (a–b) and (c–d) differ by the choice of reference energy (dashed lines) used for counting crossings.

to the trivial $\mathbb{Z}_2$-even class! Similarly, an inspection of Fig. 5.10(h) indicates that a system built from $(C_\uparrow, C_\downarrow) = (3, -3)$ is $\mathbb{Z}_2$-odd, which is topologically nontrivial.

The entire argument given to this point has been carried out in the context of hybrid Wannier center flow of the bulk Hamiltonian, as in Fig. 5.9(a) and (c), but the same can also be done for the edge state dispersions as in Fig. 5.9(b) and (d). Again, if an imaginary horizontal line is drawn at any energy in the gap, the number N_c of crossings with edge states in the half BZ will either be even or odd, a fact that cannot be changed by moving that energy up or down, or by any gradual evolution of the edge band structure. Moreover, the $\mathbb{Z}_2$ classification based on the edge band structure can readily be shown to be the same as that based on Hybrid Wannier flow using arguments similar to those applied to the Chern-insulator case in Section 5.1.1.

Figure 5.11 shows a sketch of some possible edge-state dispersions associated with a $\mathbb{Z}_2$-odd bulk insulator. It is easy to imagine a smooth path connecting the edge states of panel (a) to those of (b), or from (c) to (d), by continuously shifting and deforming the edge states (via changes to the bulk Hamiltonian, or to the edge Hamiltonian, or both). Throughout, the $\mathbb{Z}_2$ index, reflecting the number (mod 2) of curves crossing an arbitrary horizontal dashed line, remains constant, as it does if the line is shifted as in going from (b) to (c). Thus, all of these panels belong to the $\mathbb{Z}_2$-odd class; none can be continuously converted into one with an even number of crossings.

The fact that 2D TR-invariant insulators lose their $\mathbb{Z}$ classification but acquire a $\mathbb{Z}_2$ classification was first clarified in a pair of groundbreaking papers by Kane and Mele (2005a, 2005b). The term "quantum spin Hall insulator" has come into use to describe a member of the nontrivial $\mathbb{Z}_2$-odd class in 2D. This terminology is somewhat misleading, however, because the physical spin-Hall conductivity (spin-up current minus spin-down current induced transverse to $\mathcal{E}$) is *not quantized* in the presence of spin-mixing terms. A term involving σ_x, for example, can cause precession around the $\hat{x}$ spin axis, converting a spin along $+\hat{z}$ to one along $-\hat{z}$ over time. Thus, spin density is not a conserved quantity in general, and spin currents can "disappear" at the edges of samples without edge channels in a way that is impossible for charge currents. The term "quantum spin Hall" (QSH) should just be understood as referring to a $\mathbb{Z}_2$-odd topological state, and does not imply the existence of a quantized physical observable.[11]

5.2.2 Kane–Mele Model

To illustrate the distinction between the trivial and $\mathbb{Z}_2$-odd states, Kane and Mele (2005b) constructed a simple model system in which spin-up and spin-down electrons are described by a pair of complex-conjugated Haldane models with some spin-mixing terms included. Figure 5.1 still describes the model, but there are now two basis states, spin-up and spin-down, on each site. The Kane–Mele model Hamiltonian takes the form

$$H = \Delta \sum_i (-)^{\tau_i} c_i^\dagger c_i + t_1 \sum_{\langle ij \rangle} (c_i^\dagger c_j + \text{h.c.}) + \lambda_{\text{SO}} \sum_{\langle\langle ij \rangle\rangle} (i c_i^\dagger \sigma_z c_j + \text{h.c.})$$

$$+ \lambda_{\text{R}} \sum_{\langle ij \rangle} (i c_i^\dagger \hat{\mathbf{e}}_{\langle ij \rangle} \cdot \boldsymbol{\sigma} c_j + \text{h.c.}), \tag{5.17}$$

which should be compared with the Haldane model of Eq. (5.1). A double sum over spin indices s and s' is implied in each term; the spin indices are contracted over a Pauli matrix if it appears, and over the identity if it does not.[12] The notation for the first three terms is parallel to that of Eq. (5.1), except that the second-neighbor hopping t_2 has been renamed as λ_{SO} to emphasize that this represents the spin-diagonal part of the SOC. The first two terms describe the site energies and the spin-independent first-neighbor hoppings, while the third describes the spin-dependent, but still spin-diagonal, complex second-neighbor hoppings. The fourth term has been added to represent the spin-mixing part of the SOC, which in this context is usually referred to as a "Rashba" term.[13] In this term, $\hat{\mathbf{e}}_{\langle ij \rangle}$ is an in-plane unit vector defined as $\hat{\mathbf{e}}_{\langle ij \rangle} = \hat{\mathbf{d}}_{\langle ij \rangle} \times \hat{\mathbf{z}}$ where $\hat{\mathbf{d}}_{\langle ij \rangle}$, is the unit vector in the direction from site j to site i.

---

[11] For weak SOC strength, a model like that shown in Fig. 5.10(g) should have a spin-Hall conductivity σ_{SHC} close to $2e^2/h$. However, this model is in the same $\mathbb{Z}_2$-even class as that of Fig. 5.10(e). Thus, along a $\mathbb{Z}_2$-even path that connects these two insulating states, σ_{SHC} must decrease continuously, passing values close to e^2/h. This makes it obvious that the value of σ_{SHC} cannot be a measure of the $\mathbb{Z}_2$ index.

[12] For example, $c_i^\dagger \sigma_x c_j$ expands to $c_{i\uparrow}^\dagger c_{j\downarrow} + c_{i\downarrow}^\dagger c_{j\uparrow}$, while $c_i^\dagger c_j$ expands to $c_{i\uparrow}^\dagger c_{j\uparrow} + c_{i\downarrow}^\dagger c_{j\downarrow}$.

[13] Such a term is physically motivated if mirror symmetry is broken across the plane of the layer, as in the presence of a perpendicular electric field or when the layer lies on a substrate.

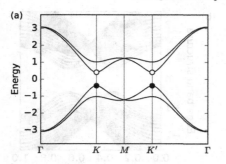
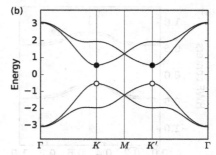

Figure 5.12 Band structures of the Kane–Mele model of Eq. (5.17), with $\Delta = 0.7$, $t_1 = -1.0$, $\lambda_R = 0.05$, and (a) $\lambda_{SO} = -0.06$; (b) $\lambda_{SO} = -0.24$. Filled (open) circles mark states of common character; band inversion at K and K' is evident in (b).

The band structure resulting from this model is illustrated in Fig. 5.12, which is the output of the PYTHTB program `kanemele_bsr.py` in Appendix D.16. The parameters are the same as for the Haldane model in Section 5.1.1 (recall that t_2 there is λ_{SO} here), but augmented by a spin-mixing Rashba coefficient $\lambda_R = 0.05$. The left–right mirror symmetry and the enforcement of Kramers degeneracies at the TRIM Γ and M are evident in the plots. At a critical value of λ_{SO} somewhere between -0.06 and -0.24, there is a simultaneous gap closure at both the K and K' points, which we expect is accompanied by a change of topology from $\mathbb{Z}_2$-even to $\mathbb{Z}_2$-odd.

To confirm that this is the case, the flow of the bulk hybrid Wannier centers and the ribbon edge band structure are plotted in Fig. 5.13 by the PYTHTB program `kanemele_topo.py` (see Appendix D.17) in a manner parallel to that of Fig. 5.4. There is a zigzag of Wannier flow in panel (c) but not (a), and a corresponding traversal of edge states across the gap in (d) but not (b), confirming the identification of topologically trivial and nontrivial states, respectively. Equivalently, the number of crossings in the half BZ with any virtual horizontal line like those in Fig. 5.10 would be even or odd in (a) or (c), respectively, and similarly for any horizontal line drawn inside the energy gap for (b) and (d).

5.2.3 *Computing the $\mathbb{Z}_2$ Invariant*

The recommended choice of method for computing the $\mathbb{Z}_2$ invariant depends on whether inversion symmetry is present. If so, it turns out that there is a simple shortcut that relies on the fact that all the Bloch states at the four TRIM in the 2D BZ have well-defined parities, which follows because the TRIM are invariant points under inversion as well as TR. Based on this observation, Fu and Kane (2007) showed that the $\mathbb{Z}_2$ index ν is just given by

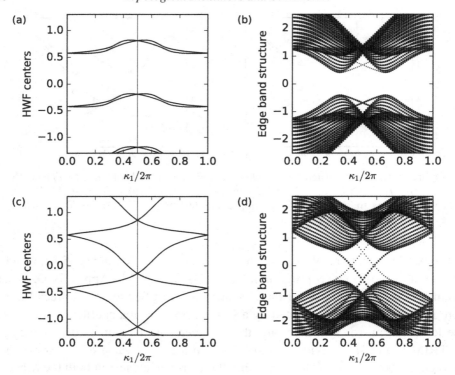

Figure 5.13 (a) Flow of hybrid Wannier centers for a Kane–Mele model in the trivial phase with $\Delta = 0.7$, $t_1 = -1.0$, $\lambda_R = 0.05$, and $\lambda_{SO} = -0.06$. (b) Edge states on a ribbon cut from the same model; those on the top and bottom edges of the ribbon are indicated by full and reduced intensity, respectively. (c–d) Same as (a–b), but in the topological phase, $\lambda_{SO} = -0.24$.

$$(-1)^{\nu} = \prod_{a=1}^{4} \prod_{m}^{N_{occ}/2} \xi_{am} \tag{5.18}$$

where index a runs over the four TRIM, m runs over the occupied pairs of Kramers-degenerate states at that TRIM (counting each pair once), and $\xi_{am} = \pm 1$ is the parity of the members of this Kramers pair.[14] That is, the system is $\mathbb{Z}_2$-odd if and only if the total number of odd-parity Kramers pairs at the four TRIM is odd.

However, when inversion symmetry is absent, as for example in the Kane–Mele model, other approaches are needed. One approach that has been widely used is to take advantage of the bulk–boundary correspondence by computing the edge band structure and inspecting the edge state dispersions as in Fig. 5.13(b) and (d). This is relatively easy to implement using standard methods for computing band structures. Moreover, for the purposes of computing the $\mathbb{Z}_2$ invariant, the details of the edge

---

[14] Since TR commutes with inversion, both members of the Kramers pair must have the same parity.

Hamiltonian do not matter; it is permissible to assume a simple truncation of a tight-binding model at the edge, or in the context of a first-principles calculation, to neglect edge structural relaxation and charge redistribution. However, the $\mathbb{Z}_2$ index is a bulk property, and it would be more natural (and probably more efficient) to compute it from a direct bulk calculation of the Bloch functions associated with the primitive unit cell.

In that spirit, a second approach, and one that will be emphasized throughout the remainder of this book, is to obtain the topological character from an inspection of the flow of the bulk hybrid Wannier centers. As we have seen in Fig. 5.4 for the TR-broken $\mathbb{Z}$ classification, and in Figs. 5.10 and 5.13 for the TR-conserving $\mathbb{Z}_2$ classification, the flow of these HWF centers reveals the topological nature of the bulk in a manner analogous to that obtained from the edge band structure, but only requires a bulk calculation. This approach was introduced by Soluyanov and Vanderbilt (2011a) and Yu et al. (2011), and can be automated fairly easily (Gresch et al., 2017).[15] As a reminder, these hybrid Wannier centers correspond to maximally localized Wannier function centers in 1D, otherwise known as multiband Berry phases or (non-Abelian) Wilson loop eigenvalues, and the method is sometimes described in these terms.

A third approach, and a sort of close cousin to the first two, is to inspect the flow of the entanglement spectrum eigenvalues as discussed for the QAH $\mathbb{Z}$ index on p. 208 and in Appendix D.14. Its extension to the QSH $\mathbb{Z}_2$ index is straightforward and is the subject of Ex. 5.11. Topologically induced features, such as the winding or criss-crossing of the curves, or Kramers-induced degeneracies at the TRIM, appear in a similar way in all three approaches. I recommend the Wannier flow approach as the most natural and straightforward, as it is based entirely on a bulk calculation with no need to introduce a boundary of any kind.

For completeness, it is worth mentioning several other approaches that have been introduced in the literature for computing the $\mathbb{Z}_2$ invariant. One, proposed by Fu and Kane (2006), involves an inspection of a certain unitary overlap matrix $\mathcal{W}_{mn}(\mathbf{k}) = \langle u_{mk} | \mathcal{T} | u_{nk} \rangle$. If the gauge is smooth on the interior of the 2D BZ, then a formula similar to Eq. (5.18), but involving the Pfaffian of $\mathcal{W}$ at the four TRIM, returns the $\mathbb{Z}_2$ invariant. Fu and Kane (2006) also describe this in terms of a concept of "time-reversal polarization," which is closely related to the hybrid Wannier flow approach, as can be seen by comparing their Fig. 3(a) with Fig. 5.13(c) here.

A related approach, also introduced by Fu and Kane (2006), is to evaluate

$$\nu = \frac{1}{2\pi} \left[\oint_{\partial B} \mathbf{A} \cdot d\mathbf{l} - \int_B \Omega \, d^2k \right] \bmod 2 \qquad (5.19)$$

---

[15] These capabilities are available in the open-source Z2PACK code package described therein.

where $\mathbf{A}$ and Ω are the Berry connection and curvature (traced over occupied states), $\mathcal{B}$ is the half BZ ($\kappa_1 \in [0, \pi]$, $\kappa_2 \in [0, 2\pi]$), and $\partial\mathcal{B}$ is the boundary of $\mathcal{B}$ traced counterclockwise. The second term involving Ω is clearly gauge-invariant, but the first is invariant only modulo 2π for an arbitrary gauge. For this equation to be meaningful, therefore, a constraint is placed on the gauge used to obtain $\mathbf{A}$ – namely, that the gauge should respect TR symmetry. That is, for each point $\mathbf{k}$ on the boundary $\partial\mathcal{B}$, the states at $-\mathbf{k}$ (which also lie on $\partial\mathcal{B}$) should be related to those at $\mathbf{k}$ by

$$|u_{2n-1}(-\mathbf{k})\rangle = \mathcal{T}|u_{2n}(\mathbf{k})\rangle \,,$$
$$|u_{2n}(-\mathbf{k})\rangle = -\mathcal{T}|u_{2n-1}(\mathbf{k})\rangle \,. \tag{5.20}$$

Fu and Kane (2006) showed that this definition conforms to the previous ones. It follows that if it is possible to choose a gauge that is continuous in the interior of $\mathcal{B}$ (and thus on its boundary), then Stokes' theorem applied in this gauge makes ν vanish in Eq. (5.19), and the system is topologically trivial. This is closely related to the issue of Wannier representability discussed in the next subsection. Finally, we note that Fukui and Hatsugai (2007) introduced a discretized method for evaluating Eq. (5.19), in which an arbitrary, not necessarily smooth, but TR-preserving gauge is chosen, and the $\mathbb{Z}_2$ invariant is evaluated by checking whether the number of nontrivial plaquettes in $\mathcal{B}$, as evaluated by a formula like Eq. (5.19) but applied to an individual plaquette, is even or odd.

5.2.4 *Wannier Representability*

In Section 5.1.5 we saw that in the case of a 2D QAH insulator, a nontrivial topological index presents a topological obstruction to the existence of Wannier functions. Is there a similar obstruction arising from the $\mathbb{Z}_2$-odd index for the case of a 2D QSH insulator?

The answer depends on how the question is formulated!

In the case of an ordinary TR-invariant insulator, we can expect to obtain a set of Wannier functions that come in Kramers pairs:

$$|w_{2n-1,\mathbf{R}}\rangle = \mathcal{T}|w_{2n,\mathbf{R}}\rangle \,,$$
$$|w_{2n,\mathbf{R}}\rangle = -\mathcal{T}|w_{2n-1,\mathbf{R}}\rangle \,. \tag{5.21}$$

If the Wannier functions are Kramers-paired according to Eq. (5.21), then the Bloch functions constructed from them necessarily obey Eq. (5.20), and vice versa. But we saw in the previous subsection that a smooth gauge obeying Eq. (5.20) is not possible when the system is $\mathbb{Z}_2$-odd.

Thus, the conclusion is that if one insists on Kramers-paired Wannier functions as in Eq. (5.21), then there *is* an obstruction for $\mathbb{Z}_2$-odd insulators, and no such set of Wannier functions is possible. Conversely, if one is willing to abandon this constraint, then a Wannier representation does become possible. In this case the Wannier functions often come out looking a bit strange, and a tight-binding Hamiltonian constructed from them in the manner of Eq. (3.94) typically lacks some of the expected symmetries.[16] The situation has been discussed, with examples provided for the Kane–Mele model, by Soluyanov and Vanderbilt (2011b).

5.2.5 Experimental Realizations

Graphene is often described as a prototypical Dirac semimetal with a zero band gap at the K and K' points in the 2D BZ. However, as discussed by Kane and Mele (2005a, 2005b), even a small SOC opens the gap, and in principle the system is converted into a QSH state that is well described at low energy by the Kane–Mele model. Nevertheless, carbon is such a light element that its SOC strength is tiny, and the resulting energy gap is only on the order of tens of microvolts (Gmitra et al., 2009). For all practical purposes, then, the system is a semimetal.

An important advance came with the prediction by Bernevig et al. (2006) that semiconductor quantum well structures in which a layer of HgTe is sandwiched between layers of a CdTe host could provide a more practical realization of the QSH state. The strong SOC in HgTe gives rise to an inverted band structure in the bulk, in which the highest zone-center Te p levels are above the Hg s levels at the zone center, in contrast to the behavior in CdTe and most other II–VI and III–V semiconductors. In the thin-film environment, Bernevig et al. (2006) predicted a transition from a trivial ($\mathbb{Z}_2$-even) to a topological ($\mathbb{Z}_2$-odd) state as the film thickness exceeds about 6 nm. Subsequent experimental work by Konig et al. (2007) dramatically confirmed this prediction, as signaled by a low-temperature conductivity quantized to within a few percent of $2e^2/h$ (consistent with two edge channels) for the best samples in the appropriate thickness range. Subsequent work on InAs/GaSb quantum wells showed similar behavior (Knez et al., 2011). In both cases, however, the energy gap was still quite small, requiring transport experiments below about 10 K with careful gating to observe clear transport evidence of the predicted edge states.

A variety of strategies have been proposed theoretically for arriving at physical realizations of the QSH state with much larger band gaps and, therefore, potentially higher-temperature operation. This search is motivated in part by the potential for applications based on the "spin filtering" of the edge states. This refers to the fact that the two edge bands crossing through the gap as in Fig. 5.9(d) have opposite

---

[16] In principle, the band structure computed from this Wannier representation should still have all the correct symmetries. However, if the Wannierized tight-binding model is truncated to some set of near-neighbor hoppings, those symmetries may be lost.

spin character; more precisely, the edge states at k and $-k$ are TR images of each other. As a result, for some given Fermi energy E_F, the right-moving electrons at k_F are not scattered into left-moving states at $-k_F$, and vice versa, except by magnetic impurities, potentially giving rise to greatly enhanced mean-free paths and low heat dissipation. The spin-filtered edge states could also prove useful in spintronics applications.

Some of the theoretically proposed strategies for achieving larger-gap QSH states are reviewed in Section IV.D of Bansil et al. (2016). For example, one potential approach is to decorate a graphene layer with adatoms or functional groups involving heavy atoms with large SOC. As of the time of this writing, however, there do not appear to have been any dramatic advances in this direction, and experimental large-gap QSH systems remain elusive.

Exercises

5.7 Consider a 2D TR-invariant insulator whose edge band structure is shown here for two different edge Hamiltonians.

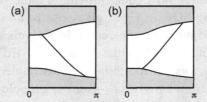

Since both are $\mathbb{Z}_2$-odd (why?), it must be possible to connect them along an adiabatic path that does not close the gap. By making a series of sketches, show how this is plausible. (*Hint:* Imagine continuously raising or lowering the energy of the edge states, remembering to enforce Kramers degeneracy at the boundaries of the half BZ.)

5.8 (a) Consider two QAH layers, each with Chern number $C = 1$, that are brought into contact gently enough that the gap is not closed. Argue that the system will have Chern number $C = 2$. Check this via a calculation of the Wannier flow for a bilayer stack of two identical Haldane layers with a moderately weak hopping connecting each vertical pair of atoms.

(b) Do the same for a TR-invariant system composed of two QSH layers brought into contact in a similar way. Argue that the system will be topologically trivial this time. Check this via a calculation of the Wannier flow for a bilayer stack of two identical Kane–Mele layers with a moderately weak spin-independent hopping connecting vertical atoms.

5.9 This exercise is intended to make the parity rule in Eq. (5.18) plausible. Consider a Kramers pair of Wannier functions as in Eq. (5.21) that generates an isolated pair of bands in a system with inversion as well as TR symmetry. From the discussion of Wannier representability, we expect this pair of bands must be topologically trivial, and they should produce an even count of odd-parity states in Eq. (5.18). Since inversion must map each of these Wannier functions onto itself with an eigenvalue $\nu = \pm 1$, the Wannier function must have its center at one of the four inversion centers $(s_1, s_2) = (0,0)$, $(1/2, 0)$, $(0, 1/2)$, or $(1/2, 1/2)$ in the 2D real-space unit cell. Show that for $\nu = -1$, each of these Wannier function contributes exactly two odd-parity states at the four TRIM.

5.10 Return to the checkerboard model of Ex. 5.3 and Appendix D.15, modifying it to make a QSH insulator.

(a) Modify program `checkerboard.py` using the spinor features of the PYTHTB program, similar to the way the Kane–Mele model was built out of the Haldane model. Use two checkerboard models with $C = +1$ and -1 to represent "spin-up" and "spin-down" electrons, with the spin-down part being the complex conjugate of the spin-up one. That is, the hoppings in diagonal directions should have phases $[1, -i, -1, i]$ instead of $[1, i, -1, -i]$ in counterclockwise rotation. For this part, the two subsystems can be uncoupled. Plot the multiband hybrid Wannier centers $\bar{y}(k_x)$ to observe the structure of the Wannier center flow.

(b) Now turn on a Rashba term that introduces σ_x and/or σ_y spin-dependent hoppings. (You can do this on one set of nearest-neighbor bonds, or both sets, or on the diagonal bonds; it shouldn't matter. However, note that the prefactors of these spin-dependent hopping terms should be pure imaginary to preserve TR symmetry.) Check that you still have Kramers degeneracies in the $\bar{y}(k_x)$ curves at the TRIM k_x values. Did the $\mathbb{Z}_2$ index change?

(c) Now change the parameters so that the underlying model is in the $C=0$ phase, and produce a third plot. What is the $\mathbb{Z}_2$ index now?

5.11 Compute and plot the entanglement spectrum (see the discussion on p. 208) for a finite-width ribbon of the Kane–Mele model following the example of the PYTHTB program in Appendix D.14. Do you see the same kind of topological crossing as in Fig. 5.13(c) and (d)?

5.3 Three-Dimensional $\mathbb{Z}_2$ Topological Insulators

Our discussion in the previous section was limited to 2D, but let us now consider 3D TR-invariant insulators, where we may expect one or more $\mathbb{Z}_2$ indices to survive. After all, if we construct a 3D insulator by making a stack of QSH layers with weak interactions between layers, we may expect it to be $\mathbb{Z}_2$-odd in some sense. Furthermore, since we can stack along any one of the three lattice directions, we might expect three indices (ν_1, ν_2, ν_3) in general, with each $\nu \in \{0, 1\}$ ($\mathbb{Z}_2$-even or -odd, respectively.) Is this correct, and are these the only indices we can define?

5.3.1 Topological Index Set

Let's first recall the discussion of TR-broken Chern insulators in 3D in Section 5.1.6. There we saw that three Chern indices (C_1, C_2, C_3) describe any given insulator, and that if any of these are nonzero, then the material can be regarded as being built from 2D QAH layers stacked in the direction of one of the reciprocal lattice vectors. For 3D TR-invariant insulators, this precedent might lead us to expect a similar behavior, such that there are only three $\mathbb{Z}_2$ indices, and that any TR-invariant insulator can be regarded as composed of QSH layers stacked in a similar way.

This turns out to be incorrect! In fact, a closer inspection suggests the possible existence of six $\mathbb{Z}_2$ indices. Consider the sketches in Fig. 5.14, each of which shows a view of the $\frac{1}{8}$ BZ obtained by letting each reduced wavevector κ_j run from 0 to π.

- The eight heavy dots are the TRIM introduced on p. 51; those points in the BZ obey $-\kappa = \kappa$ modulo a reciprocal lattice vector. At such a point the effective Hamiltonian H_κ has the same TR symmetry as that of a nonmagnetic finite system, and all Bloch eigenvalues come in Kramers pairs.
- Each straight segment is half of one of the 12 *TR-invariant lines*, each of which gets mapped onto itself by $-\kappa \rightarrow \kappa$ under TR. The effective Hamiltonian H_κ on such a line has the same TR symmetry as that of a nonmagnetic chain or polymer.

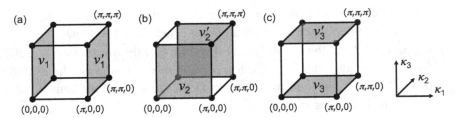

Figure 5.14 Sketch of one-eighth of the BZ of a 3D TR-invariant insulator. Heavy dots indicate TRIM; shaded planes indicate TR-invariant planes. Topological indices ν_1 and ν_1' are defined as the $\mathbb{Z}_2$ indices associated with the labeled TR-invariant planes in (a), and similarly for ν_2 and ν_2' in (b) and ν_3 and ν_3' in (c).

• Each shaded plane is one-fourth of one of the six *TR-invariant planes*, each of which gets mapped onto itself by $-\kappa \rightarrow \kappa$ under TR. The effective Hamiltonian H_κ on such a plane has the same TR symmetry as that of a nonmagnetic 2D system.

Since each of these latter planes is effectively TR-invariant, a $\mathbb{Z}_2$ index can be associated with each, with the labeling chosen as shown in the figure. That is, ν_1 and ν_1' are the $\mathbb{Z}_2$ indices of the planes at $\kappa_1 = 0$ and π, respectively, and similarly for ν_2, ν_2', ν_3, and ν_3'.

So now we seem to have six $\mathbb{Z}_2$ indices, not three! But are they all independent? The answer to this question, which is central to the theory of 3D TIs, was provided independently by Fu et al. (2007), Moore and Balents (2007), and Roy (2009). It turns out that there are four topological $\mathbb{Z}_2$ indices, not three or six, and that one of these, the "strong index," has a special status.

To see that this should be the case, we follow a line of argument similar to that given by Fu et al. (2007). Consider a pair of TR-invariant planes that meet at an edge of the $\frac{1}{8}$ BZ, as, for example, the $\kappa_2 = 0$ and the $\kappa_1 = \pi$ planes with indices ν_2 and ν_1'; these are shown as shaded in Fig. 5.15(a). Further, consider the mapping of the hybrid Wannier flow $\bar{s}_3(\kappa_1, \kappa_2)$ along the path (κ_1, κ_2) from $(0,0) \rightarrow (\pi, 0) \rightarrow (\pi, \pi)$, as shown in Fig. 5.15(e). Each segment covers a half-BZ of a 2D TR-invariant system, so the HWF centers are Kramers-degenerate at the end

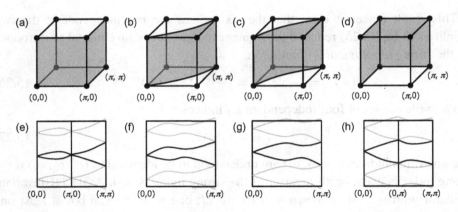

Figure 5.15 (a–d) Gradual deformation of a 2D manifold in the $\frac{1}{8}$ BZ of a TR-invariant insulator with two occupied bands; locations (κ_1, κ_2) in the 2D projected BZ are indicated at bottom. (e–h) Corresponding evolution of the HWF center flow $\bar{s}_3(\kappa_1, \kappa_2)$ (vertical axis) obtained from Berry phases computed on the loop $\kappa_3 \in [0, 2\pi]$. A representative set of HWF centers is drawn as dark curves; periodic images are lightly shaded. Panels (e) and (h) reveal topological indices $(\nu_2, \nu_1') = (0, 1)$ and $(\nu_1, \nu_2') = (1, 0)$ respectively.

points and the manner of connection of the curves (simple vs. zigzag) determines the $\mathbb{Z}_2$ index (0 or 1). In the example shown, we find $\nu_2 = 0$ and $\nu_1' = 1$.

Next consider a gradual deformation of the path in the (κ_1, κ_2) plane, starting from $(0,0) \rightarrow (\pi, 0) \rightarrow (\pi, \pi)$ as in panel (a), proceeding smoothly through some intermediate configurations as shown in (b–c), and ending with $(0,0) \rightarrow (0, \pi) \rightarrow (\pi, \pi)$ as in (d). Panels (e–h) show a possible set of HWF center flow diagrams corresponding to (a–d), respectively. For the intermediate paths the Kramers degeneracy of the HWF centers occurs only at the two end points, but we can still define a $\mathbb{Z}_2$ index $\bar{\nu}$ for the path as a whole. However, a continuous evolution from (e) to (f) clearly requires that the sum of $\mathbb{Z}_2$ indices of the subpanels in (e) must equal that of (f):

$$\bar{\nu} = \nu_2 + \nu_1' \quad \text{mod } 2. \tag{5.22}$$

By the same token, the progress from (g) to (h) requires

$$\bar{\nu} = \nu_1 + \nu_2' \quad \text{mod } 2. \tag{5.23}$$

It follows that

$$\nu_1 + \nu_1' = \nu_2 + \nu_2' \quad \text{mod } 2. \tag{5.24}$$

Similar arguments apply in relating the indices of the other cube faces. Switching to a parity-like notation, we conclude that[17]

$$(-)^{\nu_1 + \nu_1'} = (-)^{\nu_2 + \nu_2'} = (-)^{\nu_3 + \nu_3'}. \tag{5.25}$$

This result makes it clear that the six indices are *not* independent; the two equalities in Eq. (5.25) reduced the number to *four*. It is conventional to introduce ν_0, the *strong topological $\mathbb{Z}_2$ index*, via

$$(-)^{\nu_0} = (-)^{\nu_1 + \nu_1'} = (-)^{\nu_2 + \nu_2'} = (-)^{\nu_3 + \nu_3'} \tag{5.26}$$

and to write the set of four independent $\mathbb{Z}_2$ indices as[18]

$$(\nu_0; \nu_1' \nu_2' \nu_3'). \tag{5.27}$$

The unwritten indices (ν_1, ν_2, ν_3) are understood to be identical to (ν_1', ν_2', ν_3') if the strong index is even, or all reversed if the strong index is odd. A 3D TR-invariant insulator with ν_0 odd is known as a *strong TI*; one with ν_0 even but at least one of the ν_j' odd is a *weak TI*; and one described by index set (0; 000) is said to be topologically trivial.

---

[17] This result implicitly assumes the absence of a bulk band closure anywhere in the 3D BZ. See Section 5.4 for a counterexample when this assumption is not valid.

[18] The choice of the primed indices in the convention has the consequence that if a strong TI has a band inversion at only one of the eight TRIM, then $(\nu_1' \nu_2' \nu_3')$ specifies its location.

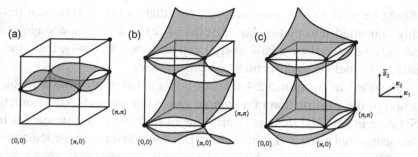

Figure 5.16 Sketches of possible flows of the hybrid Wannier center sheets $\bar{s}_3(\kappa_1, \kappa_2)$ versus in-plane wavevector for a TR-invariant insulator. One real-space repeat unit in the s_3 direction is shown. (a) Normal insulator with $\nu_1 = \nu_1' = \nu_2 = \nu_2' = 0$. (b) Weak TI with $\nu_1 = 1$, $\nu_1' = 1$, $\nu_2 = 0$, $\nu_2' = 0$. (c) Strong TI with $\nu_1 = 1$, $\nu_1' = 0$, $\nu_2 = 0$, $\nu_2' = 1$.

It may also be useful to visualize the topology of weak and strong TIs in terms of the hybrid Wannier sheet structure plotted throughout the one-fourth projected BZ, rather than just along line segments or curves as was done earlier. Calculations of HWF sheet structures of this kind were presented for a variety of tight-binding models, and also for some first-principles calculations of known topological materials, by Taherinejad et al. (2014).

Three possible outcomes for an insulator with two occupied bands are sketched in Fig. 5.16. Panel (a) shows a topologically trivial sheet structure in which there is no connection of the Wannier sheets with the ones in neighboring cells along $\hat{z}$. The situation is similar to the sheet structure shown in Fig. 4.17 for a single-occupied-band insulator as discussed in connection with electric polarization, except that here for spinors we have two sheets glued together by Kramers degeneracies at the four projected TRIM. The indices are $\nu_1 = \nu_1' = \nu_2 = \nu_2' = 0$. From Fig. 5.16(a) we cannot quite conclude that the system is topologically normal, since it remains consistent with both $(0; 000)$ and $(0; 001)$ index sets. A calculation of the hybrid Wannier flow $\bar{s}_2$ versus (κ_1, κ_3) can be used to obtain the values of $\nu_1, \nu_1', \nu_3, \nu_3'$, thereby providing some redundant information about ν_0 and ν_1' while also determining ν_3'. [Of course, a calculation of $\bar{s}_1$ versus (κ_2, κ_3) would have worked just as well.]

Figure 5.16(b) shows a case in which the sheets connect normally along κ_1 but zigzag along κ_2; the nontrivial flow in the κ_2–s_3 plane implies that $\nu_1 = \nu_1' = 1$. This is therefore a weak TI with index set $(0; 100)$ or $(0; 101)$. In Fig. 5.16(c), by contrast, we find $\nu_1 = 1$, $\nu_1' = 0$, $\nu_2 = 0$, and $\nu_2' = 1$. This identifies the state as a strong TI, consistent with both $(1; 010)$ and $(1; 011)$ index sets. If the last index is needed, it can always be determined as before via an additional calculation of Wannier flow along a different set of axes.

It should be clear that a sheet structure like that in Fig. 5.16(a) can never be smoothly converted into that in Fig. 5.16(b) or (c) without some singular event that can only be caused by a bulk energy gap closure. Obviously, then, the three materials shown belong to different topological classes.

We concluded in Section 5.2.4 that an odd $\mathbb{Z}_2$ index presents a topological obstruction to the construction of a smooth gauge that respects TR symmetry for the QSH case in 2D. The same is clearly true of both weak and strong TIs in 3D, since any gauge that respects TR in the full 3D BZ would do so in particular on each of the six TR-invariant planes, which we know to be impossible if any are $\mathbb{Z}_2$-odd.

5.3.2 Strong and Weak Topological Insulators

One way to think of the distinction between weak and strong TIs is that the former can be regarded as a system of stacked QSH layers, while the latter cannot. Consider, for example, an infinite stack of QSH layers normal to $\mathbf{b}_3$ that are initially well separated, and suppose the lattice constant in this third direction is decreased until the layers come into contact. Initially, when the layers are far enough apart, the band dispersions in the κ_3 direction will be negligible, and the Bloch states at $(\kappa_1, \kappa_2, 0)$ and $(\kappa_1, \kappa_2, \pi)$ will be essentially identical to each other, and to a single isolated layer, in their in-plane character. That is, the lack of dispersion with respect to κ_3 implies that the Wannier flow must be trivial for $\bar{s}_1(\kappa_3)$ and $\bar{s}_2(\kappa_3)$, leading to $\nu_2 = \nu_2' = 0$ and $\nu_1 = \nu_1' = 0$, respectively. The QSH character of the layers implies $\nu_3 = \nu_3' = 1$, so the topological index set is $(0; 001)$. If the layers are gradually brought into contact until they reach their final configuration, and if no bulk gap closure occurs during this process, then the index set cannot change; we must have a weak TI with indices $(0; 001)$.

Other weak index sets can be understood in terms of QSH layers stacked in other directions. For example, an insulator with index set $(0; 101)$ can be regarded as being constructed out of stacked QSH layers normal to reciprocal lattice vector $\mathbf{b}_1 + \mathbf{b}_3$. More generally, the weak indices of a weak TI can be thought of as giving the Miller indices of the QSH stacking direction.[19] A strong TI, by contrast, can never be adiabatically connected to a stacked set of weakly coupled QSH layers.

The terms "weak" and "strong" also refer to the degree of robustness of the corresponding topological state against various kinds of perturbations. The most straightforward sense in which a weak TI is not robust has to do with pairing of layers. The point of Ex. 5.8(b) was to show that, when two QSH layers are brought into contact, the result (assuming no gap closure along the way) is a topologically trivial layer. Thus, if we have a weak TI composed of stacks of QSH layers and we

---

[19] See also the related discussion of the QAH case on p. 216.

double the periodicity along the stacking direction, as, for example, by dimerizing the layers in pairs, the index set in the enlarged cell is just (0; 000) and the system becomes trivial. This is explored further in Ex. 5.16.

5.3.3 Surface States

Yet another sense in which strong TIs are "stronger" than weak TIs is connected with the implications for surface states. Consider, for example, a weak TI with indices (0; 001), which we may regard as composed of QSH layers stacked along the third axis. Each layer is fully gapped in the bulk, so there is no reason to expect surface states on the (001) surface. In contrast, on a (100) or (010) surface, the topologically required edge channels still exist, but acquire some dispersion in the κ_3 direction, turning into Fermi loops[20] in the surface band structure. Thus, weak TIs require metallic surface states on some, but not all, exposed surfaces. By contrast, we claim that all surfaces must be metallic for a strong TI.

Before providing a justification of this claim it is important to clarify an implicit assumption – namely, that TR is conserved at the surface as well as in the bulk. If TR symmetry is broken at the surface, either spontaneously, by an applied magnetic field, or by proximity to an adjacent magnetic insulator, the presence of Kramers degeneracy at the four projected TRIM in the surface band structure can no longer be justified, and the arguments given in the following discussion can fail. We shall return to this point in Chapter 6 (see, for example, Fig. 6.8).

Our claim can be justified by considering the bulk–boundary correspondence on each of the four TR-invariant planes intersecting the surface. Consider an (001) surface of the material shown in Fig. 5.15, for example, where we found the four $\mathbb{Z}_2$ indices $\nu_1 = \nu_1' = 1$ and $\nu_2 = \nu_2' = 0$. These must be reflected not only in the Wannier flows shown in Fig. 5.15(e) and (h), but also in the connectivity of surface states, as illustrated in Fig. 5.13. That is, the fact that $\nu_1 = \nu_1' = 1$ means that the surface-BZ segments defined by $\kappa_2 \in [0, \pi]$ at $\kappa_1 = 0$ and $\kappa_1 = \pi$ must both exhibit an odd number of surface states crossing the Fermi level. Similarly, $\nu_2 = \nu_2' = 0$ means that the segments $\kappa_1 \in [0, \pi]$ at $\kappa_2 = 0$ and $\kappa_2 = \pi$ must show an even number of such crossings.

Some of the surface Fermi loop topologies consistent with these rules are shown, for six different sets of topological indices, in Fig. 5.17. In these figures, solid line segments indicate that the $\mathbb{Z}_2$ index of the TR-invariant plane projecting to this line is even, while a dashed line indicates that it is odd. Thus, every dashed line segment of length π must have an odd number of Fermi loops crossing it, while

---

[20] As mentioned on p. 217, the term "Fermi loop" is used for 2D metals, in analogy to the term "Fermi surface" for 3D metals, to describe a constant-energy band contour at E_F.

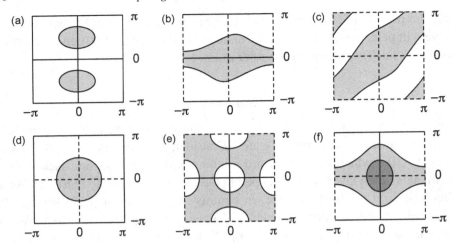

Figure 5.17 Possible configurations of surface Fermi loops in the projected surface BZ for the (001) surface of 3D TR-invariant insulators having different bulk topologies. Shaded (heavily shaded) regions have one (two) more surface bands occupied than white regions. (a) $\nu_1 = \nu_1' = \nu_2 = \nu_2' = 0$. (b) $\nu_1 = \nu_1' = 1$, $\nu_2 = \nu_2' = 0$. (c) $\nu_1 = \nu_1' = \nu_2 = \nu_2' = 1$. (d) $\nu_1 = 0$, $\nu_1' = 1$, $\nu_2 = 0$, $\nu_2' = 1$. (e) $\nu_1 = 1$, $\nu_1' = 0$, $\nu_2 = 1$, $\nu_2' = 0$. (f) $\nu_1 = 1$, $\nu_1' = 0$, $\nu_2 = 0$, $\nu_2' = 1$.

every solid line segment must have an even number of crossings. Each panel shows one possible arrangement of surface Fermi loops consistent with these rules. Each figure also has two-fold rotational symmetry in the plane of the paper, $(\kappa_1, \kappa_2) \rightarrow (-\kappa_1, -\kappa_2)$, as required by TR symmetry; some have other symmetries in addition, while others do not.

The simplest case is that of $\nu_1 = \nu_1' = \nu_2 = \nu_2' = 0$, which could represent either a trivial insulator with index set (0; 000) or a weak TI with indices (0; 001). This case is consistent with the surface BZ shown in Fig. 5.17(a) in which there are two trivial electron pockets; each Fermi loop always crosses a line segment twice. It would also be consistent with a fully gapped surface band structure with no Fermi loops at all. All other panels have at least one dashed line, meaning that an (001) surface of any of these materials is necessarily metallic. Panels (b)–(c) show two different weak TIs, with the former being consistent with the topology of the material in Fig. 5.15. In all cases, many other arrangements of Fermi loops are consistent with the rules; we have chosen to illustrate just a single representative for each.

The remaining panels (d–f) in Fig. 5.17 show possible surface Fermi structures for strong TIs. Panel (d) illustrates the simplest case of a single electron pocket surrounding the projected TRIM at the BZ center. Panel (e) illustrates a configuration of three hole pockets surrounding three of the four projected TRIM. Panel (f) reminds us that nested Fermi loops are also possible.

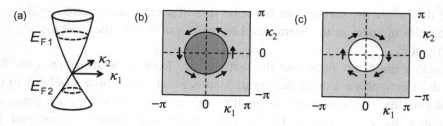

Figure 5.18 (a) Sketch of Dirac cone in the surface band structure of a strong TI. Possible Fermi levels are indicated by dashed circles. (b–c) Fermi loop structure in the surface BZ for the Fermi level positions E_{F1} and E_{F1} respectively. Increasing shading denotes increasing surface band occupation. Arrows indicate spin chirality.

Note that the Fermi loop structure of any *isolated* TR-invariant 2D system, such as an atomically thin layer or a slab cut from the crystalline bulk, can only follow the rules indicated in Fig. 5.17(a). That is, if there are any surface states, they must appear in such a way that the Fermi loops cut every segment of length π an even number of times. All of the structures shown in Fig. 5.17(b–f) are forbidden to occur for an isolated 2D system. The fact that such structures *can* appear at the surface of a 3D insulator is again an example of an *anomaly*, a concept introduced on p. 218 in connection with 3D Chern insulators. In the present case, this anomaly is resolved by noting that the topological index set $(\nu_1, \nu_1', \nu_2, \nu_2')$ governs both the top and bottom surfaces of any thick but finite slab cut from the 3D material in question. Counting both surfaces, then, one always finds an even number of Fermi loops crossing any line segment, as required. There is a strong analogy here to the discussion of Fig. 5.6 in Section 5.1.6.

The configuration shown in Fig. 5.17(d) is typical of materials in the Bi_2Se_3 family, which have a band inversion near the BZ center. In fact, this kind of paradigmatic configuration has formed the basis for much of the discussion of strong TIs in the literature. The situation is illustrated further in Fig. 5.18, where panel (a) shows a sketch of the surface band structure in the bulk band gap in such a TI. The axis in this figure coincides with one of the TRIM in the projected 2D band structure, so the crossing of bands at this point is protected by Kramers degeneracy. The structure shown in the figure is known as a *Dirac cone*.[21] The signature feature of the Dirac cone is that the dispersion is linear at leading order in k_x and k_y at the crossing, which we take to occur at energy E_c. This leads to the presence of an electron pocket in the surface Fermi structure for a Fermi energy $E_{F1} > E_c$, or a hole pocket if $E_{F2} < E_c$, as shown in Fig. 5.18(b–c), respectively.

---

[21] This might be better described as a "Weyl cone" (c.f. Section 5.4), but in 2D the "Dirac cone" usage is by now well established.

Note that while the surface can never be free of surface states crossing the gap, the density of states at the Fermi level formally vanishes in the limit that E_F falls exactly at E_c.

Another important feature of the surface-state Dirac cone structure is the fact that the Berry phase around the Fermi loop is exactly π, which is related to the "spin chirality" illustrated by the small arrows in Fig. 5.18(b–c). As a reminder, the entire discussion here is for the case of spinor electron systems with TR invariance. Since the states at (k_x, k_y) and $(-k_x, -k_y)$ are related by TR symmetry, their spin expectation values must be in opposite directions. This means that as one traverses the loop, the spin orientation vector traces out a path on the Bloch sphere that divides it exactly in half. We learned in Section 3.2.1 that the Berry phase for a path on the Bloch sphere is minus one-half of the traced-out solid angle, or exactly π (modulo 2π) in the present case.[22]

In the paradigmatic case of the (111) surface of Bi_2Se_3-class materials with E_F close to the crossing, the spins lie in the x–y plane and rotate through a 2π angle as one cycles around the Fermi loop. When E_F is close to the Dirac crossing, the loop is almost circular and the spin orientations stay very close to the x–y plane, but the structure becomes more warped and the spins show increased canting above and below the plane as the Fermi level gets farther from the Dirac crossing.

Apparently similar Dirac cones appear in the band structure of an isolated graphene sheet at the K and K' points in the 2D BZ, but there are crucial differences. First, the crossings in graphene occur only in the absence of SOC, in which case two cones are superimposed on top of one another, one for spin up and one for spin down (or four cones counting those at K and K'). Here, however, we find only a single cone in the entire 2D BZ. This is a situation that can occur *only* on the surface of a strong TI. In fact, any isolated 2D sheet, or any 2D surface of a 3D insulator whose strong index ν_0 is even, must have an even number of surface Dirac cones, while any surface of a strong TI must have an odd number.[23]

Note that while we have been interpreting Figs. 5.17 and 5.18 as a visualization of the surface energy bands $E_{surf}(\kappa_1, \kappa_2)$, similar figures would result from a

---

[22] One might think of applying Stokes' theorem to the Fermi pocket in Fig. 5.18(b) and, recalling that TR symmetry implies $\Omega(\mathbf{k}) = -\Omega(-\mathbf{k})$, conclude that the Berry phase should vanish. The flaw in this reasoning is that there is a singularity exactly at the Dirac point that hides a delta-function concentration of π Berry flux, as can be confirmed by computing the Berry phase of a surrounding circular loop of infinitesimal radius in a linearized limit. See also the discussion of Fig. 5.3 on p. 206.

[23] Actually, the counting of Dirac cones at the four TRIM becomes rather subtle for general surface band structures. The basic idea is that one chooses a "reference gap" and then declares that a Dirac cone occurs at a given TRIM if and only if the pattern of Kramers degeneracies closes the reference gap at that TRIM. For example, in Fig. 5.11(a), one would probably assign the large gap at $\kappa = \pi$ as the reference gap, in which case there is a Dirac cone at $\kappa = 0$ but not at π. In Fig. 5.11(d), one might instead choose the gap at $\kappa = 0$ as the reference gap, in which case there is a Dirac cone at $\kappa = \pi$ but not at 0. A change in the choice of reference gap may reverse the list of TRIM carrying Dirac cones, but does not change the evenness or oddness of the total number of Dirac cones.

visualization of the HWF center flow $\bar{s}_3(\kappa_1, \kappa_2)$. For example, if we would plot the intersection of one of the Wannier sheet structures shown in Fig. 5.16 with a constant-$\bar{s}_3$ plane, the resulting curves would generate figures very much like those shown in Fig. 5.17. In particular, the same topological rules would apply; each solid (dashed) line segment has to have an even (odd) number of constant-$\bar{s}_3$ curves crossing it. Moreover, the points of Kramers degeneracy at the TRIM are essentially Dirac cones in the HWF sheet structure, a feature that is evident in Fig. 5.16, although somewhat obscured by the fact that the sheets are plotted in only a quarter of the BZ.

5.3.4 Fu–Kane–Mele Model

A tight-binding model that exhibits some of the topological phases discussed previously for TR-invariant insulators in 3D is the Fu–Kane–Mele model proposed by Fu et al. (2007). It consists of s-like orbitals (two counting spin) on each site of a diamond lattice with a Hamiltonian of the form

$$H = \sum_{\langle ij \rangle} t_{ij} \left(c_i^\dagger c_j + \text{h.c.} \right) + \lambda_{\text{SO}} \sum_{\langle\langle ij \rangle\rangle} \left(i c_i^\dagger \mathbf{g}_{ij} \cdot \boldsymbol{\sigma} c_j + \text{h.c.} \right). \tag{5.28}$$

In the first term the t_{ij} are real first-neighbor hoppings, taken equal to t_0 for all bonds except those aligned along (111), which have hoppings $t_0 + \Delta t$. The second term represents spin-dependent second-neighbor hoppings incorporating the effects of SOC. Here $\mathbf{g}_{ij}$ is a vector of length $\sqrt{2}$ in the direction of $\mathbf{d}_{il} \times \mathbf{d}_{lj}$, where $\mathbf{d}_{il}$ and $\mathbf{d}_{lj}$ are the bond vectors of the first-neighbor pairs making up the second-neighbor bond. The model has inversion and TR symmetry, so that the bulk bands are doubly degenerate throughout the BZ (see Section 2.1.6). They also respect three-fold rotational symmetry around the (111) rhombohedral axis, which implies that the $\mathbb{Z}_2$ indices of the $\kappa_1 = 0$, $\kappa_2 = 0$, and $\kappa_3 = 0$ planes are equal, and similarly for those at $\kappa_j = \pi$. It thus suffices to compute the $\mathbb{Z}_2$ invariant on just two of these planes, which we take to be the ones at $\kappa_3 = 0$ and $\kappa_3 = \pi$.

This model is implemented by the PYTHTB program fkm.py provided in Appendix D.18. This program calculates the band structure (see Fig. D.5), and also computes the Wannier flow needed to identify the invariants as shown in Fig. 5.19. For the chosen parameters $t_0 = 1$, $\Delta t = 0.4$, and $\lambda_{\text{SO}} = 0.125$, it is evident that the flow is trivial in the $\kappa_3 = 0$ plane but nontrivial at $\kappa_3 = \pi$. In view of the three-fold symmetry, it follows that $\nu_1 = \nu_2 = \nu_3 = 0$ and $\nu_1' = \nu_2' = \nu_3' = 1$, so we have a strong TI with index set (1; 111). By modifying the parameter Δt that controls the hopping strength along the (111)-oriented bonds, the model passes through trivial and weak topological as well as strong topological phases, as explored in Ex. 5.14.

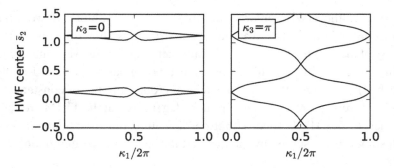

Figure 5.19 Hybrid Wannier center flow in the $\kappa_3 = 0$ and $\kappa_3 = \pi$ planes in the BZ of the Fu–Kane–Mele model of Eq. (5.28). Parameter values are $t_0 = 1$, $\Delta t = 0.4$, and $\lambda_{SO} = 0.125$.

The model can also be used as a basis for illustrating the existence of topologically protected surface states, which again is left to Ex. 5.15.

5.3.5 Experimental Realizations

Soon after the appearance of the first theory papers about the topological classification of 3D TR-invariant insulators (Fu and Kane, 2007; Fu et al., 2007; Moore and Balents, 2007; Roy, 2009), events moved quickly. Hsieh et al. (2008) identified $Bi_{1-x}Sb_x$ as a strong TI in the vicinity of $x \simeq 0.1$ by using high-resolution angle-resolved photoemission spectroscopy (ARPES) to map the surface state dispersions. These dispersions are not simple in this material; five crossings were detected, for example, between the zone center and the zone boundary of the surface BZ, indicating a nontrivial topology of some kind. Closer study reveals this to be a strong TI with index set $(1; 111)$ in the rhombohedral setting. However, the relatively small band gap (approximately 40 meV) and the complexity of the surface band structure make this a somewhat unattractive candidate for a practical TI material.

Shortly afterward, a seminal ARPES study by Xia et al. (2009) identified Bi_2Se_3 as a strong TI with a single Dirac cone centered at Γ in the 2D surface BZ, and a large band gap of approximately 0.3 eV. A pioneering theory paper published almost simultaneously by Xia et al. (2009) provided a detailed understanding of the role of relativistic effects[24] in driving the band inversion at Γ that leads to the topological state in this material, and placed it in context in a family of materials in which Bi_2Te_3 and Sb_2Te_3 were also predicted to be topological while Sb_2Se_3 (being composed of lighter atoms) was not. The topological index set of

---

[24] The band inversion is often described as driven by SOC, but scalar-relativistic shifts of orbital energies, which scale in a similar way with atomic number, also play an important role.

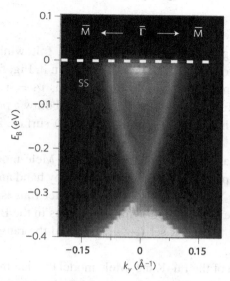

Figure 5.20 Angle-resolved photoemission spectroscopy image of the Dirac cone on the surface of Bi_2Se_3. The brightest intensities indicate regions of occupied bulk states (including an electron pocket at the bottom of the conduction band, barely visible at top center); lesser intensities trace the location of the Dirac-cone surface states, with the crossing point located just above the valence-band maximum. Reprinted by permission from Macmillan Publishers Ltd.: Xia et al. (2009), copyright 2009.

this material was found to be $(1; 000)$ in the rhombohedral setting, which differs in its weak indices from that of $Bi_{1-x}Sb_x$. Numerous subsequent experimental studies characterized the surface bands of these materials in exquisite detail and confirmed many of the theoretical predictions about them, including the spin-chiral nature of the Dirac-cone surface states. Hasan and Kane (2010) and Qi and Zhang (2011) provide excellent reviews of these developments on both experimental and theoretical fronts. Surface transport measurements have proved more difficult in view of sample purity issues, but considerable progress has been made here, too (see, for example, Wu et al., 2016).

Since then, a great many other TIs have been identified. Bansil et al. (2016) offer an excellent review of these more recent developments, in which theory and experiment have worked hand in hand to reveal other classes of TI compounds. A common profile is that the materials are nonmagnetic, are insulating, and incorporate heavy atoms so as to provide strong SOC. Despite this vigorous activity, as of this writing Bi_2Se_3 and members of its class remain the workhorse materials of choice for most investigations and potential applications of TIs because of their ease of preparation, relatively large band gap, and the presence of just a single Dirac cone surrounding the Γ point in the surface BZ.

Exercises

5.12 Sketch what Fig. 5.15(e–g) might look like, following the same paths in (κ_1, κ_2) space, for each of the materials shown in Fig. 5.16.

5.13 Suppose each material shown in Fig. 5.17 has $\nu_3 = 1$. For each of the six cases, give the full $(\nu_0; \nu_1' \nu_2' \nu_3')$ index set, and draw a possible diagram like the one shown in Fig. 5.17, but for the (100) surface instead of the (001) surface.

5.14 Explore the phase diagram of the Fu–Kane–Mele model empirically as a function of the parameter Δt by adjusting it by hand and watching how the results change. You should find five parameter ranges, delineated by four band-touching events at high-symmetry points in the BZ. For each of these phases, specify the index set $(\nu_0; \nu_1' \nu_2' \nu_3')$ and the range of Δt values over which it occurs.

5.15 Consider a slab of the Fu–Kane–Mele model that is finite in the third lattice direction (i.e., the surface is normal to $\mathbf{b}_3$). In the manner of Fig. 5.13(b) and (d), plot the surface energy bands versus κ_1 at $\kappa_2 = 0$ and $\kappa_2 = \pi$, and versus κ_2 at $\kappa_1 = 0$ and $\kappa_1 = \pi$. Are your results consistent with expectations based on the topological index set?

5.16 Based on the discussion at the end of Section 5.3.2, we expect that a weak TI, which can be thought of as composed of stacked QSH layers, can be made trivial if the periodicity is doubled along the stacking direction.

    (a) Set up two 3D PYTHTB models similar to that of Ex. 5.8(b) with weak hoppings between layers, but stacked infinitely along the $\hat{\mathbf{z}}$ direction. One model should have the primitive periodicity along $\hat{\mathbf{z}}$, while the other has doubled periodicity and allows for two distinct vertical hopping strengths in alternating layers of vertical bonds. (The latter can be constructed from the former by using the PYTHTB `make_supercell` method and then modifying some of the hoppings.) To study the topological properties, plot the Wannier flow $\bar{s}_2(\kappa_1)$ at $\kappa_3 = 0$ or $\kappa_3 = \pi$ for three cases: the single-periodicity model; the doubled model with equal vertical hoppings; and again with unequal hoppings. Discuss.

    (b) By contrast, a period-doubled strong TI remains a strong TI. Demonstrate this by doubling the Fu–Kane–Mele model along some direction and plot the Wannier flow before and after doubling. Discuss.

5.4 Weyl Semimetals

While we have been focusing up to now on topological aspects of the band structures of insulators, it is also possible to find topological characteristics in the band structures of some metallic systems. In particular, systems in which the Fermi level lies at or close to isolated degeneracies between the valence and conduction bands in the 3D BZ, known as *Weyl semimetals*, have received a great deal of attention recently, as have some of their cousins (e.g., *Dirac semimetals* and *nodal-loop semimetals*). To approach this topic, it is best to start with a discussion of *conical intersections* in arbitrary parameter spaces.

5.4.1 Conical Intersections

In Chapter 3 we introduced a model of a triangular molecule undergoing a pseudorotation as sketched in Fig. 3.2 and solved by the program `trimer.py` in Appendix D.6. The point of Ex. 3.7 was to show that in the 3D space of parameters (u, v, α), there is a special isolated point, at $u = v = 0$ and $\alpha = \pi$, at which the two lowest eigenenergies become degenerate. A bit of numerical investigation using a program like `trimer.py` will show that the splitting of the degeneracy is linear in the distance from the touching point in any given direction in parameter space. Such an intersection is known as a "conical intersection" for this reason. That is, if any one of the three parameters is held fixed at the touching value while the other two are varied, the energy eigenvalues exhibit a conical structure like that shown in Fig. 5.18(a).

These conical intersections are topological in themselves, as we can see by the analogy with the behavior of a free spinor with gyromagnetic ratio $\gamma > 0$ in a magnetic field as described by Eq. (3.23) in Section 3.1.3. In this ideal case the energy eigenvalues are just $E(\mathbf{B}) = \pm E_0 \sqrt{B_x^2 + B_y^2 + B_z^2}$, where $E_0 = \gamma \hbar / 2$ in Eq. (3.23), so that we have a perfectly isotropic conical behavior in the parameter space (B_x, B_y, B_z). We saw in Section 3.2.2 that the Chern number on a sphere in $\mathbf{B}$ space surrounding the origin is -1 for the lower-energy (spin "up" along $+\hat{\mathbf{B}}$) solution, and $+1$ for the higher-energy (spin "down") one. Thus, such an intersection point can be regarded as a source or sink of a 2π quantum of Berry flux in the surrounding parameter space.

The same applies to the trimer model. As discussed on p. 96, Fig. 3.11 implies a Chern number $C^{(\varphi\alpha)} = +1$ for the torus enclosing the conical intersection point. Further investigation shows that the ground-state Chern number on any sphere surrounding the conical intersection point in (u, v, α) space is also $C = +1$, where the connection with Eq. (3.44) is given by assigning $u = s \cos \varphi$ and $v = \sin \varphi$.[25] After all, we can imagine deforming the surface from the initial torus to any final sphere, first by pinching off the torus on the side far from the conical intersection,

---

[25] The sign convention is such that (u, v, α) is a right-handed coordinate system and C is the integral of $\hat{\mathbf{n}} \cdot \boldsymbol{\Omega}$ with $\hat{\mathbf{n}}$ outward-directed on the sphere.

Figure 5.21 Sketch of 2D manifold enclosing a conical intersection (star) as it is deformed from an initial torus, representing (φ, α) space, to a surrounding sphere in (u, v, α) space (see text).

and then shrinking it to a sphere, as sketched in Fig. 5.21; the Chern number cannot change during this process if no other gap closure is encountered.

To discuss these kinds of band touching events in more generality, consider an N-dimensional parameter space $\boldsymbol{\lambda} = (\lambda_1, \ldots, \lambda_N)$. Let the origin $\boldsymbol{\lambda} = \mathbf{0}$ in parameter space be chosen close to the point where two eigenvalues are found close in energy, while other states are far away. In this case we may use the upper and lower states at $\boldsymbol{\lambda} = \mathbf{0}$ (or any linear combination of them, if the states are degenerate) as a basis for an approximate description of the dispersions for nearby $\boldsymbol{\lambda}$. The effective Hamiltonian in this two-state space then takes the form

$$H_{2\times 2}(\boldsymbol{\lambda}) = f_0(\boldsymbol{\lambda}) I + f_1(\boldsymbol{\lambda}) \sigma_1 + f_2(\boldsymbol{\lambda}) \sigma_2 + f_3(\boldsymbol{\lambda}) \sigma_3 \tag{5.29}$$

where the four f_j are some smooth functions of $\boldsymbol{\lambda}$, I is the 2×2 identity, and the σ_j are the Pauli matrices of Eq. (2.20), relabeled with indices $\{1, 2, 3\}$ instead of $\{x, y, z\}$ since they they have nothing to do with spin here. (In fact, $\{I, \sigma_1, \sigma_2, \sigma_3\}$ can just be thought of as a basis for writing any possible 2×2 Hermitian matrix.) We encountered this form in Eq. (2.86), and its eigenvalues are just $E = f_0 \pm \sqrt{f_1^2 + f_2^2 + f_3^2}$.

What this means is that $f_0(\boldsymbol{\lambda})$ is irrelevant for the purposes of finding a degeneracy, which will occur if and only if $f_1(\boldsymbol{\lambda}) = f_2(\boldsymbol{\lambda}) = f_3(\boldsymbol{\lambda}) = 0$. Since this requires three equations to be satisfied simultaneously, it can be satisfied generically in a parameter space of dimension $N \geq 3$, but not if $N \leq 2$. That is, by varying $(\lambda_1, \lambda_2, \lambda_3)$ in a three-dimensional parameter space, we can generically find special points at which the two energy bands touch, which are our conical intersection points. The term "generically" implies that while we are not guaranteed to find any such touching events, we should not be surprised to find them, either; no special fine-tuning or extra symmetry is needed to arrive at one. If some additional parameters of the model are adjusted, the conical intersection points cannot immediately disappear; to first order they will simply be shifted such that they occur at a slightly different point in the 3D parameter space. In dimension $N \leq 2$, by contrast, such conical intersections will occur "with probability zero" unless tuned by hand or in the presence of special symmetries.

This is an example of what is known as a *codimension argument*.[26] If we linearize in the vicinity of a degeneracy and let W (of dimension N) be the full vector space while V (of dimension M) is the subspace in which degeneracies occur, then $\dim W - \dim V = N - M$ is known as the "codimension of V in W." Because the degeneracies require the coefficients of three Pauli matrices to vanish, we expect a codimension of 3 in the present case. That is, we generically expect no degeneracies in dimension $N \leq 2$; points in a parameter space of $N = 3$; lines (loops) in a parameter space of $N = 4$; surfaces in $N = 5$; and so on.

If we do find an isolated degeneracy in a parameter space of dimension $N = 3$, we can expand Eq. (5.29) in λ in the vicinity of the crossing. After resetting the origin of λ to coincide with the crossing and dropping the $f_0 I$ term of Eq. (5.29), which has no influence on the existence of a degeneracy, we obtain

$$H_{2\times 2}(\lambda) = \sum_{\mu\nu} A_{\mu\nu} \lambda_\mu \sigma_\nu \tag{5.30}$$

where $A_{\mu\nu} = \partial_\mu f_\nu$ is a 3×3 Jacobian matrix evaluated at $\lambda = 0$. Here we have dropped higher than linear terms in λ, an approximation that is valid sufficiently close to the degeneracy point provided that A is not singular ($\det A \neq 0$). In that case, $\lambda_1 = \lambda_2 = \lambda_3 = 0$ is the unique solution in this vicinity.

We can go one step further and assign a *chirality* $\chi = \pm 1$ to the isolated degeneracy, with the convention that the degeneracy point acts as a source of 2π Berry flux for the lower-energy state and a sink of 2π for the upper state if $\chi = +1$, and vice versa if $\chi = -1$. In other words, the Chern number of the lower state on any small sphere surrounding the degeneracy point in λ space (computed from the outward-directed flux in a right-handed coordinate system) is $2\pi\chi$. The choice of $A_{\mu\nu} = c\delta_{\mu\nu}$ for real c exemplifies the two possibilities: We obtain $\chi = -1$ according to the discussion in Sections 3.1.3 and 3.2.2 when $c < 0$ in Eq. (3.23) or, conversely, we obtain $\chi = +1$ if $c > 0$. More generally, we can imagine gradually modifying the conditions giving rise to the conical intersection in such a way that the matrix A smoothly changes. In this process the Chern number on a small sphere surrounding the crossing point cannot suddenly change unless a singularity is encountered. The relevant singularity here is the case $\det A = 0$, at which point the neglect of higher-than-linear terms in Eq. (5.30) is no longer justified. From such continuity arguments, it follows that the chirality of a conical intersection is given by[27]

$$\chi = \text{sgn}(\det A). \tag{5.31}$$

---

[26] A nice pedagogical introduction to this style of argument can be found in Berry (1985).
[27] Note that the sign convention in Eq. (5.31) depends on the order in which the three parameters are assigned to a right-handed coordinate system.

From a topological point of view, then, conical intersections come in two flavors, one with positive and one with negative chirality, depending on the sign of the determinant of A.

Up to now we have been considering an arbitrary parameter space, but we can also consider the case in which one or more of the parameters are wavevectors components of a crystal Hamiltonian. We encountered one such case in Section 5.1.1, where Fig. 5.2 shows the evolution of the band structure of the 2D Haldane model in the vicinity of the phase boundary between the trivial and topological states. Figure 5.2(c) clearly shows the presence of a conical intersection in (k_x, k_y, t_2) space (two wavevectors and one model parameter) at the K' point in the 2D BZ. From the point of view of the 2D space (k_x, k_y), the codimension argument implies that a band touching will *not* generically occur, a statement that is supported by the other panels in Fig. 5.2; external tuning of some third parameter is needed, as in panel (c).[28] In contrast, from the viewpoint of the extended (k_x, k_y, t_2) space, a conical intersection *can* occur generically. The transfer of a 2π flux of Berry curvature from the lower to the higher band as t_2 passes through its critical value at the normal-to-topological transition is intimately connected with the chirality-determined Chern number of a 2D surface surrounding the touching point in the extended 3D space, as explored in Ex. 5.17.

5.4.2 Weyl Points

We now specialize to the case that the parameter space in question is just the manifold of Bloch wavevectors $\mathbf{k}$ in the BZ of some 3D crystal. According to the codimension argument given earlier, we expect that conical intersections can then occur generically, where the wavevector components (k_x, k_y, k_z) have enough freedom to force two bands to become degenerate at a point $\mathbf{k}_0$ in wavevector space. Letting $\mathbf{q} = \mathbf{k} - \mathbf{k}_0$, the leading-order effective Hamiltonian in the space of crossing bands generically takes the form

$$H_{2\times2}(\mathbf{q}) = (E_0 + \hbar\,\mathbf{v}_0 \cdot \mathbf{q})\,I + \sum_{\mu\nu} A_{\mu\nu}\,q_\mu\sigma_\nu \qquad (5.32)$$

where E_0 and $\mathbf{v}_0$ are the energy and average Fermi velocity at the crossing point, respectively, and $A_{\mu\nu}$ is the 3×3 Jacobian matrix as before. (Here the σ_ν are "pseudospin" operators acting in the band space, not the spin space.) The terms involving the 2×2 identity matrix I that were dropped from Eq. (5.30) have been restored here, since the shape of the energy bands may play an important role.

---

[28] This statement assumes the absence of additional symmetries that, if present, may allow (or even require) the presence of band touchings in 2D (see also p. 258).

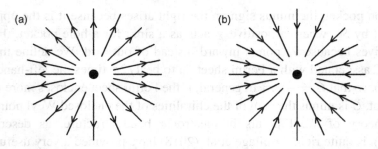

Figure 5.22 Sketch showing flux of Berry curvature of the lower-energy band emerging from or vanishing into the Weyl point at the center. (a) Negative chirality, $\chi = -1$. (b) Positive chirality, $\chi = +1$.

A band touching of the kind described by Eq. (5.32) is known as a *Weyl point* because of the resemblance of Eq. (5.32) to the famous Weyl Hamiltonian

$$H(\mathbf{p}) = \chi \, c \, \mathbf{p} \cdot \boldsymbol{\sigma} \qquad (5.33)$$

introduced by Herman Weyl (1929) in the context of elementary particle physics, where $\chi = \pm 1$ is the chirality, c is the speed of light, $\mathbf{p}$ is momentum, and $\boldsymbol{\sigma}$ is spin. The sign of χ determines whether the equation describes particles whose spin is parallel or antiparallel to their momentum. Equation (5.32) reduces to Eq. (5.33) with $\mathbf{p} = \hbar \mathbf{q}$ and $E_0 = 0$ on the assumptions of spatial isotropy and Lorentz invariance. The Weyl fermions described by this equation behave just like massless Dirac fermions, but of just one handedness. That is, the massless Dirac equation can be regarded as a composition of two Weyl Hamiltonians with opposite chirality. Weyl proposed that a species of particle could exist in only one state of handedness as described by Eq. (5.33). For some time it was thought that neutrinos might be Weyl fermions, but this position became untenable when neutrinos were found to have mass. More recently, the book by Volovik (2003) has been influential in drawing the attention of theorists to the phenomenology of Weyl fermions in the condensed-matter context.

A word about the sign of χ is in order. The sign is still given by Eq. (5.31) using the A matrix from Eq. (5.30), and we repeat that a positive χ corresponds to a source and sink of Berry curvature in the lower and upper bands, respectively. This means that the Berry flux integrated over a Fermi sheet surrounding an electron or hole pocket containing a Weyl point is

$$\int_{S_F} \boldsymbol{\Omega} \cdot \hat{\mathbf{v}}_F \, dS = -2\pi \chi \,. \qquad (5.34)$$

Here $\hat{\mathbf{v}}_F$ is a unit vector in the direction of the Fermi velocity; that is, it is normal to the Fermi sheet and points in the direction of unoccupied states. For the case of an electron pocket, the minus sign on the right arises because it is the upper band that is cut by S_F, where a positive χ acts as a sink; for a hole pocket, the minus sign survives because $\hat{v}_F$ points inward instead of outward. We define the Chern number C associated with a Fermi sheet S_F to be $1/2\pi$ times the left-hand side of Eq. (5.34), so that $C = -\chi$.[29] In general, if the Fermi sheet encloses more than one Weyl point, C is minus the sum of the chiralities of the enclosed Weyl points.[30]

The theory of Weyl points in electronic band structure, as described by Eq. (5.32), is quite rich. Armitage et al. (2018) have provided a very useful review focusing on the theory underlying such systems, while the reviews by Hasan et al. (2017) and Yan and Felser (2017) provide more coverage of experimental developments. As a point of interest, Gosálbez-Martínez et al. (2015) surveyed all of the Weyl points in the occupied states of bcc Fe and found that they are remarkably plentiful, numbering more than 100. For the present purposes, we shall specialize our discussion by focusing our attention as follows.

1. Weyl points may occur between any adjacent pair of bands, including deep in the valence or conduction bands. Here, we restrict our attention to Weyl points at or near the Fermi energy – that is, connecting the highest valence and lowest conduction band.[31]

2. TR and inversion symmetries play an important role. We assume spinful electrons throughout the present discussion. If the Hamiltonian commutes with both $\mathcal{T}$ and $\mathcal{I}$ (or, for that matter, if it commutes with $\mathcal{I} * \mathcal{T}$), then all bands are Kramers degenerate already (see Section 2.1.6), and any additional touchings would generate points of fourfold degeneracy known as *Dirac points*. We defer consideration of this possibility until Section 5.5.2, excluding it from our discussion until then.

3. Crystallographic symmetries may also play an important role. For example, in a crystal with a rotational symmetry axis, Weyl points can occur generically on the symmetry axis in **k**-space because states belonging to different irreducible representations can cross without mixing at some point on the symmetry line. In some cases this can even result in higher-order nodes with chiral charge $|\chi| > 1$, such that they act as a source or sink of 4π or 6π Berry flux (see, for example, Tsirkin et al., 2017). We exclude this possibility as well in the following discussion, concentrating just on generic locations in the 3D

---

[29] The minus sign in this relation can be traced back to our choice of sign conventions in Eqs. (3.1) and (3.13) for the Berry phase and connection. The reader should be aware that other conventions are commonly used in the literature.

[30] This statement is sufficient in simple situations but needs to be generalized for more complex ones, such as when multiple nested Fermi sheets are generated by several bands crossing E_F. A careful treatment can be found in Section III of Gosálbez-Martínez et al. (2015).

[31] In Eq. (5.34), Ω refers to the Berry curvature of the single band crossing E_F, but the result also holds if we replace it with the total Berry curvature of the occupied-band manifold as a whole as expressed in Eq. (3.136).

BZ that are unaffected by any symmetries other than possible inversion and TR symmetries.

4. As pointed out by Soluyanov et al. (2015), Weyl points can be classified into so-called *Type I* and *Type II* varieties according to the strength of the $\mathbf{v}_0$ term in Eq. (5.32). To clarify the situation, we choose a direction $\hat{\mathbf{q}}$ and consider the form of the dispersion along a ray in this direction in $\mathbf{k}$ space. Letting $\alpha_\nu = \sum_\mu A_{\mu\nu}\hat{q}_\mu$, the linearized-in-$q$ Hamiltonian of Eq. (5.32) simplifies to

$$H_{2\times2}(q) = E_0 I + (\beta I + \alpha\sigma_\alpha)\,q \qquad (5.35)$$

along this ray, where $\beta = \hbar\hat{\mathbf{q}} \cdot \mathbf{v}_0$, α is the norm of $\boldsymbol{\alpha}$, and σ_α is the Pauli matrix $\hat{\boldsymbol{\alpha}} \cdot \boldsymbol{\sigma}$. Its eigenvalues are then $E = E_0 + (\beta \pm \alpha)q$. When $\beta = 0$, we get a symmetric Weyl cone like that shown in Fig. 5.23(a), or as encountered earlier in Fig. 5.18(a). For nonzero β, however, the dispersion develops an asymmetry. For small but positive β, for example, the slope of the down-going lower band steepens to the left and becomes more gradual to the right of the crossing point, and vice versa for the upper band, as shown in Fig. 5.23(b). When $|\beta| > \alpha$, both bands increase in energy to the right, corresponding the the Type-II scenario shown in Fig. 5.23(c). More precisely, a Weyl point is classified as being of Type II if $|\beta| > \alpha$ for some direction $\hat{\mathbf{q}}$, and of Type I otherwise. This distinction is important because when E_F is close to to E_0, a pair of hyperbolic Fermi sheets extends away from the Weyl point in the Type II case, instead of a small ellipsoidal pocket being present as in the Type I case. Thus, the density of states does not vanish as E_F tends to E_0 in the Type II case. We shall restrict our considerations to Type-I Weyl points in the following discussion.

We have been focusing on the local properties of Weyl points in reciprocal space, but a crucial *global* property of the population of Weyl points also needs to be discussed. Namely, the Nielsen–Ninomiya "Fermion doubling" theorem (Nielsen and Ninomiya, 1983) states that the sum of the chiralities of the Weyl points in the 3D BZ must vanish. That is,

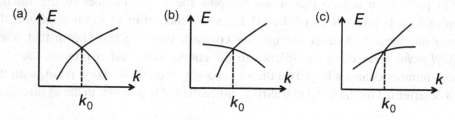

Figure 5.23 Sketch of energy dispersions in the vicinity of a Weyl point at k_0. (a) Isotropic Type-I Weyl point. (b) Anisotropic Type-I Weyl point. (c) Type-II Weyl point.

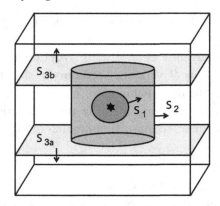

Figure 5.24 Sketch of three surfaces surrounding a Weyl point (star) in a region that is otherwise free of band touchings in the 3D BZ. Arrows show surface normal directions $\hat{\mathbf{n}}$ used for computing the Chern index on each surface.

$$\sum_j \chi_j = 0 \qquad (5.36)$$

where index j runs over the Weyl points in the BZ.[32] We can see why this must be so by considering the sketch in Fig. 5.24, which shows a Weyl point (denoted by the star) in a region of the BZ that is otherwise free of band touchings. As discussed earlier, the Chern index of the lower band on the inner sphere S_1 is $-\chi$. Now imagine gradually deforming this surface until it adopts the shape of S_2 (a cylinder including its end caps), and then increasing the radius of the cylinder until it merges at the side boundaries of the BZ and fills the entire region between planes S_{3a} and S_{3b}. If no other conical intersections were encountered during these deformations, the Chern number on surface S_3, computed as the sum of contributions from S_{3a} and S_{3b} with downward and upward unit normals $\hat{\mathbf{n}}$, respectively, as shown in the sketch, must still equal $-\chi$. Actually, each of these "planes" is a 2D BZ, a torus. It then follows that $\chi = C_a - C_b$, where the slice Chern numbers C_a and C_b are now both defined with $\hat{\mathbf{n}}$ pointing upward.

On p. 215 we argued that if we compute the Chern number of the valence manifold on (κ_1, κ_2) slices of the 3D BZ as a function of κ_3 in an *insulator*, the Chern number could never change. By contrast, we have just shown that when a *Weyl point* connecting the valence to the conduction band is present, the slice Chern number changes by ± 1 as the slice passes through the Weyl point, with the sign determined by the Weyl chirality χ. Since the BZ is periodic in the κ_3 direction,

---

[32] Here we are discussing touchings between the highest valence and lowest conduction bands, but the theorem also applies to any adjacent pair of bands.

the slice Chern number has to return to itself after a full loop, proving the Nielsen–Ninomiya theorem of Eq. (5.36).

Thus, a single Weyl point can never appear in isolation; it always has to be part of a population of Weyl points whose chiralities sum to zero. We can say more about this population by returning for a moment to the implications of inversion and TR symmetry.

If either of these symmetries is present, then $E(\mathbf{k}) = E(-\mathbf{k})$, and it follows that a Weyl point at $\mathbf{k}_0$ necessarily has a partner at $-\mathbf{k}_0$. Moreover, we saw on p. 105 that inversion and TR impose the constraints $\mathbf{\Omega}(\mathbf{k}) = \mathbf{\Omega}(-\mathbf{k})$ and $\mathbf{\Omega}(\mathbf{k}) = -\mathbf{\Omega}(-\mathbf{k})$, respectively, on the Berry curvature. However, if we imagine that small spheres are constructed around these Weyl points, if a given vector is outward-pointing at point $\mathbf{k}$ on one of these spheres, the same vector will be inward-pointing at $-\mathbf{k}$ on the partner sphere. This amounts to an additional sign reversal on the Chern number computed on the surrounding spheres, and it follows that the two chiralities are related by

$$\chi(\mathbf{k}_0) = -\chi(-\mathbf{k}_0) \qquad \text{(inversion symmetry)},$$
$$\chi(\mathbf{k}_0) = \chi(-\mathbf{k}_0) \qquad \text{(TR symmetry)}. \qquad (5.37)$$

If neither inversion nor TR is present, then a Weyl point at $\mathbf{k}_0$ generically has no partner at $-\mathbf{k}_0$.

An *ideal Weyl semimetal* is a 3D crystal in which there is a population of Weyl points exactly at the Fermi energy E_F, with no other Fermi surfaces present. In other words, it is almost an insulator, but with isolated band touchings between the top valence and lowest conduction band occurring exactly at E_F. Such a material would have a formally vanishing density of states at the Fermi energy (see Ex. 5.18). More broadly, the term *Weyl semimetal* is applied to metals in which this is almost true, so that the global Fermi surface consists of a collection of small electron or hole pockets enclosing Weyl points.[33]

The symmetry considerations allow us to discuss the following cases:

- TR is present but not inversion ("nonmagnetic Weyl semimetals"). In this case the presence of a Weyl point at $\mathbf{k}_0$ requires the existence of a partner at $-\mathbf{k}_0$ with the *same* chirality, so the Nielsen–Ninomiya theorem implies that these cannot be the only Weyl points. At a minimum, another pair of Weyl points must be present elsewhere in the BZ with the opposite chirality. Generically these will occur at a

---

[33] Some authors also use the term "topological metal" to denote a system in which Fermi sheets with nonzero Chern numbers are present. In addition to Weyl semimetals, this category includes ordinary metals such as bcc Fe in which some of the Fermi sheets enclose a Weyl point. See, for example, Haldane (2014) and Gosálbez-Martínez et al. (2015).

different energy, making an ideal Weyl semimetal impossible, although if mirror or other symmetries are present they may be pinned to the same energy.

- Inversion symmetry is present, but not TR ("magnetic Weyl semimetals"). In this case a minimal Weyl semimetal with E_F coinciding with the energy of a pair of opposite-chirality Weyl points at $\pm\mathbf{k}_0$ is possible.
- Neither inversion nor TR is present ("generic Weyl semimetals"). In this case Weyl points generically occur with an arbitrary distribution, although mirror or other symmetries may induce relations between them.

5.4.3 Fermi Arcs and the Chiral Anomaly

Much of the discussion of Weyl semimetals in the literature has focused on three physical consequences arising from the presence of Weyl points. The first is the occurrence of "Fermi arcs" in the surface band structure, and the second and third are the "chiral anomaly" and "chiral magnetic effect" that arise in the context of transport theory.

The presence of *Fermi arcs*, as discussed by Wan et al. (2011), is yet another example of a bulk–boundary correspondence. The principle is illustrated in Fig. 5.25, which shows an example of a magnetic Weyl semimetal in which Weyl points of opposite chirality are located at positions $\pm\mathbf{k}_0$ in the 3D BZ. To characterize the Fermi loop structure on the (001) surface, it is instructive to construct a cylinder like that shown in Fig. 5.25(a) enclosing one Weyl point. Its base is a circle in the (κ_1, κ_2) plane parametrized by λ going from 0 to 1, and it extends fully across the BZ in the κ_3 direction. Since the top and bottom are identified, the cylinder is really a closed manifold (i.e., a torus), and carries a nonzero Chern number given by the chirality of the enclosed Weyl point. Fig. 5.25(b) shows a sketch of the surface band structure plotted versus λ; because

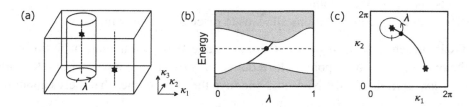

Figure 5.25   (a) Sketch of 3D BZ of a Weyl semimetal with two Weyl points shown by stars. The cylinder indicates a manifold enclosing one Weyl point, following a path in the (κ_1, κ_2) projected plane parametrized by λ running from 0 to 1. (b) Surface band structure on the (001) surface plotted versus λ. The existence of an up-crossing surface state implies a Chern number of $+1$ on the cylinder. The filled circle shows a crossing of the surface state with the Fermi energy (dashed line). (c) Illustration of the surface Fermi loop structure, showing a Fermi arc connecting the projected positions of the Weyl points.

of the nonzero Chern number, a surface state is topologically required to cross the gap as shown. The crossing of this state with E_F is shown as the heavy dot in panel (b) and in the view of the surface BZ in panel (c), where the λ loop is again indicated. Since we can repeat this construction for cylinders of different radii, and since we are required to find a crossing with E_F at any radius, the dot in Fig. 5.25(c) must actually belong to a "Fermi arc" extending out of the projected location of the Weyl point as shown.

It follows, then, that each Weyl point acts as a source from which a Fermi arc is emitted. But the arc has to terminate, and can do so only at the surface projection of another Weyl point of opposite chirality, as illustrated in Fig. 5.25(c). The Nielsen–Ninomiya theorem guarantees that there is an equal number of "sources" and "sinks" for the Fermi arcs, allowing a globally consistent configuration to emerge.

Here we have another example of an anomaly in the surface band structure, a concept that we encountered previously on p. 218 and again on p. 247. That is, we have a Fermi loop structure that could never exist in an isolated 2D system, or at the surface of a 3D insulator (even that of a TI of any kind). The anomaly is resolved for the case of a thick but finite slab by noting that while one must find a Fermi arc connecting the two projected Weyl points on the top surface, there must be another arc on the bottom surface, giving rise to an "ordinary" closed electron or hole pocket for the slab as a whole.

If multiple Weyl points are present in the 3D BZ, then multiple Fermi arcs will appear in the surface band structure. (As an exception, if two Weyl points of opposite chirality project to the same point in the surface BZ, no Fermi arc is needed.) The picture that arises when the Weyl points fall somewhat above or below the Fermi energy is discussed by Haldane (2014), where Fermi arcs are found to emerge out of, and then merge into, the pockets of surface-projected bulk states at the Fermi energy.

A second topological consequence of the presence of a Weyl point near the Fermi energy is the *chiral anomaly*. We have in mind that the global Fermi surface is composed of some number of electron or hole pockets, each surrounding a Weyl point and carrying a Chern number C_i equal to minus the chirality χ_i of the enclosed Weyl node. We also assign an electron density n_i to each pocket (or "valley"), positive for electron pockets and negative for holes. Then, in the presence of simultaneously applied electric and magnetic fields, one finds that

$$\frac{dn_i}{dt} = \frac{e^2}{h^2 c}\,(\boldsymbol{\mathcal{E}} \cdot \mathbf{B})\,\chi_i\,, \tag{5.38}$$

as we shall do shortly. That is, when focusing on the neighborhood of a single valley in isolation, one concludes that particle population grows or shrinks with time, so that particle number is not conserved!

Of course, particle number really *is* conserved. This becomes self-evident when we remember the Nielsen–Ninomiya theorem, which ensures that the sum of chiralities vanishes. Thus, the total dn/dt is zero due to cancellation of the contributions from Weyl points of positive and negative chirality. In fact, the chiral anomaly can be regarded as providing an alternative proof of the Nielson–Ninomiya theorem via a charge-conservation argument.

The chiral anomaly is often derived via a consideration of the properties of the Landau-level spectrum in the presence of a finite magnetic field; this argument is sketched, for example, in Section II.C.2 of the review by Armitage et al. (2018). As an alternative, we will sketch a semiclassical argument along the lines of one given by Son and Spivak (2013). We simply apply the semiclassical equations of motion given in Eq. (5.11), crucially including the anomalous velocity term in Eq. (5.11a), to the case of spatially uniform $\mathcal{E}$ and $\mathbf{B}$ fields. After several manipulations designed to move all terms involving $\dot{\mathbf{r}}$ and $\dot{\mathbf{k}}$ to the left, we obtain (see Ex. 5.19)

$$\left(1 + \frac{e}{\hbar c}\mathbf{B}\cdot\boldsymbol{\Omega}\right)\dot{\mathbf{r}} = \mathbf{v}_g + \frac{e}{\hbar}\mathcal{E}\times\boldsymbol{\Omega} + \frac{e}{\hbar c}\left(\mathbf{v}_g\cdot\boldsymbol{\Omega}\right)\mathbf{B}, \tag{5.39a}$$

$$\left(1 + \frac{e}{\hbar c}\mathbf{B}\cdot\boldsymbol{\Omega}\right)\dot{\mathbf{k}} = -\frac{e}{\hbar}\mathcal{E} - \frac{e}{\hbar c}\mathbf{v}_g\times\mathbf{B} - \frac{e^2}{\hbar^2 c}(\mathcal{E}\cdot\mathbf{B})\boldsymbol{\Omega}. \tag{5.39b}$$

The factor $(1 + e\mathbf{B}\cdot\boldsymbol{\Omega}/\hbar c)$ appearing on the left-hand sides is a density-of-states enhancement factor that arises when Berry curvature and magnetic fields are both present. As explained by Xiao et al. (2005), it can be derived based on a reconsideration of Liouville's theorem on the conservation of phase-space volume in light of Eq. (5.11). The physics is basically the same as that of the Středa formula discussed on p. 33. There we worked in 2D; in that context, we can combine Eq. (1.34) with Eq. (5.8) for a single band to get

$$\Delta n = \frac{1}{(2\pi)^2}\int\left(\frac{e}{\hbar c}B_\perp\Omega_{xy}\right)f(\mathbf{k})\,d^2k \tag{5.40}$$

where $f(\mathbf{k})$ is an occupation factor. The zero-field density n_0 is given by Eq. (5.40) but without the factor in parentheses, so $\Delta n/n_0 = eB_\perp\Omega_{xy}/\hbar c$, giving an enhancement ratio of $1 + e\mathbf{B}\cdot\boldsymbol{\Omega}/\hbar c$ when generalized to 3D.

To derive the chiral anomaly, we track the flow of occupied wave packets in $\mathbf{k}$-space as expressed by Eq. (5.39b). The $\mathbf{k}$-space flux density $\mathcal{F}$, defined as $d\mathbf{k}/dt$ times the density of states per "phase space" volume $d^3r\,d^3k$, is just $(2\pi)^{-3}$ times the left-hand side of Eq. (5.39b). If there is an electron pocket surrounding the Weyl point, then the density n of electrons associated with the pocket grows at a rate given by the total flux passing through the surrounding Fermi sheet[34] as expressed by

$$\frac{dn}{dt} = \int_{S_F} \mathcal{F} \cdot \hat{\mathbf{v}}_F \, d^2k,$$

(5.41)

where $\hat{\mathbf{v}}_F$ is again the unit vector normal to the Fermi surface. Only the last term on the right side of Eq. (5.39b) contributes to the net flux, giving

$$\frac{dn}{dt} = \frac{-1}{(2\pi)^3} \frac{e^2}{\hbar^2 c} (\mathcal{E} \cdot \mathbf{B}) \int_{S_F} \mathbf{\Omega} \cdot \hat{\mathbf{v}}_F \, d^2k.$$

(5.42)

But the integral on the right is just 2π times the Fermi-sheet Chern number, and comparing with Eq. (5.34), we obtain Eq. (5.38). The Nielsen–Ninomiya theorem guarantees that the sum of Chern numbers of all Fermi sheets must vanish, so that overall electron number is conserved. However, from a **k**-space viewpoint, electron density seems to be "teleported" from one electron or hole pocket to another via Weyl "portals" when $\mathcal{E} \cdot \mathbf{B}$ is nonzero.

A third phenomenon associated with the presence of Weyl points is the *chiral magnetic effect*, which can also be derived from Eq. (5.39). Let's take a look at the other terms appearing there. The first two terms on the right side of Eq. (5.39b) describe the acceleration of electrons resulting from electric and magnetic forces, while the first two on the right side of Eq. (5.39a) correspond to the group velocity and anomalous velocity, respectively. We now focus on the third term in Eq. (5.39a), which gives rise to a current

$$\mathbf{J} = \frac{-e^2 \mathbf{B}}{(2\pi)^3 \hbar^2 c} \sum_n \int_{BZ} f_{n\mathbf{k}} (\nabla_{\mathbf{k}} E_{n\mathbf{k}}) \cdot \mathbf{\Omega}_{n\mathbf{k}} \, d^3k$$

(5.43)

where $f_{n\mathbf{k}}$ is the occupation function for band n.[35] Applying an integration by parts, this becomes

$$\mathbf{J} = \frac{e^2 \mathbf{B}}{(2\pi)^3 \hbar^2 c} \sum_n \int_{BZ} E_{n\mathbf{k}} (f_{n\mathbf{k}} \nabla_{\mathbf{k}} \cdot \mathbf{\Omega}_{n\mathbf{k}} + \mathbf{\Omega}_{n\mathbf{k}} \nabla_{\mathbf{k}} \cdot f_{n\mathbf{k}}) \, d^3k.$$

(5.44)

(There are no boundary terms because the BZ is a closed manifold.) The first term can be dropped since $\nabla_{\mathbf{k}} \cdot \mathbf{\Omega}_{n\mathbf{k}} = 0$ except possibly for delta-function singularities at Weyl points, but when these do occur the two bands that meet at a Weyl point give equal and opposite contributions that cancel. As for the second term, at low

---

[34] After occupied electrons pass through and equilibrate, a new E_F will be established, but here we compute the rate at which states cross the fixed original E_F.

[35] Our semiclassical derivation here loosely follows Son and Yamamoto (2012), Son and Spivak (2013), and Zhong et al. (2016).

temperature $\nabla_{\mathbf{k}} \cdot f_{n\mathbf{k}}$ turns into a delta function on the Fermi surface, so that Eq. (5.44) becomes

$$\mathbf{J} = \frac{-e^2 \mathbf{B}}{2\pi \hbar^2 c} \sum_i \mu_i \int_{S_i} \mathbf{\Omega}_{n\mathbf{k}} \cdot \hat{\mathbf{v}}_F \, dS \qquad (5.45)$$

where $\hat{\mathbf{v}}_F$ is again a unit normal to the Fermi surface. Here we have assumed for simplicity that bands up to $n - 1$ are fully occupied, band n consists of several electron pockets surrounding Weyl points, and higher bands are empty. The sum is over pockets i with Fermi surfaces S_i, and we have replaced $E_{n\mathbf{k}}$ in each surface integral by a local Fermi level μ_i associated with that pocket. This is meaningful provided that scattering between pockets is slow compared to equilibration within a pocket. Finally, using Eq. (5.34) we obtain

$$\mathbf{J} = \left(\frac{e^2}{\hbar^2 c} \sum_i \mu_i \chi_i \right) \mathbf{B} . \qquad (5.46)$$

This strange-looking result expresses the chiral magnetic effect, also discussed in Section II.C.2 of Armitage et al. (2018). It suggests that an electron pocket surrounding a Weyl point contributes a steady ground-state current in the presence of a $\mathbf{B}$ field. Can this result in a net $\mathbf{J}$ parallel to $\mathbf{B}$? On the one hand, in global equilibrium the answer is clearly "no" since then all the μ_i are identical, and $\sum_i \chi_i$ vanishes by the Nielsen-Ninomiya theorem, Eq. (5.36). On the other hand, if scattering between pockets is slow enough that local Fermi levels μ_i can be established and driven out of equilibrium with each other, a net current can result. For example, Eq. (5.38) suggests that this could be accomplished by applying an $\mathcal{E}$-field pulse parallel to a preexisting static $\mathbf{B}$ field.

The presence of surface Fermi arcs, the chiral anomaly of Eq. (5.38), and the chiral magnetic effect of Eq. (5.46) all have fascinating potential consequences for the transport (especially magnetotransport) and optical properties of Weyl semimetals. These will not be discussed further here, but the interested reader is referred to the excellent reviews by Hasan et al. (2017), Yan and Felser (2017), and Armitage et al. (2018) mentioned earlier.

5.4.4 Experimental Realizations

Much theoretical and experimental effort has gone into the identification of Weyl semimetals in recent years. An early discussion by Murakami (2007) focused on an intermediate Weyl semimetal phase that could arise in the context of a transition between strong TI and normal phases in systems with TR but broken inversion. Interest switched to magnetic Weyl semimetals following the influential paper of

Wan et al. (2011) suggesting the possible existence of a Weyl semimetal phase and associated Fermi arcs in rare-earth pyrochlore iridates. Unfortunately, this proposal does not seem to have been borne out by subsequent work. Attention soon switched back to nonmagnetic semimetals with the discovery of the TaAs family of materials, also including TaP, NbAs, and NbP, in which ARPES measurements clearly revealed the presence of 24 Weyl points and associated surface Fermi arcs (Huang et al., 2015; Lv et al., 2015; Weng et al., 2015; Xu et al., 2015b). Searches for simpler nonmagnetic Weyl semimetals with fewer (ideally just four) Weyl points at or near E_F are ongoing. The distinct properties of Type-II Weyl points (see p. 259) were first pointed out by Soluyanov et al. (2015) and proposed to exist in WTe_2. Soon afterward, $MoTe_2$ was suggested as a platform where the Weyl points might be more accessible (Sun et al., 2015). Interest has also been returning to the search for magnetic Weyl semimetals, with the Heusler class of materials having been put forth as attractive candidates for hosting such states (Yan and Felser, 2017). The field is still developing rapidly at the time of this writing.

Exercises

5.17 Reread the discussion at the very end of Section 5.4.1, and write a PYTHTB program to compute the Chern number on a small sphere surrounding the conical intersection point at K' in (k_x, k_y, t_2) space. It is probably easiest to do this by solving on a 2D mesh of (θ, ϕ) values using the `wf_array` method, where $\theta \in [0, \pi]$ and $\phi \in [0, 2\pi]$ represent the polar and azimuthal angles in (k_x, k_y, t_2) as viewed from the conical intersection point. (Don't worry about the redundancy associated with extremal values $\theta = 0$ and π; it shouldn't matter.) Repeat for the conical intersection at K, and check that you get the opposite Chern number.

5.18 Show that the contribution of states near a 3D Weyl point at energy E_0 to the density of states $\rho(E)$ of Eq. (2.38) is quadratic in $|E - E_0|$.

5.19 Carry out the algebra to convert Eq. (5.11) into the form given in Eq. (5.39).

5.20 Consider a tetragonal crystal with a cell of dimensions $a \times a \times c$ that hosts an ideal TR-broken Weyl semimetal state with just two Weyl points on the k_z axis with chiralities ± 1 at $k_z = \pm k_0$. Starting from Eq. (5.15), noting the discussion about slice Chern numbers on p. 260, and assuming the Chern number vanishes on the slice at $k_z = \pi/c$, show that the AHC is $\sigma_{AHC} = e^2 k_0 / hc$, or just proportional to the separation between the Weyl points.

5.21 At the beginning of Section 5.4.4 we mentioned that a Weyl semimetal phase generically intervenes at a topological phase transition between a strong TI and a normal insulator. To explore this, begin by reading Section III of Murakami and Kuga (2008), referred to here as MK. Modify the Fu–Kane–Mele PᴇTHTB program accordingly, setting the parameters as in the caption of MK Fig. 5 so that your results are consistent with MK Fig. 6. To "see" the location of the Weyl points, it useful to plot the Berry curvature computed on a nearby plane, say at $k_x = 1.02$, over the region k_y, $k_z \in [-0.15, 0.15]$; the Berry flux emerging from the Weyl points should appear as "bright spots" whose sign reflects the chirality. Generate a series of such plots as the parameters are varied as in MK Fig. 6(b) and explain what you see.

5.5 Other Topological Systems

In the preceding sections of this chapter, we have considered TR-broken QAH (Chern) insulators in 2D and 3D, TR-invariant QSH insulators in 2D, TR-invariant weak and strong TIs in 3D, and Weyl semimetals. This covers some of the most-studied classes of topological systems, but does not begin to exhaust the growing zoo of topological states currently under investigation. While it is beyond the scope of this book to go into these in any detail, this section gives a very brief overview of some of these systems.

5.5.1 Topological Crystalline Insulators

The QAH state was defined in terms of a $\mathbb{Z}$ topological index without reference to any symmetry, but this is a somewhat unusual case. Another topological invariant that relies on nothing other than translational symmetry – so obvious that it hardly needs mentioning – is the number N_{occ} of occupied bands in an insulator. More generally, topological invariants are often defined in the context of some specified additional symmetries. For example, the entire discussion of the QSH and TI states of nonmagnetic insulators relied heavily on the presence of TR symmetry. In particular, our starting point was to classify a pair of TR-invariant insulators as belonging to the same class if and only if they can be connected along an adiabatic path without gap closure and while preserving the TR symmetry everywhere along the path.

This suggests that we can play the same game using other symmetries. For example, suppose we forget about TR symmetry and focus instead on inversion symmetry. Then we will consider two centrosymmetric insulators to belong to the same class if and only if they can be connected without gap closure along a path on which inversion symmetry is always present. The resulting topological classification of 3D insulators is described elegantly by Turner et al. (2012) based on the parity eigenvalues of the occupied states at the eight TRIM. For example, the number of occupied odd-parity states at the BZ center Γ is clearly a topological invariant.[36] The parities of the states at the other TRIM depend in part on the choice of origin (there are eight possible inversion centers to choose from) and of crystallographic axes, but some interesting global invariants emerge. For example, let N_{odd} be the total number of occupied odd-parity states at the eight TRIM. Turner et al. (2012) showed that if the system is insulating, has no nonzero Chern indices, and N_{odd} is twice an odd integer, then the system is $\mathbb{Z}_2$-odd in an inversion-symmetric sense that cannot be removed by a shift of origin or rotation of axes. Such an insulator is sometimes referred to as an *axion insulator* (see, for example, Wan et al., 2011) for reasons that will become clear in Chapter 6.[37]

But inversion is just one of many crystallographic symmetries. Consider, for example, a crystal of spinless electrons with M_z mirror symmetry, such that $k_z = 0$ and $k_z = \pi/c$ are special planes in the BZ on which Bloch states can be classified as mirror-even or mirror-odd. Assuming the crystal is not a 3D QAH insulator, the total Chern number on each of these planes must vanish, but the *mirror Chern number* defined as $(C_{odd} - C_{even})/2$, reflecting the difference of the Chern numbers of the mirror-odd and mirror-even manifolds, is clearly a topological invariant on each of the special k_z planes. Rocksalt-structure SnTe was proposed to be a mirror-enforced TI of this kind by Hsieh et al. (2012), with experimental confirmation following soon after (Tanaka et al., 2012). Since then, $Pb_{1-x}Sn_xTe$ alloys have been the focus of most investigation (Dziawa et al., 2012; Xu et al., 2012).

These systems have come to be known collectively as *topological crystalline insulators,* an admittedly imperfect name intended to convey that the topology is enforced by crystalline symmetries. Note that a bulk–boundary correspondence does not always exist in such cases. For example, if the bulk classification relies on an M_x mirror symmetry, then surfaces with unit normals lying in the y–z plane may respect the M_x symmetry; if they do, Dirac-cone surface states similar to those on the surfaces of TR-protected TIs can be expected. However, other surfaces cannot possibly respect this symmetry, and such surfaces will generically be gapped. An extreme case is that of axion insulators: In general, no metallic surfaces are required, since inversion symmetry is always broken at all surfaces.

Many other crystallographic symmetries can also be the basis for a topological crystalline insulator classification. In fact, the initial paper of Fu (2011) introduced the concept in terms of a model system with a four-fold rotational symmetry. Indeed, for any space group one can imagine setting up a topological classification using a definition of equivalence based on connection along an insulating path

---

[36] Here we assume that we are comparing only insulators with the same N_{occ}, so this could just as well have been expressed in terms of the number of even-parity states.

[37] As shown by Alexandradinata et al. (2014), a hybrid Wannier flow (or Wilson loop) analysis, similar to that employed in Section 5.3.1 for TR-invariant insulators, can also profitably be applied to the inversion-symmetric case.

that everywhere respects the specified space-group symmetry. There are 230 space groups in the crystallographic classification of nonmagnetic materials alone, so that this might seem like a herculean task, but surprisingly there has been significant recent progress in this direction (Bradlyn et al., 2017; Freed and Moore, 2013; Po et al., 2017). Magnetic insulators are less explored, although an important benchmark is the paper of Mong et al. (2010) on antiferromagnetic TIs.

5.5.2 Nodal Semimetals

The Weyl semimetals that we considered in Section 5.4 belong to a larger class of materials that are almost insulators in the sense that the highest valence band is gapped from the lowest conduction band everywhere in the BZ except at one or a few isolated points ("nodes") or lines ("nodal loops"). In the case of Weyl semimetals, these nodal points are touchings of just two bands; they carry a chirality as in Eq. (5.31) and can occur generically in the 3D BZ without any special symmetry (other than translational symmetry).

When other crystallographic symmetries are present, bands can be "glued together" into band groups, as we saw for Si in Section 3.6.1. Looking back at Fig. 3.16, we are reminded that bands 1–2 and 3–4 are degenerate at the X point and bands 2–4 meet at Γ; this was why we argued that the four-band group needed to be treated as a unit when constructing Wannier functions. Actually each band is Kramers-degenerate, so four bands meet at X and six at Γ.

A *Dirac semimetal* is a crystal with inversion and TR symmetry in which the doubly degenerate valence and conduction bands meet at an isolated point of fourfold degeneracy. Such a degeneracy point is known as a *Dirac point* in analogy with the massless Dirac equation, where a similar Dirac cone structure arises in the dispersion of four states that meet at the origin of momentum space. In 3D crystal band structures, Dirac points do not occur generically unless some additional symmetries (beyond TR and inversion) are present. The graphene band structure plotted in Fig. 2.8 shows a 2D Dirac point at K; here the degeneracy can be traced to the presence of a three-fold rotational symmetry, but this degeneracy is not robust against the introduction of spin-orbit effects. The inclusion of SOC opens a tiny gap (on the order of tens of microvolts), so strictly speaking this is not a "true Dirac point." For all practical purposes it is one, however, and graphene is often taken as exemplifying a system in which the Fermi level lies at (or, with inevitable doping, near) a pair of 2D Dirac points.

An interesting question is whether it is possible to find 3D materials in which the valence and conduction bands touch only at one or more true Dirac points. As reviewed by Armitage et al. (2018), there are three known scenarios by which this can happen. One is by external fine-tuning, as by pressure or composition, to the boundary between strong TI and normal behavior of a material like Bi_2Se_3. However, this does not count as generic behavior; for that one needs to turn to the second scenario, involving a band crossing along a symmetry axis in the BZ, or the third, which involves a degeneracy pinned to a high-symmetry point in the BZ. The second scenario is exemplified by materials in the Na_3Bi and Cd_3As_2 classes (Wang et al., 2012, 2013). Search principles for the third scenario have been identified by Young et al. (2012) and Steinberg et al. (2014). Here again, interesting transport properties have been predicted to occur for such materials.

In some cases it can also happen that the highest occupied and lowest unoccupied bands become degenerate along an entire line (i.e., loop) in the 3D BZ, giving rise to a *nodal-loop semimetal*. One scenario in which this can occur arises in the presence of an additional symmetry such as a mirror symmetry. In this case there are planes (e.g., at $k_z = 0$ for mirror symmetry M_z) on which all the Hamiltonian eigensolutions have definite mirror parity. The states of opposite parity can cross without mixing, and the locus of crossing point forms a loop of degeneracy. This may happen both for spinless and spinful electrons. As an example of the latter, a mirror-protected nodal loop occurs on the $k_z = 0$ plane of bcc Fe, although it is somewhat overshadowed by the presence of several other Fermi sheets (Gosálbez-Martínez et al., 2015).

The second scenario occurs in spinless systems when $\mathcal{I} * \mathcal{T}$ (inversion composed with TR) is a symmetry. In this case the Hamiltonian $H_\mathbf{k}$ is effectively real everywhere in the BZ, and the 2×2 effective Hamiltonian in the subspace of crossing bands takes a form like that in Eq. (5.29) but excluding the imaginary matrix σ_2, so that only the real matrices σ_1 and σ_3 appear on the right-hand side. In other words, it is possible to find a basis in which the Hamiltonian and all its eigenfunctions are real, and the codimension is now 2 instead of 3. In this case, nodal loops can generically form in 3D (and nodal points can appear in 2D).

Moreover, there is a nice connection with Berry phases in the presence of nodal loops. Since the eigenfunctions are effectively real, the Berry phase on any gapped closed path must be 0 or π. If a path like P_1 in Fig. 5.26 can be contracted to a point without encountering a gap closure, then clearly its Berry phase must have been zero all along, since it cannot change during the contraction. However, if a path like P_2 in the figure is linked with the nodal loop, then by a similar argument its Berry phase must be the same as that of a tiny circle around the nodal loop. But this is just π, since the effective Hamiltonian very close to the loop will always take a form that can be reduced (after some linear transformations) to

$$H_{2\times2}(\mathbf{q}) = f_0(\mathbf{q})\, I + q_1\sigma_1 + q_2\sigma_3 , \tag{5.47}$$

where q_1 and q_2 are independent coordinates in directions perpendicular to the loop. Then circling around the loop induces a 2π rotation of the pseudospin solution in

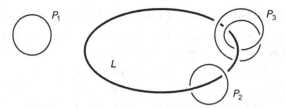

Figure 5.26 Sketch of a nodal loop L (heavy line) in the BZ of a 3D semimetal protected by $\mathcal{I} * \mathcal{T}$ symmetry. Gapped paths P_1, P_2, and P_3 (light lines) encircle the nodal loop 0, 1, and 2 times, and thus have Berry phases of 0, π, and 0, respectively.

the x–z plane, giving a Berry phase of π, as discussed at the end of Section 3.2.1. These considerations lead to the conclusion that the Berry phase of any gapped closed path has a $\mathbb{Z}_2$ index that is determined by the evenness or oddness of its winding number around the nodal loop, yielding Berry phases of 0, π, and 0 for P_1, P_2, and P_3 in Fig. 5.26, respectively.

Generically there is always some variation of the energy along any nodal loop. If E_F falls above the maximum energy or below the minimum energy along the loop, one finds a tubular Fermi electron or hole pocket surrounding the nodal loop. If E_F lies in the range of energies of the loop, then a series of cigar-like electron and hole pockets are strung together like a necklace around the loop.

These and other features of nodal-loop semimetals, as well as possible consequences for transport and other experiments, are covered in a recent review by Fang et al. (2016).

5.5.3 Broader Contexts

The topological states discussed thus far have all been ones that arise in the context of the electronic band theory of crystals described at the single-particle ("free-Fermion") level. In recent years, however, the investigation of topological states has broadened to cover other kinds of physical systems. These are mentioned briefly in this subsection.

Strongly Correlated Topological states. The topological states discussed previously, including the QAH and QSH states in 2D and the $\mathbb{Z}_2$ TIs in 3D, are robust against interactions in a perturbative sense, that is, as long as the interactions are sufficiently weak that the system remains adiabatically connected to a noninteracting one of the given topology. However, when interactions become strong, new kinds of topological phases become possible, and much effort has recently gone into categorizing these. Most closely related to the noninteracting phases are the "symmetry-protected topological phases." As reviewed by Senthil (2015) and Wen (2017), these fermionic and bosonic systems are characterized by a unique ground state that preserves the symmetry of the Hamiltonian and in which all excitations are gapped. A classic example is the Haldane spin chain in 1D (Haldane, 1983). A more exotic class of states that goes under the name of "topologically ordered phases" includes the fractional quantum Hall state, potential fractional QAH states, and other states with multiple degenerate ground states and excitations with fractional charges and/or statistics. These states also exhibit long-range topological entanglement, as discussed, for example, in Kitaev and Preskill (2006), Chapter 8 of Wen (2007), and Wen (2017).

Topological Superconductors. Superconductors are materials that are metals in their normal state, but that become gapped over all or almost all of the Fermi surface in the superconducting state. The Bardeen–Cooper–Schrieffer ("BCS") theory describes the gap opening as resulting from an electron pairing instability mediated by phonons, although other mechanisms (such as magnon-mediated interactions) are under investigation to explain more exotic (e.g., high-temperature) forms of superconductivity. A topological classification can also be applied to these superconducting states. The "s-wave" state described by the standard form of BCS theory plays the role of a "normal" or "trivial" superconductor, and other topological classes are defined if it is impossible to adiabatically transform a member of one class into a member of another. For example, "$p_x + ip_y$ superconductors" in 2D form a topologically distinct class, closely related to the "A_1 phase" of superfluid ^{3}He (Volovik, 2003). A superconducting state can be characterized by a "Bogoliubov–de Gennes" (BdG) equation that plays a role analogous to the Bloch Hamiltonian of an ordinary insulator, and the topological classification corresponds to the topological invariants governing the BdG equations and their solutions. For a recent review, see Sato and Ando (2017).

Photonic Crystals. Just as Bloch's theorem can be applied to find solutions of the Schrödinger equation in the presence of a 3D-periodic potential $V(\mathbf{r}) = V(\mathbf{r} + \mathbf{R})$, one can find Bloch-like solutions of Maxwell's equations for electromagnetic waves propagating in a dielectric medium whose index of refraction is a 3D-periodic function $n(\mathbf{r}) = n(\mathbf{r} + \mathbf{R})$ of space. Such *photonic crystals* may occasionally occur naturally, but most interest has focused on artificially engineered structures with periodicities spanning from microns to centimeters, many of which have important optoelectronic applications. The book by Joannopoulos et al. (2011) provides a good introduction. With the rapid growth of interest in topological phases, it is only natural that analogues of various topological states have been explored in photonic crystal systems. Since these typically respect TR symmetry, most of the focus has been on analogues of 2D QSH and 3D $\mathbb{Z}_2$ TI states and their topologically protected edge/surface modes, and on Weyl points and other nodal degeneracies between "photon bands." Despite the similarities with electron crystals, however, there are important differences. For one thing, photons are bosons, and while Berry phases, connections, and curvatures can be defined in the BZ in full analogy with electronic systems, concepts related to an occupied set of valence bands, as governed by the Pauli exclusion principle, have no analog in the photonic arena. A second important difference is that Maxwell's equations are linear, so interactions typically play no role. A short review is provided by Lu et al. (2014); the interested reader can find further references therein.

Cold-Atom Lattices.   In recent years it has become possible to create and confine ultracold dilute gases of neutral atoms, and to subject these gases to designed periodic potentials created by standing-wave laser fields. These *cold-atom lattice* systems offer an attractive platform for the artificial realization of crystalline phases of matter, using either bosonic or fermionic constituents. Because it is possible to tune not only the potential energy landscape, but also the form and strength of the interparticle interactions, these systems present a unique opportunity for realizing exotic, strongly correlated phases of matter. In particular, researchers are now creating and probing topological states in these fascinating synthetic crystalline systems. For a recent review, see Goldman et al. (2016).

Floquet States.   Here we consider a time-dependent but temporally periodic Hamiltonian $H(t + T) = H(t)$, and focus on the eigenstates of the unitary evolution operator U that propagates states forward in time by one period, from $t = 0$ to $t = T$, according to the time-dependent Schrödinger equation. The eigenvectors of U are quasi-stationary states $\psi_{n\omega}(t)$ known as *Floquet states*, and the eigenvalues $e^{-i\omega_n t}$ define "quasi-energies" $\epsilon_n = \hbar\omega_n$, providing a Bloch-like formalism in the time domain. If the system is also spatially periodic in N dimensions, the quasi-energies form bands $\epsilon_{n\mathbf{k}}$ in analogy to the energy bands $E_{n\mathbf{k}}$ of an ordinary static crystalline Hamiltonian. Berry connections and curvatures can be defined for the eigenfunctions associated with these bands, opening the door to a topological classification of Floquet crystals in much the same way as for ordinary crystals. Special interest has focused on narrow-gap crystalline insulators subjected to time-periodic laser or microwave fields (see, for example, Cayssol et al., 2013; Lindner et al., 2011; Roy and Harper, 2017). Floquet states are also being investigated in cold-atom lattices, as in the work of Jotzu et al. (2014). A new wrinkle was added by Wilczek (2012), who proposed that some interacting systems might spontaneously break continuous time-translational symmetry to form Floquet-like states, and the topological properties of these are also under active investigation. Bukov et al. (2015) and Roy and Harper (2017) provide recent perspectives on this field.

Clearly, there is much room for overlap among the various types of topological states, and new types of quantum systems that are amenable to topological classification are sure to emerge in the coming years.

Exercises

5.22 Look up the paper by Fu (2011) and program the tight-binding model of a topological crystalline insulator described there. Referring to the equations and figures therein, set up your model following Eq. (1) using the parameters following Eq. (2). Check that you can reproduce the bulk and surface band structures shown in Fig. 2, and then compute the hybrid Wannier flow along $\hat{z}$ in the same projected BZ as in Fig. 2(b). Comment on any similarity to the surface band structure.

5.23 The context of the following figure is the same as that of Fig. 5.26.

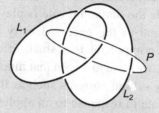

Here paths L_1 and L_2 denote nodal loops of touching of valence and conduction bands in a spinless model with inversion and TR symmetry. What is the Berry phase (traced over all valence-band states) on path P?

6

Orbital Magnetization and Axion Coupling

Recall that the orbital magnetization $\mathbf{M}_{orb}$ introduced in Section 1.2 describes the density of circulating orbital currents in a solid, as shown schematically in Fig. 1.9. In most magnetic materials the dominant role is played by the spin magnetization $\mathbf{M}_{spin}$, which is proportional to the excess density of spin-up versus spin-down electrons, and the orbital component is a small effect induced by the spin-orbit coupling (SOC). However, we have just seen that much of the physics of topological insulators is dominated by orbital currents such as those that flow in the quantum anomalous Hall (QAH) state in response to an applied electric field. It is therefore worth taking a closer look at the orbital magnetization, which, it turns out, can be expressed as a Brillouin zone (BZ) integral of a quantity that looks very much like a Berry curvature. Moreover, the orbital magnetoelectric coupling, which measures the change in $\mathbf{M}_{orb}$ induced by an electric field, will be of particular interest, especially a "Chern–Simons" contribution taking the form of a combination of Berry connections and curvatures. This Chern–Simons coupling has an important connection to the theory of topological insulators and contributes to the isotropic or "axion" magnetoelectric response, which in turn is closely related to the surface anomalous Hall conductivity. The purpose of this last chapter is to introduce these concepts and to trace some of the interesting connections among them.

6.1 Orbital Magnetization

For a finite system such as an atom or a molecule, the orbital magnetic dipole moment is given by

$$\mathbf{m} = \frac{1}{2c} \int \mathbf{r} \times \mathbf{j}(\mathbf{r}) \, d^3r \tag{6.1}$$

where $\mathbf{j}(\mathbf{r})$ is the local current density and the speed of light in the denominator comes with the Gaussian units used here. A naive first guess at a formula

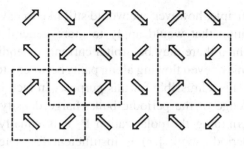

Figure 6.1 Model 2D crystal with broken time-reversal symmetry in which arrows indicate the local current flow pattern $\mathbf{j}(\mathbf{r})$. The three dashed squares indicate three different choices of unit cell boundaries that would give rise to very different definitions of orbital magnetization according to Eq. (6.2).

for the bulk orbital magnetization in a magnetic crystal could be, in analogy with Eq. (4.2),

$$\mathbf{M} = \frac{1}{2cV_{\text{cell}}} \int_{\text{cell}} \mathbf{r} \times \mathbf{j} \, d^3r \qquad (6.2)$$

where the integral is over the interior of the unit cell of volume V_{cell}. In general, however, the result will depend on the choice of unit cell, as illustrated for a model current flow in Fig. 6.1. In analogy with the discussion of Fig. 4.1, where we concluded that Eq. (4.2) is hopelessly flawed as a definition of electric polarization, we come to a similar conclusion here regarding Eq. (6.2) for the orbital magnetization.

In practice the difficulties are not nearly as severe as they were for the case of electric polarization. In almost all ferromagnetic materials,[1] the orbital currents are associated with partially filled d or f shells of magnetic atoms. Since these orbitals are fairly localized, it is a common practice to construct a set of nonoverlapping atomic spheres around the magnetic ions and approximate $\mathbf{M}$ as the sum of contributions from inside these spheres:

$$\mathbf{M} = \frac{1}{V_{\text{cell}}} \sum_s \mathbf{m}_s . \qquad (6.3)$$

Here s runs over magnetic ions located at sites $\boldsymbol{\tau}_s$, and $\mathbf{m}_s$ is the orbital magnetic moment obtained by integrating $(1/2c)(\mathbf{r} - \boldsymbol{\tau}_s) \times \mathbf{j}$ over the interior of the atomic sphere centered on $\boldsymbol{\tau}_s$. This approach has been widely used in the literature and is often quite accurate (see, for example, Ceresoli et al., 2010).

---

[1] The present discussion applies more generally to any material with a net magnetization, including ferrimagnets and canted antiferromagnets.

As a matter of principle, however, we would still like to have a proper definition of orbital magnetization that would apply to more general time-reversal (TR)–broken systems in which there may be orbital currents extending into the bonding regions between atoms, or even flowing along paths between atoms. Clearly, neither Eq. (6.2) nor Eq. (6.3) is admissible for this purpose. In fact, just as we concluded that a perfect knowledge of the periodic bulk charge density $\rho(\mathbf{r})$ is insufficient in principle for determining the polarization $\mathbf{P}$, it is equally true that a perfect knowledge of the periodic bulk $\mathbf{j}(\mathbf{r})$ is insufficient for determining the orbital magnetization $\mathbf{M}$ (Hirst, 1997).

Instead, we need a quantum-mechanical expression that accesses some information about the Bloch eigenstates. For a finite system like an atom or molecule described at the single-particle level, the quantum expression corresponding to Eq. (6.1) is

$$\mathbf{m} = \frac{-e}{2c} \sum_n^{\text{occ}} \langle \psi_n | \mathbf{r} \times \mathbf{v} | \psi_n \rangle \tag{6.4}$$

where $\mathbf{v}$ is the velocity operator and the ψ_n are the occupied eigenstates. For a crystal, however, we have to be careful how to proceed. Just as the expectation value $\langle \psi_{n\mathbf{k}} | \mathbf{r} | \psi_{n\mathbf{k}} \rangle$ is ill defined in the Bloch representation, so too is $\langle \psi_{n\mathbf{k}} | \mathbf{r} \times \mathbf{v} | \psi_{n\mathbf{k}} \rangle$. In the case of the electric polarization, we found a way of expressing the polarization as a BZ integral of the Berry connection; we can now hope to do something similar for the orbital magnetization.

Several developments led to a more or less simultaneous solution of this problem coming from different points of view. An approach based on the Wannier representation of an insulator was put forth by Thonhauser et al. (2005) and Ceresoli et al. (2006), while a formulation based on the semiclassical theory of Eqs. (5.11) and (5.39) was proposed by Xiao et al. (2005). An additional derivation based on a long-wave perturbation analysis was later provided by Shi et al. (2007). Happily, all these derivations arrived at the same formula for the ground-state orbital magnetization, namely (in 2D)

$$M = \frac{e}{\hbar c} \frac{1}{(2\pi)^2} \sum_n \int_{\text{BZ}} \text{Im} \, \langle \partial_x u_{n\mathbf{k}} | H_{\mathbf{k}} + E_{n\mathbf{k}} - 2E_{\text{F}} | \partial_y u_{n\mathbf{k}} \rangle f_{n\mathbf{k}} \, d^2k \tag{6.5}$$

where $\partial_\mu = \partial/\partial k_\mu$, $E_{n\mathbf{k}}$ is the band energy, and the occupation factor $f_{n\mathbf{k}}$ limits the integral to occupied states up to the Fermi energy E_{F}.[2] Note that the term involving E_{F} is proportional to the total Chern number for an insulator, as described in Eqs. (3.67) and (3.68), and can be dropped entirely for a trivial insulator.

---

[2] We work here at zero temperature so that $f_{n\mathbf{k}} \in \{0, 1\}$.

Equation (6.5) expresses what is sometimes referred to as the "modern theory of orbital magnetization." We shall sketch two derivations of this result in the following discussion. A good review of the theory has been provided by Thonhauser (2011) and an update on recent applications appears in Hanke et al. (2016).

6.1.1 Dimensionality

There are some interesting parallels between the electric polarization **P** and orbital magnetization **M**, but also some important differences. The concept of polarization arises first in 1D, where we have seen that it constrains the surface charge Q_{surf} modulo the charge quantum e. If we want to know the P_x component in a 2D or 3D system, connected to edge or surface charges, then we use the 1D Berry-phase theory in the x direction and carry out a simple average over wavevectors in the other "spectator" directions.

In a similar way, the orbital magnetization arises first in 2D, since there is no such thing as a circulating current in 1D. In 2D, M_{orb} is a pseudoscalar quantity (invariant under inversion but odd under TR) that determines the edge current I as in Fig. 1.9. This time there is no question of any ambiguity modulo a quantum; the edge current is completely determined to be $I = M_{orb}$ independent of edge preparation, as argued on physical grounds in Section 1.2 based on charge conservation. In 3D, M_{orb} is a pseudovector, giving the 2D surface sheet current to be $\mathbf{K}_{surf} = \mathbf{M}_{orb} \times \hat{\mathbf{n}}$ for surface normal $\hat{\mathbf{n}}$. Generalizing what we said a moment ago, once we have the theory in hand in 1D (for polarization) or 2D (for magnetization), we can carry out a simple average over wavevectors in the spectator direction(s). For the magnetization, this leads to the 3D formula

$$M_\alpha = \frac{e}{2\hbar c} \frac{1}{(2\pi)^3} \sum_n \int_{BZ} \mathrm{Im}\, \varepsilon_{\alpha\mu\nu} \langle \partial_\mu u_{n\mathbf{k}} | H_\mathbf{k} + E_{n\mathbf{k}} - 2E_F | \partial_\nu u_{n\mathbf{k}} \rangle\, d^3k \qquad (6.6)$$

where $\varepsilon_{\alpha\mu\nu}$ is the fully antisymmetric tensor. This is sometimes written as

$$\mathbf{M} = \frac{e}{2\hbar c} \frac{1}{(2\pi)^3} \sum_n \int_{BZ} \mathrm{Im}\, \langle \nabla_\mathbf{k} u_{n\mathbf{k}} | \times (H_\mathbf{k} + E_{n\mathbf{k}} - 2E_F) | \nabla_\mathbf{k} u_{n\mathbf{k}} \rangle\, d^3k. \qquad (6.7)$$

6.1.2 Derivation in the Wannier Representation

We begin by briefly sketching the derivation given by Thonhauser et al. (2005), based on a Wannier-function picture, of a formula for the contribution to M_{orb} from a single occupied band in a 2D insulator. We assume that TR is broken in such a way that $M_{orb} \neq 0$, but that the band is topologically trivial ($C = 0$). We imagine

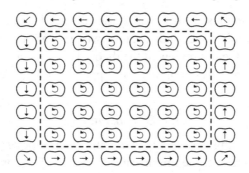

Figure 6.2 Sketch of current flow (arrows) associated with the Wannier functions in a crystallite cut from a magnetic 2D insulator. The dashed rectangle indicates the dividing line between the interior "bulk" region and the exterior "skin" region (see text).

cutting a rectangular crystallite from the bulk as shown in Fig. 6.2. Then the orbital magnetic moment of the crystallite is

$$m = \frac{-e}{2c} \sum_i \langle \psi_i | \mathbf{r} \times \mathbf{v} | \psi_i \rangle \tag{6.8}$$

where $|\psi_i\rangle$ are the occupied eigenstates of the sample. Since we are in 2D, magnetic moments m and magnetizations M are written as pseudoscalars (understood as along $\hat{\mathbf{z}}$), and $\mathbf{r} \times \mathbf{v}$ is understood to mean $xv_y - yv_x$. Now imagine that we have carried out a unitary rotation among the occupied states to arrive at a set of localized Wannier-like functions $|w_s\rangle$ spanning the same occupied subspace, as indicated by the "blobs" in Fig. 6.2. We can just as well express m as a trace over these orbitals instead, obtaining

$$m = \frac{-e}{2c} \sum_s \langle w_s | \mathbf{r} \times \mathbf{v} | w_s \rangle . \tag{6.9}$$

Letting $\mathbf{r} = (\mathbf{r} - \bar{\mathbf{r}}_s) + \bar{\mathbf{r}}_s$ in each matrix element, where $\bar{\mathbf{r}}_s = \langle w_s | \mathbf{r} | w_s \rangle$ is the Wannier center, we can write m as the sum $m_{\mathrm{LC}} + m_{\mathrm{IC}}$ of two terms: a "local circulation" (LC) piece

$$m_{\mathrm{LC}} = \frac{-e}{2c} \sum_s \langle w_s | (\mathbf{r} - \bar{\mathbf{r}}_s) \times \mathbf{v} | w_s \rangle \tag{6.10}$$

representing the sum of the circulations of each Wannier function about its own center, and an "itinerant circulation" (IC) contribution

$$m_{\mathrm{IC}} = \frac{-e}{2c} \sum_s \bar{\mathbf{r}}_s \times \bar{\mathbf{v}}_s \tag{6.11}$$

where $\bar{\mathbf{v}}_s = \langle w_s | \mathbf{v} | w_s \rangle$ is the velocity expectation in Wannier function s.

Now consider the limit that the size of the crystallite grows toward infinity, and let

$$M = M_{LC} + M_{IC} \tag{6.12}$$

where the two terms correspond to the magnetic moment per unit area associated with Eqs. (6.10) and (6.11), respectively. The dashed line in Fig. 6.2 indicates a division into a "bulk" interior region and a "skin" exterior region; in the limit of increasing sample size, we scale the thickness of the skin region such that it is a vanishing fraction of the sample size but much greater than a lattice constant.

We focus first on M_{LC}. Here our limiting procedure ensures that the skin region makes a negligible contribution, since it has a negligible area. In the bulk region, by contrast, each Wannier function tends toward a corresponding bulk Wannier function $|w_{\mathbf{R}}\rangle$ (where $\mathbf{R}$ is the cell label). Since the expectation value of the velocity $\langle w_{\mathbf{R}}|\mathbf{v}|w_{\mathbf{R}}\rangle$ vanishes for a bulk Wannier function, Eq. (6.10) becomes

$$M_{LC} = \frac{-e}{2cA_{cell}} \langle w_0|\mathbf{r} \times \mathbf{v}|w_0\rangle \tag{6.13}$$

where $|w_0\rangle$ is the bulk Wannier function in the home unit cell and A_{cell} is the unit cell area. Using Eq. (2.47) and carrying out a few lines of algebra, this can also be written as

$$M_{LC} = \frac{e}{\hbar cA_{cell}} \, \mathrm{Im} \, \langle w_0|xHy|w_0\rangle \,. \tag{6.14}$$

The IC term is a little trickier. Only the skin region contributes to this term, since $\bar{\mathbf{v}}_s$ vanishes for bulk-like Wannier functions. However, this time the skin region gives a finite contribution to $\mathbf{M}$ in the limit of large size, in spite of its negligible area. We understand this by looking at Fig. 6.2, where the IC term corresponds to the circulation of a current around the periphery of the the sample as shown by the straight arrows. We know that a circulating current corresponds to a bulk orbital magnetization, so the M_{IC} term is just given by this net edge current. Remarkably, Thonhauser et al. (2005) were able to show that this current can also be written in terms of bulk Wannier functions as

$$M_{IC} = \frac{e}{2\hbar cA_{cell}} \, \mathrm{Im} \, \sum_{\mathbf{R}} \langle w_0|H|w_{\mathbf{R}}\rangle \left(R_x \langle w_{\mathbf{R}}|y|w_0\rangle - R_y \langle w_{\mathbf{R}}|x|w_0\rangle \right). \tag{6.15}$$

Equation (6.13) for M_{LC} is quite intuitive. Just as the contribution of this band to the polarization P is given by the electric dipole expectation $-e\langle w_0|\mathbf{r}|w_0\rangle$, so is its contribution to M_{LC} given by its magnetic dipole expectation $(-e/2c)\langle w_0|\mathbf{r}\times\mathbf{v}|w_0\rangle$. Equation (6.15) for M_{IC} is much less intuitive; indeed, its presence is somewhat surprising.

Equations (6.14) and (6.15) provide the final expressions for the two contributions to the orbital magnetization in the Wannier representation. These can then be converted back to the Bloch representation using expressions such as Eqs. (3.99) and (3.100) in Section 3.5.2 to obtain

$$M_{LC} = \frac{e}{(2\pi)^2 \hbar c} \, \text{Im} \int_{BZ} \langle \partial_x u_\mathbf{k} | H_\mathbf{k} | \partial_y u_\mathbf{k} \rangle \, d^2 k \tag{6.16}$$

and

$$M_{IC} = \frac{e}{(2\pi)^2 \hbar c} \, \text{Im} \int_{BZ} \langle \partial_x u_\mathbf{k} | E_\mathbf{k} | \partial_y u_\mathbf{k} \rangle \, d^2 k . \tag{6.17}$$

The itinerant term can also be written more simply as

$$M_{IC} = \frac{-e}{(2\pi)^2 2\hbar c} \int_{BZ} E_\mathbf{k} \Omega_\mathbf{k} , \tag{6.18}$$

which is just the BZ integral of the band energy times the Berry curvature. Adding Eqs. (6.16) and (6.17) and restoring the sum over occupied bands, we recover the previously claimed expression, Eq. (6.5), for the orbital magnetization M of a topologically trivial insulator.

6.1.3 Semiclassical Theory

A very different derivation of Eq. (6.5) was given by Xiao et al. (2005) based on the semiclassical theory introduced in Eq. (5.11) and discussed in Sections 5.1.7 and 5.4.3. First, citing earlier work of Sundaram and Niu (1999), they observe that the orbital magnetic moment of a wave packet centered at $\mathbf{k}$ in band n of a 3D crystal (insulator or metal) can be written as

$$\mathbf{m}_{n\mathbf{k}} = \frac{-ie}{2\hbar c} \, \langle \nabla_\mathbf{k} u_{n\mathbf{k}} | \times (H_\mathbf{k} - E_{n\mathbf{k}}) | \nabla_\mathbf{k} u_{n\mathbf{k}} \rangle . \tag{6.19}$$

As a result, they argued, the total band energy takes the form

$$E_{tot} = \frac{1}{(2\pi)^3} \sum_n \int \left(1 + \frac{e}{\hbar c} \mathbf{B} \cdot \mathbf{\Omega}_{n\mathbf{k}} \right) (E_{n\mathbf{k}} - \mathbf{m}_{n\mathbf{k}} \cdot \mathbf{B}) \, f_{n\mathbf{k}} \, d^3 k \tag{6.20}$$

in the presence of a magnetic field $\mathbf{B}$, where $f_{n\mathbf{k}} = 1$ for states with $E_{n\mathbf{k}} < E_F$ and $= 0$ otherwise, and $E_{n\mathbf{k}}$ is the band energy. The first factor in parentheses on the right-hand side is the density-of-states factor discussed in Eq. (5.39), while the second is the renormalized band energy after taking into account the coupling with the external $\mathbf{B}$ field. The orbital magnetization can be defined as the derivative

$M_\alpha = -\partial E_{\text{tot}}/\partial B_\alpha$ taken at fixed particle density and evaluated at $\mathbf{B} = 0$, which gives two contributions. The one involving the magnetic moment $\mathbf{m}_{nk}$ is

$$\mathbf{M}_{\text{mom}} = \frac{e}{2\hbar c} \frac{1}{(2\pi)^3} \sum_n \int \text{Im} \langle \nabla_{\mathbf{k}} u_{n\mathbf{k}} | \times (H_{\mathbf{k}} - E_{n\mathbf{k}}) | \nabla_{\mathbf{k}} u_{n\mathbf{k}} \rangle f_{n\mathbf{k}} \, d^3 k, \quad (6.21)$$

while the one coming from $\mathbf{B}$ in the density-of-states factor is

$$M_{\text{DOS}} = \frac{e}{\hbar c} \frac{1}{(2\pi)^3} \sum_n \int \text{Im} \langle \nabla_{\mathbf{k}} u_{n\mathbf{k}} | \times (E_{n\mathbf{k}} - E_{\text{F}}) | \nabla_{\mathbf{k}} u_{n\mathbf{k}} \rangle f_{n\mathbf{k}} \, d^3 k. \quad (6.22)$$

The appearance of E_{F} in Eq. (6.22) arises because of a subtlety. When $\mathbf{B}$ is turned on while working at fixed particle density, the states added or subtracted at $E_{n\mathbf{k}} < E_{\text{F}}$ inside the Fermi sea as a result of the density-of-states factor $(1 + e\mathbf{B} \cdot \mathbf{\Omega}_{n\mathbf{k}}/\hbar c)$ must be compensated by subtracting or adding occupation of states at the Fermi energy. In other words, new states are not really just created at $E_{n\mathbf{k}}$; they are moved there from E_{F}. This accounts for the fact that the energies appear in the combination $E_{n\mathbf{k}} - E_{\text{F}}$.

Once again, Eqs. (6.21) and (6.22) sum up to give the previously claimed result in Eq. (6.7). This derivation applies equally to metals and insulators, even including topological insulators.

6.1.4 Discussion

The expressions derived in the preceding subsections clearly show that the orbital magnetization depends on Berry-derived quantities. In fact, the $E_{n\mathbf{k}}$ and E_{F} terms in, for example, Eq. (6.5) just multiply the Berry curvature $\Omega = -2\text{Im} \langle \partial_x u_{\mathbf{k}} | \partial_y u_{\mathbf{k}} \rangle$, while the remaining term inserts a factor of the Hamiltonian inside the matrix element between wave function derivatives.

Unlike the expressions for the electric polarization and the anomalous Hall conductivity (AHC), which involved only the Bloch functions $|u_{n\mathbf{k}}\rangle$, here we also need the Hamiltonian $H_{n\mathbf{k}}$ (and corresponding eigenenergy $E_{n\mathbf{k}}$). Thus, unlike $\mathbf{P}$ and σ_{AHC}, $\mathbf{M}_{\text{orb}}$ is not a function of the occupied-state manifold alone, but also depends on the Hamiltonian. For example, if H is scaled by a multiplicative factor, $\mathbf{M}_{\text{orb}}$ gets scaled by the same factor, while $\mathbf{P}$ and σ_{AHC} remain invariant.

Note that while both the Wannier-based derivation and the semiclassical derivation express the final result for $\mathbf{M}_{\text{orb}}$ as the sum of two terms, the decompositions are different. That is, neither the "local circulation" nor the "itinerant circulation" term in the former derivation corresponds directly with the "orbital moment" or "density-of-states" term in the latter one.

In Section 1.3 we argued on physical grounds that M_{orb} would be expected to vary linearly with electrostatic potential at fixed E_{F} according to Eq. (1.28) or (1.32) in a 2D Chern insulator. Recall that while it may seem strange that a "bulk quantity" such as M_{orb} should depend on the Fermi energy position in the gap of an insulator, the resolution comes by considering the change in filling of the edge channel as the position of E_F in the gap is modified. We can derive Eq. (1.32) in the present context by taking the derivative with respect to E_F in Eq. (6.5), yielding

$$\frac{\partial M_{orb}}{\partial E_F} = \frac{1}{ec}\sigma_{AHC}, \tag{6.23}$$

which is consistent with Eq. (1.32) in view of Eq. (1.24).

Since symmetry-induced degeneracies are often found to connect the occupied bands into band groups, it is useful to derive a multiband version of the theory that is gauge-invariant in the multiband sense (i.e., invariant to any mixing among the occupied bands). This has been done in Ceresoli et al. (2006), where Eqs. (44–45) provide expressions for two separate gauge-invariant contributions connected, respectively, with the "local" and "itinerant" terms of the single-band formalism. The gauge-invariance properties are discussed further in Souza and Vanderbilt (2008), where it is shown that the difference of these two terms, which corresponds with the magnetic moment term in the semiclassical derivation, is related to a quantity that is measurable in principle and takes the form of an f-sum rule in the magneto-optical spectrum (see also Yao et al., 2008).

These "modern theory" formulas for $\mathbf{M}_{orb}$ have been implemented in several electronic-structure codes, and the applications to date are nicely summarized in the reviews by Thonhauser (2011) and Hanke et al. (2016). Applications to simple ferromagnetic metals and more complex magnetic materials show differences, and often improvements, relative to conventional calculations based on integrating currents inside atomic spheres (Ceresoli et al., 2010; Hanke et al., 2016; Lopez et al., 2012; Nikolaev and Solovyev, 2014). The work of Hanke et al. (2016), in particular, provides evidence that the modern theory is often crucial for a proper understanding of orbital magnetization in noncollinear, topological, and heterogeneous systems.

In principle, once an expression for $\mathbf{M}$ is in hand, we can also gain access to derivatives of $\mathbf{M}$, such as the magnetic susceptibility tensor $\chi_{\mu\nu} = \partial M_\mu / \partial B_\nu$. The needed derivations have been carried out by Gonze and Zwanziger (2011), but are quite tricky because the application of an orbital magnetic field breaks ordinary translational symmetry. Their derivation is based on a perturbation theory of the single-particle density matrix that borrows some ideas from Essin et al. (2010). We shall not pursue this topic further here.

Instead, the cross-term response of the orbital magnetization to an applied *electric* field is something that we can hope to derive more easily, since the electric-field perturbation has already been discussed in Section 4.6. We turn our attention to this problem next.

Exercises

6.1 Carry out the algebra leading to Eq. (6.14).

6.2 Show that Eqs. (6.16) and (6.17) are gauge-invariant (in fact, the integrands are gauge-invariant), and also invariant against a shift of the zero of energy (recall that these formulas assume a trivial 2D insulator).

6.3 After summing over occupied bands n, the expressions in Eqs. (6.16) and (6.17) are applicable to multiband insulators consisting of isolated occupied bands, but the presence of degeneracies at high-symmetry points in the BZ often forces us to treat the occupied band manifold as a group. For this purpose, corresponding multiband gauge-invariant expressions were derived by Ceresoli et al. (2006), where they appear as Eqs. (44–45). Show that these equations reduce to Eqs. (6.16) and (6.17) in the case of a Hamiltonian gauge (i.e., one for which the $|u_k\rangle$ are eigenfunctions of H_k) for a system with isolated bands.

6.4 Verify that $dM_{orb}/dE_F = eC/hc$ for a 2D QAH insulator as claimed on p. 284.

6.2 Magnetoelectric Coupling and Surface Anomalous Hall Conductivity

An insulating material is said to show *magnetoelectric* (ME) behavior if an applied electric field induces changes in the magnetism or, conversely, if an applied magnetic field induces changes in the electric polarization. In some materials with low enough symmetry, it may happen that this cross-coupling occurs at linear order. In other cases, one may have to go to a higher order (e.g., magnetization changes at second order in electric field). Fiebig (2005) and Rivera (2009) provide useful reviews of ME effects. It turns out that one piece of the story of the linear ME coupling relates closely to Berry phases and topological insulators, as will be explained in the remainder of this chapter.

6.2.1 Linear Magnetoelectric Coupling

We confine our attention to the linear ME effect, which is described by a 3×3 matrix defined by

$$\alpha_{\mu\nu} = \left(\frac{\partial P_\mu}{\partial B_\nu}\right)_{\mathcal{E}} = \left(\frac{\partial M_\nu}{\partial \mathcal{E}_\mu}\right)_{\mathbf{B}}. \tag{6.24}$$

The subscript $\mathcal{E}$ on the $\partial P/\partial B$ term indicates that the derivative is taken at fixed electric field, and similarly for the subscript $\mathbf{B}$ on the last expression. These are necessarily equal by the Maxwell relation

$$\frac{\partial}{\partial B_\nu}\left(\frac{\partial E}{\partial \mathcal{E}_\mu}\right) = \frac{\partial}{\partial \mathcal{E}_\mu}\left(\frac{\partial E}{\partial B_\nu}\right), \tag{6.25}$$

where $P_\mu = -\partial E/\partial \mathcal{E}_\mu$ is the polarization, $M_\nu = -\partial E/\partial B_\nu$ is the magnetization, and $E(\mathcal{E}, \mathbf{B})$ is the electromagnetic enthalpy density expressed in terms of field variables $\mathcal{E}$ and $\mathbf{B}$, consistent with $E = E_0 - \mathcal{E} \cdot \mathbf{P} - \mathbf{B} \cdot \mathbf{M}$ at linear order.

Since $\mathbf{P}$ and $\mathcal{E}$ are inversion-odd but TR-even, while $\mathbf{M}$ and $\mathbf{B}$ are inversion-even but TR-odd, the ME coupling is odd under either $\mathcal{I}$ or $\mathcal{T}$. For ordinary purposes, then, we would conclude that α has to vanish for any insulating crystal having either inversion or TR symmetry. Since there are only a few noncentrosymmetric insulating ferromagnets, this would seem to be a fairly severe symmetry restriction. However, there are some antiferromagnets, such as Cr_2O_3, for which neither $\mathcal{I}$ nor $\mathcal{T}$ is a symmetry by itself, although $\mathcal{I} * \mathcal{T}$ *is* a symmetry. There are also some noncollinear magnets and canted antiferromagnets in which the spin pattern breaks both inversion and TR symmetries. In such cases the ME tensor α need not vanish, although its entries are often strongly constrained by other (e.g., rotational or mirror) symmetries.

In systems where it is allowed by symmetry, the ME response can have several different physical origins. On the one hand:

- There is a *purely electronic* or *frozen-lattice* response, defined as the $\mathbf{M}$ that arises in response to $\mathcal{E}$ (or the $\mathbf{P}$ that arises in response to $\mathbf{B}$) if the atomic coordinates and lattice vectors are clamped.
- The remainder constitutes the *lattice-mediated* response. From the $\partial M/\partial \mathcal{E}$ viewpoint, a first-order $\mathcal{E}$ induces first-order displacements of atomic coordinates, which in turn induce a first-order change in magnetization. A similar argument applies to the $\partial P/\partial B$ viewpoint. In crystals that are both piezoelectric and piezomagnetic, the coupling can be mediated in a similar way but with induced lattice strain replacing internal atomic displacements as the intermediate variable. The computational treatment of these responses is discussed by Íñiguez (2008) and Wojdeł and Íñiguez (2009).

On the other hand:

- There is a *spin response*, corresponding to the change of spin magnetization $\partial M_{\mathrm{spin}}/\partial \mathcal{E}$. This is equivalent to $\partial P/\partial B$ when taking into account only the Zeeman term in the Hamiltonian, which comprises the $\mathbf{B} \cdot \sigma$ term in Eq. (2.23).
- There is an *orbital response*, corresponding to the change of orbital magnetization $\partial M_{\mathrm{orb}}/\partial \mathcal{E}$. This is equivalent to $\partial P/\partial B$ considering only the orbital-current term in the Hamiltonian, which consists of the $(\mathbf{p} + e\mathbf{A}/c)^2$ term in Eq. (2.23).

In general, the spin-electronic, spin-lattice, orbital-electronic, and orbital-lattice pieces all contribute to the ME coupling. Since the spin magnetization usually dominates over the orbital part in most magnetic materials, especially systems whose magnetism arises on $3d$ transition-metal ions for which SOC is weak, the spin contribution is usually assumed to dominate. Experimentally it is not easy to distinguish between the frozen-lattice and lattice-mediated responses, and phenomenological models typically include both implicitly. Recent developments in first-principles methodologies now allow for a systematic calculation of all contributions, as has been done for Cr_2O_3 by Malashevich et al. (2012), although such calculations are not yet commonplace.

Before proceeding, a few words about conventions and units are in order. It is common in the ME literature to work in terms of the $(\mathcal{E}, \mathbf{H})$ frame, as opposed to the $(\mathcal{E}, \mathbf{B})$ frame of Eq. (6.24), since it is easier to control $\mathbf{H} = \mathbf{B} - 4\pi\mathbf{M}$ than $\mathbf{B}$ in the laboratory setting.[3] There is also a choice of SI or Gaussian units; we have chosen to use Gaussian units in our discussion. Even then, however, there is an additional choice of where to place a factor of 4π. Some authors prefer $\alpha_{\mu\nu} = 4\pi \, \partial M_\nu/\partial\mathcal{E}_\mu = 4\pi \, \partial P_\mu/\partial H_\nu$, while others prefer $\alpha_{\mu\nu} = \partial M_\nu/\partial\mathcal{E}_\mu = \partial P_\mu/\partial H_\nu$; we denote these as Conventions I and II, respectively. Note that α is dimensionless in either case, but a commonly used practice is to quote ME couplings in units of g.u. ("Gaussian units" using Convention I) following Rivera (1994). Here we use Convention II instead, for which we introduce the notation g.u.$'$ as the corresponding unit, with the conversion $1\,\text{g.u.}' = 4\pi$ g.u. In the common case that the magnetic susceptibility is negligible, we can set $\mathbf{H} \simeq \mathbf{B}$, so that $\alpha_{\mu\nu} = 4\pi \, \partial P_\mu/\partial B_\nu$ and $\partial P_\mu/\partial B_\nu$ in the two conventions, respectively. The same two conventions are available when using SI units, $\alpha_{\mu\nu} = \mu_0 \, \partial M_\nu/\partial\mathcal{E}_\mu = \partial P_\mu/\partial H_\nu \simeq \mu_0 \, \partial P_\mu/\partial B_\nu$ in Convention I and $(\mu_0/4\pi) \, \partial M_\nu/\partial\mathcal{E}_\mu = (1/4\pi)\partial P_\mu/\partial H_\nu \simeq (\mu_0/4\pi) \, \partial P_\mu/\partial B_\nu$ in Convention II, with α having units of inverse velocity. Another common practice is to quote α in units of ps/m using Convention I, with the conversion being $1\,\text{g.u.} = 1/c = 3.34 \times 10^3$ ps/m. In view of all this, the reader is encouraged to check conventions carefully when reading the literature on ME effects.

From here on we shall be interested only in the frozen-lattice orbital ME response, since this is the only piece that has a contribution of topological character. This contribution is sometimes called the "orbital magnetoelectric polarizability" in the literature. We shall see that this piece is, in principle, only well defined modulo a quantum in a way that parallels the theory of electric polarization, while the lattice-mediated and spin contributions are all uniquely defined in the usual sense. Henceforth we drop the 'orb' subscript from $\mathbf{M}_{\text{orb}}$, and it is understood that we are always talking about frozen-ion orbital ME responses unless otherwise indicated.

---

[3] In this respect, there is an asymmetry between magnetostatics and electrostatics. For the former, the paradigmatic case is the control of the free current passing through a solenoidal winding, which determines $\mathbf{H}$, not $\mathbf{B}$; in the latter, one controls the potential difference across a capacitor, which determines $\mathcal{E}$, not $\mathbf{D}$.

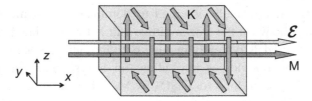

Figure 6.3 Block of material exhibiting an isotropic magnetoelectric coupling. The electric field (top long arrow) induces an orbital magnetization (bottom long arrow), which in turn generates circulating currents on the surfaces (shorter arrows). The surface currents can equally well be ascribed to a surface AHC.

6.2.2 Relation of Magnetoelectric Coupling to the Surface Anomalous Hall Conductivity

A link to topological properties can be anticipated by thinking about the connection between the (orbital) ME coupling and the *surface anomalous Hall conductivity* (surface AHC). For simplicity, consider an insulating block of material with insulating surfaces and with an isotropic linear ME coupling $\alpha_{\mu\nu} = \alpha_{\mathrm{iso}}\delta_{\mu\nu}$. As shown in Fig. 6.3, an electric field applied along the $\hat{\mathbf{x}}$ direction then induces a magnetization along the same direction, which corresponds to a surface current $\mathbf{K} = c\mathbf{M} \times \hat{\mathbf{n}}$ (Gaussian units) on a surface with unit normal $\hat{\mathbf{n}}$. On the top surface, for example, we get a current in the $-\hat{\mathbf{y}}$ direction in response to $\mathcal{E}_x$, which can be interpreted as a surface anomalous Hall response $\sigma_{\mathrm{AHC}}^{\mathrm{surf}} = \sigma_{yx} = -c\alpha_{\mathrm{iso}}$. In other words, from the point of view of the surface current response, an isotropic bulk ME coupling and a surface AHC are indistinguishable phenomena. We have here another example of an apparent surface property that is actually fixed by a bulk property; we encountered this most recently for the surface current, which is actually fixed by the bulk M_{orb} as also discussed in Section 1.2.

An even more germane example is that of the surface charge of an insulator, which is fixed – not absolutely, but modulo a quantum – by the bulk polarization. As a reminder, one way to understand the presence of this "modulo" is with reference to Fig. 1.3, in which we imagined an entire isolated surface band in the bulk gap. If this surface band goes from being empty to fully occupied, the surface charge density changes by a quantized amount $-e/A_{\mathrm{surf}}$; if we ascribe this to a change of bulk polarization, we would conclude that $\mathbf{P}$ has changed by this same amount. Thus, the bulk polarization should be regarded as well defined only modulo the quantum e/A_{surf}, and a knowledge of any one of its values constrains the surface charge of an insulating surface to take on one of a lattice of possible values as given by Eq. (1.6).

We can apply a similar argument here. Consider an insulating bulk with isotropic ME coupling α_{iso}, and again suppose there is an isolated surface band in the bulk

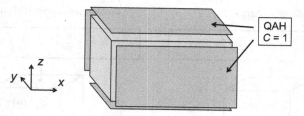

Figure 6.4 The block of material from Fig. 6.3 has now been dressed by attaching four pieces of a QAH insulator with Chern index $C = 1$ (relative to an outward-directed normal) to its four sides. The effective bulk ME coupling has been reduced by e^2/hc.

gap on its surface as in Fig. 1.3. Suppose, moreover, that this surface band has a nonzero Chern number C, so that in isolation it would act like a 2D QAH system. Then $\sigma_{\text{AHC}}^{\text{surf}}$ obviously changes by Ce^2/h when this surface band goes from being empty to being filled. If we then ascribe this to a change in the bulk ME coupling, we conclude that α_{iso} has changed by $-Ce^2/hc$. Thus, we conclude on physical grounds that the bulk orbital ME coupling is only well defined modulo a quantum e^2/hc, in the same sense that the bulk polarization is only well defined modulo a quantum e/A_{surf}. In analogy to the relation between polarization and surface charge anticipated in Eq. (1.6) and demonstrated in Section 4.5.4 leading to Eq. (4.101), we can write

$$\sigma_{\text{AHC}}^{\text{surf}} := -c\alpha_{\text{iso}} \tag{6.26}$$

where ':=' now indicates indeterminacy modulo the quantum e^2/hc.

Another argument to the same effect is sketched in Fig. 6.4. We imagine a block of material having some initial isotropic ME coupling α_{iso} when four sheets cut from a 2D QAH insulator with Chern index $C = 1$ are attached to its surfaces. Since the surface and the QAH sheet are both insulating, it is plausible that this can be done without closing the surface gap. Now the surface AHC has been increased by e^2/h, so the effective α_{iso} has been reduced by e^2/hc.

In view of this quantum of indeterminacy, it is natural to express α_{iso} in terms of a phase angle θ defined via

$$\alpha_{\text{iso}} = \frac{e^2}{hc}\frac{\theta}{2\pi} = \left(\frac{\alpha}{4\pi^2}\right)\theta \tag{6.27}$$

where α is again the fine structure constant. Then an uncertainty in α_{iso} modulo e^2/hc just corresponds to an uncertainty in θ modulo 2π.

If the surface is metallic, there will be an additional contribution arising from the partially occupied surface band. For the case of surface charge, this was expressed

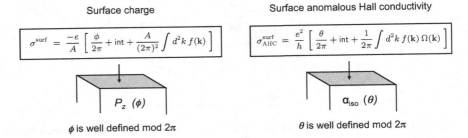

Figure 6.5 Left: Relation of surface charge σ^{surf} to the (k_x, k_y)-averaged Berry phase ϕ determining the polarization P_z. Right: Relation of the surface anomalous Hall conductivity $\sigma^{\text{surf}}_{\text{AHC}}$ to the axion angle θ determining the isotropic magneto-electric response α_{iso}. In each case, 'int' is an arbitrary integer and the last term accounts for the possibility of a metallic surface.

by Eq. (4.102) as an integral over the occupied 2D Fermi sea. The corresponding result here is, in view of Eq. (5.13),

$$\sigma^{\text{surf}}_{\text{AHC}} := -c\alpha_{\text{iso}} + \frac{e^2}{(2\pi)^2\hbar} \int_{\text{BZ}} f(\mathbf{k})\,\Omega(\mathbf{k})\,d^2k \qquad (6.28)$$

in terms of the Berry curvature Ω of the surface band crossing E_{F}. Using Stokes' theorem, the integral in the second term is just the Berry phase ϕ taken along the surface Fermi loop. Eq. (6.28) can then be rewritten rather elegantly in terms of the two phase angles ϕ and θ:

$$\sigma^{\text{surf}}_{\text{AHC}} := \frac{e^2}{h}\frac{\phi - \theta}{2\pi}. \qquad (6.29)$$

The correspondence between the formulas for surface charge and surface AHC in terms of Berry phases ϕ and axion angles θ is illustrated in Fig. 6.5 to emphasize their similarity.

For ordinary ME materials, the ME response is so weak that we can always make the branch choice such that θ is close to zero. We will not be populating entire surface bands or wrapping materials in Chern insulators. With the conversions given on p. 287, we find that $\theta \simeq 0.01$ ps/m for Cr_2O_3, a typical ME material, as reported by Malashevich et al. (2012),[4] which can be compared with $e^2/hc = (2\alpha)$ g.u. $= (\alpha/2\pi)$ g.u.$'$ $= 48.7$ ps/m. The 2π indeterminacy of θ is nevertheless there in principle, and as we shall see, topological materials can exhibit θ values on the order of π.

---

[4] In keeping with the context established on p. 287, this includes only frozen-ion orbital contributions. The total isotropic part of α for Cr_2O_3 was computed to be $\theta \simeq 1$ ps/m by the same authors, in rough agreement with measured values. For more context on the range of ME coefficients of known materials, see Fiebig (2005) and Rivera (2009).

In summary, we have concluded based on physical arguments that the orbital ME coupling may best be regarded as a multivalued quantity, and that it determines the surface AHC at an insulating surface modulo a quantum, in much the same way as occurs with electric polarization and surface charge. In Chapter 4 we were able to understand this indeterminacy of **P** modulo a quantum after deriving an expression for **P** in terms of Berry phases, Eq. (4.40) in 1D or Eq. (4.56) in 3D, or equivalently as integrals of the Berry connection **A** over the BZ, Eq. (4.41) in 1D or Eq. (4.44) in 3D. The essential observation was that while **A** itself is not gauge-invariant, its integral over the BZ *is* gauge-invariant, but only modulo a quantum. In 1D, where extraneous dimensions are omitted, this just corresponds to the fact that some gauge transformations change the Berry phase by 2π.

This begs the following questions: If we derive a formal expression for the orbital ME coupling, will it also involve Berry connections in a similar way? How will it behave under gauge transformations? Can we confirm the expected indeterminacy in α_{iso} and explain it in similar terms? These questions will occupy us for the remainder of this chapter.

Exercises

6.5 The 3×3 magnetoelectric tensor can be decomposed into a symmetric part $\alpha_{\mu\nu}^{\text{symm}} = (\alpha_{\mu\nu}+\alpha_{\nu\mu})/2$ (six independent components) and an antisymmetric part $\alpha_{\mu\nu}^{\text{anti}} = (\alpha_{\mu\nu}-\alpha_{\nu\mu})/2$ (three independent components). Show that the latter can be repackaged into a vector $\mathcal{A}$ such that the antisymmetric component of the responses are give by $\mathbf{P} = -\mathcal{A} \times \mathbf{B}$ and $\mathbf{M} = \mathcal{A} \times \mathcal{E}$.

6.6 On p. 289 we implicitly assumed that each edge where facets meet, such as between the top ($+\hat{\mathbf{z}}$) and front ($-\hat{\mathbf{y}}$) facets in Fig. 6.4, can be insulating – that is, free of any chiral edge channel. Justify this assumption by using the concepts of Section 1.3.2 in the case of a weak electric field applied parallel to the edge.

6.7 The sketch below shows five stages in the evolution of the Dirac-cone surface-state structure of a strong topological insulator as an applied TR-breaking perturbation at the surface is reversed, with $\sigma_{\text{AHC}}^{\text{surf}}$ changing from $-e^2/2h$ to $+e^2/2h$ during this process.

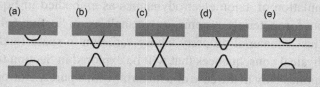

Panels (a–e) correspond to an amplitude B of the surface TR-breaking perturbation of $1.2B_c$, $0.5B_c$, 0, $-0.5B_c$, and $-1.2B_c$, respectively, where

B_c is a critical value at which the cone first touches the Fermi energy (dashed line). Make a series of sketches showing how you expect ϕ, θ, and $\sigma_{\text{AHC}}^{\text{surf}}$ of Eq. (6.29) to vary with B in this process.

6.3 Chern–Simons Axion Coupling

The notion that there is a topological aspect to the orbital ME coupling was first clearly expounded by Qi et al. (2008). They adopted a field-theoretic approach in which a Chern–Simons effective field theory was obtained in 4D and used, via a kind of dimensional reduction, to describe TIs in 3D and 2D. In particular, they showed that a given TR-invariant insulator in 3D could be characterized by a "Chern–Simons" contribution θ_{CS} to Eq. (6.27) defined as a BZ integral over the occupied bands of the form

$$\theta_{\text{CS}} = -\frac{1}{4\pi} \int_{\text{BZ}} \varepsilon_{\mu\nu\sigma} \, \text{Tr} \left[A_\mu \partial_\nu A_\sigma - \frac{2i}{3} A_\mu A_\nu A_\sigma \right] d^3k \qquad (6.30)$$

where $\varepsilon_{\mu\nu\sigma}$ is the third-rank antisymmetric tensor, the trace is over occupied bands, and the A_μ are Berry connection matrices. We shall derive this equation in Section 6.3.1. The subscript 'CS' is attached to θ to emphasize that θ_{CS} is not necessarily the only contribution to the total θ in Eq. (6.27) in the materials context.

An isotropic ME coupling as in Eq. (6.27) corresponds to a term in the electromagnetic Lagrangian of the form

$$\mathcal{L}_{\text{EM}} = \frac{\theta}{2\pi} \frac{e^2}{hc} \boldsymbol{\mathcal{E}} \cdot \mathbf{B} = \left(\frac{\mathfrak{a}}{4\pi^2} \right) \theta \, \boldsymbol{\mathcal{E}} \cdot \mathbf{B}. \qquad (6.31)$$

In the context of elementary particle theory, Eq. (6.31) is called an "axion" term, and A_μ in Eq. (6.30) has the interpretation of a fundamental gauge potential. Even if θ is treated as a fixed background property of the vacuum, a nonzero θ could have some observable consequences, such as endowing magnetic monopoles (if they exist) with a fractional electric charge (Wilczek, 1987; Witten, 1979). If instead $\theta(\mathbf{r}, t)$ is treated as a dynamical quantum field, its quantized excitations, known as "axions," have been considered as candidates for dark-matter particles.

The formulation of axion electrodynamics as embodied in Eq. (6.31) can also be carried out for materials systems, as we shall see in Section 6.4. In this context, θ becomes just a material property whose variation in space (and perhaps also in time) has physical consequences that will be explored in Section 6.4.2. As written, Eq. (6.31) applies only in the case of a material with a purely isotropic ME coupling α_{iso}. The Chern–Simons θ_{CS} is not necessarily the only contribution to the isotropic response, but as we shall see, it is the most interesting one from the point of connections to topology and the potential for colossal responses.

Taking a rather different approach from that of Qi et al., Essin et al. (2009) carried out a derivation of the orbital ME coupling based on a formulation by Xiao et al. (2009) of the response of the Berry-phase polarization to a slow spatial variation of the crystal Hamiltonian, here taken as a slow variation of the electromagnetic vector potential so as to represent a magnetic field. They obtained conclusions that were consistent with those of Qi et al. regarding TR-invariant systems, but also clarified that θ_{CS} remains well defined and characterizes an isotropic contribution to the ME response for general crystalline systems with broken TR and inversion symmetries.

Subsequently, two groups carried out a careful derivation of all frozen-ion contributions to the orbital ME response of an insulating crystal, finding that other "nontopological" terms are present in addition to the topological Chern–Simons one. On the one hand, Essin et al. (2010) adopted the dP/dB framework, considering an insulating crystal in the presence of a weak uniform magnetic field. Since the Hamiltonian eigenstates do not take the Bloch form in this case, the formulation was based instead on a perturbation theory of the one-particle density matrix. These authors also revisited the case of a slow spatial variation of the electromagnetic vector potential, obtaining consistent results. On the other hand, Malashevich et al. (2010) took the $dM/d\mathcal{E}$ approach, obtaining an expression for the ME coupling by deriving an expression for $\mathbf{M}_{orb}$ in the presence of a perturbing electric field. Both groups obtained identical formulas. More recently, consistent results were obtained by Gao et al. (2014) by extending the semiclassical theory outlined in Sections 5.1.7 and 5.4.3 to the second order in applied fields.

The next section briefly outlines the derivation by Malashevich et al. In the remaining sections, we shall focus on the Chern–Simons axion contribution α^{CS} and discuss its physical consequences.

6.3.1 Derivation of the Orbital Magnetoelectric Tensor

Here we briefly review the derivation of Malashevich et al. (2010), who obtained a complete expression for the (frozen-ion orbital) ME tensor of an arbitrary insulating crystal in the single-particle context. Our main goal is to obtain Eq. (6.30) for the Chern–Simons component, but along the way we shall see that other contributions of nontopological character appear as well.

Malashevich et al. (2010) began by generalizing the derivation of the orbital magnetization expression of Ceresoli et al. (2006) to the case in which a small electric field $\mathcal{E}$ is present. This could be done at a finite $\mathcal{E}$-field using the methods of Section 4.6, but for our purposes it is sufficient to work in terms of the first-order field-perturbed cell-periodic Bloch functions

$$|u_{n\mathbf{k}}\rangle = (1 + i\hbar e\, T_{n\mathbf{k}}^2 v_{\mathbf{k},\nu})\, |u_{n\mathbf{k}}^0\rangle, \tag{6.32}$$

where the superscript '0' denotes a zero-field quantity and the perturbation contribution from Eq. (4.21) has been included inside the parentheses. As a reminder, for an insulator at zero field $\mathbf{M}_{\text{orb}}$ is given by

$$M_\alpha^0 = \frac{e}{2\hbar c} \frac{1}{(2\pi)^3} \sum_n \int_{\text{BZ}} \text{Im}\; \varepsilon_{\alpha\mu\nu} \langle \partial_\mu u_{n\mathbf{k}}^0 | H_\mathbf{k}^0 + E_{n\mathbf{k}}^0 | \partial_\nu u_{n\mathbf{k}}^0 \rangle \, d^3k \qquad (6.33)$$

where the sum runs over the J occupied states only. The term involving E_F in Eq. (6.6) has been dropped on the assumption that the material is not a Chern insulator, as discussed following Eq. (6.5). The $H_\mathbf{k}^0$ and $E_{n\mathbf{k}}^0$ terms correspond to the LC and IC contributions, respectively.

The next step is to generalize to the field-perturbed case. We shall do this in the context of an arbitrary multiband gauge following the framework of of Section 3.6.4. That is, in general the states $|u_{n\mathbf{k}}\rangle$ have been transformed from the Hamiltonian eigenstate basis by a $\mathbf{k}$-dependent $J{\times}J$ unitary matrix as in Eq. (3.15). As a consequence, the unperturbed Hamiltonian $H_{mn\mathbf{k}}^0$ is not necessarily diagonal as it would be ($H_{mn\mathbf{k}}^0 = \delta_{mn} E_{n\mathbf{k}}^0$) in the Hamiltonian gauge.

Malashevich et al. then found that the local- and itinerant-circulation pieces become

$$M_\alpha^{\text{LC}} = \frac{e}{2\hbar c} \frac{1}{(2\pi)^3} \sum_n \int_{\text{BZ}} \text{Im}\; \varepsilon_{\alpha\mu\nu} \langle \partial_\mu u_{n\mathbf{k}} | H_\mathbf{k}^0 | \partial_\nu u_{n\mathbf{k}} \rangle \, d^3k \qquad (6.34)$$

and

$$M_\alpha^{\text{IC}} = \frac{e}{2\hbar c} \frac{1}{(2\pi)^3} \sum_{mn} \int_{\text{BZ}} \text{Im}\; \varepsilon_{\alpha\mu\nu} \langle \partial_\mu u_{m\mathbf{k}} | \partial_\nu u_{n\mathbf{k}} \rangle H_{nm\mathbf{k}} \, d^3k . \qquad (6.35)$$

In the second equation $H_\mathbf{k} = H_\mathbf{k}^0 + e\boldsymbol{\mathcal{E}} \cdot \mathbf{A}_\mathbf{k}$ which, from a heuristic standpoint, is a translation of $H_\mathbf{k} = H_\mathbf{k}^0 + e\boldsymbol{\mathcal{E}} \cdot \mathbf{r}$ into the Bloch representation. Separating the two contributions coming from $H_\mathbf{k}$ as $\mathbf{M}^{\text{IC}} = \mathbf{M}^{\text{IC},0} + \mathbf{M}^{\text{IC},\mathcal{E}}$, they obtained

$$M_\alpha^{\text{IC},0} = \frac{e}{2\hbar c} \frac{1}{(2\pi)^3} \sum_{mn} \int_{\text{BZ}} \text{Im}\; \varepsilon_{\alpha\mu\nu} \langle \partial_\mu u_{m\mathbf{k}} | \partial_\nu u_{n\mathbf{k}} \rangle H_{nm\mathbf{k}}^0 \, d^3k \qquad (6.36)$$

and

$$M_\alpha^{\text{IC},\mathcal{E}} = \frac{e}{2\hbar c} \frac{1}{(2\pi)^3} \sum_{mn} \int_{\text{BZ}} \text{Im}\; \varepsilon_{\alpha\mu\nu} \langle \partial_\mu u_{m\mathbf{k}}^0 | \partial_\nu u_{n\mathbf{k}}^0 \rangle \, e\boldsymbol{\mathcal{E}} \cdot \mathbf{A}_{nm\mathbf{k}} \, d^3k . \qquad (6.37)$$

The replacement of the field-perturbed wave functions $|u_{n\mathbf{k}}\rangle$ by the zero-field ones $|u_{n\mathbf{k}}^0\rangle$ in the last term in Eq. (6.37) is justified because there is already an explicit factor of field $\mathcal{E}$ and we are only interested in the first-order response overall. The field-perturbed orbital magnetization is then the sum of three pieces: $\mathbf{M} = \mathbf{M}^{\text{LC}} + \mathbf{M}^{\text{IC},0} + \mathbf{M}^{\text{IC},\mathcal{E}}$.

The authors then reorganized these expressions in such a way that each term would be individually multiband gauge-invariant. They made use of the gauge-covariant derivative $|\check{\partial}_\nu u_{n\mathbf{k}}\rangle$ of Eq. (3.145) to define two gauge-covariant objects

$$F_{mn\mathbf{k},\mu\nu} = \langle\check{\partial}_\mu u_{m\mathbf{k}}|\check{\partial}_\nu u_{n\mathbf{k}}\rangle \tag{6.38}$$

and

$$\Gamma_{mn\mathbf{k},\mu\nu} = \langle\check{\partial}_\mu u_{m\mathbf{k}}|H_\mathbf{k}^0|\check{\partial}_\nu u_{n\mathbf{k}}\rangle \tag{6.39}$$

so as to write $\mathbf{M} = \widetilde{\mathbf{M}}^{LC} + \widetilde{\mathbf{M}}^{IC,0} + \widetilde{\mathbf{M}}^{IC,\mathcal{E}}$ with

$$\widetilde{M}_\alpha^{LC} = \frac{e}{2\hbar c}\frac{1}{(2\pi)^3}\int_{BZ}\varepsilon_{\alpha\mu\nu}\,\text{Im}\,\text{Tr}\,[\Gamma_{\mu\nu}]\,d^3k\,, \tag{6.40}$$

$$\widetilde{M}_\alpha^{IC,0} = \frac{e}{2\hbar c}\frac{1}{(2\pi)^3}\int_{BZ}\varepsilon_{\alpha\mu\nu}\,\text{Im}\,\text{Tr}\,[H^0 F_{\mu\nu}]\,d^3k\,, \tag{6.41}$$

and, after some algebra,

$$\widetilde{M}_\alpha^{IC,\mathcal{E}} = \frac{-e}{2\hbar c}\mathcal{E}_\alpha\frac{1}{(2\pi)^3}\int_{BZ}\varepsilon_{\mu\nu\sigma}\,\text{Tr}\,\left[A_\mu^0\partial_\nu A_\sigma^0 - \frac{2i}{3}A_\mu^0 A_\nu^0 A_\sigma^0\right]d^3k\,. \tag{6.42}$$

The matrices $H^0, A_\mu, F_{\mu\nu}$, and $\Gamma_{\mu\nu}$ are all $J\times J$ Hermitian matrices (for J occupied bands), 'Tr' indicates a trace over the band index, and the $\mathbf{k}$ subscripts have now been suppressed.

The first two terms $\widetilde{\mathbf{M}}^{LC}$ and $\widetilde{\mathbf{M}}^{IC,0}$ depend linearly on $\mathcal{E}$ through the field dependence of the wave functions in Eq. (6.32). Explicit linear response expressions for these terms were derived in Malashevich et al. (2010), and are often referred to as "Kubo terms" or "cross-gap terms" since they take a form typical of linear response formulations of the Kubo–Greenwood type. In these terms, the integrand itself (inside the $\mathbf{k}$ integral) is multiband gauge-invariant. Thus, the Kubo terms' values are completely unambiguous, and they can be shown to vanish whenever inversion or TR symmetry is present. As a result, they are not capable of carrying any of the topological flavor that we are anticipating from our discussion in Section 6.2.2.

Our interest here is in the third term, $\widetilde{\mathbf{M}}^{IC,\mathcal{E}}$, which is just proportional to $\mathcal{E}$, giving rise to a perfectly isotropic contribution to the ME tensor. We now drop the '0' subscript in Eq. (6.42) for brevity, and arrive at a contribution to α_{iso} as in Eq. (6.27) with $\theta = \theta_{CS}$ given by Eq. (6.30). We repeat the result here for reference:

$$\theta_{CS} = -\frac{1}{4\pi}\int_{BZ}\varepsilon_{\mu\nu\sigma}\,\text{Tr}\,\left[A_\mu\partial_\nu A_\sigma - \frac{2i}{3}A_\mu A_\nu A_\sigma\right]d^3k\,. \tag{6.43}$$

This is just the Chern–Simons contribution that we were after. Note that the Kubo terms may also contain isotropic components, so in general we have $\theta = \theta_{\text{Kubo}} + \theta_{\text{CS}}$, but the Chern–Simons contribution will be of more interest to us going forward.

6.3.2 Gauge Dependence of the Axion Coupling

Equation (6.43) is the desired formula for the Chern–Simons axion coupling θ_{CS}. Note that θ_{CS} is a functional of the ground-state wave functions alone. As a reminder, this is *not* the case for the orbital magnetization of Eq. (6.33) or for the Kubo terms in Eqs. (6.40) and (6.41), which require a knowledge of the Hamiltonian as well. Moreover, θ_{CS} can easily be seen to be dimensionless (A and dk have inverse dimensions).

In all these respects, the expression in Eq. (6.43) for θ_{CS} is strongly analogous to the one for the Berry phase of a 1D insulator, which we can write for optimal parallelism as

$$\phi = \int_{\text{BZ}} \text{Tr}\,[A_\mu]\, dk_\mu\,. \tag{6.44}$$

Again, this is a dimensionless function of the ground-state wave functions alone.

Another similarity appears when we consider the gauge-transformation properties of Eqs. (6.43) and (6.44). In the multiband context with J occupied bands, a general gauge transformation is given by Eq. (3.140), which we repeat here for reference:

$$|\tilde{u}_{n\mathbf{k}}\rangle = \sum_m U_{mn}(\mathbf{k})\,|u_{m\mathbf{k}}\rangle\,. \tag{6.45}$$

Here $U_{mn}(\mathbf{k})$ is a $\mathbf{k}$-dependent $J \times J$ unitary matrix that mixes the occupied states. The corresponding change to the Berry connection is given by Eq. (3.143); dropping the band and $\mathbf{k}$ labels for conciseness, this is

$$\tilde{A} = U^\dagger A_\mu U + U^\dagger i\partial_\mu U\,. \tag{6.46}$$

Then the change in Berry phase $\Delta\phi = \tilde{\phi} - \phi$ is given by

$$\Delta\phi = \int_{\text{BZ}} \text{Tr}\,[U^\dagger i\partial_\mu U]\, dk_\mu\,. \tag{6.47}$$

In the single-band case, U reduces to a scalar $e^{-i\beta(\mathbf{k})}$ with $\tilde{A}_\mu = A_\mu + \partial_\mu\beta$; for the multiband case, we can do something similar and write

$$U(\mathbf{k}) = e^{-iB(\mathbf{k})} \tag{6.48}$$

where B is a $J \times J$ Hermitian matrix that varies smoothly with $\mathbf{k}$. There is some arbitrariness in doing so: Just as β can be incremented by $2\pi m$ for integer m without changing $e^{-i\beta}$, so can any eigenvalue of B be incremented by $2\pi m$ in the diagonal representation of B without changing U. Note that B and U commute, so that we obtain

$$\tilde{A}_\mu = U^\dagger (A_\mu + \partial_\mu B) U \tag{6.49}$$

and

$$\begin{aligned}
\operatorname{Tr}[\tilde{A}_\mu] &= \operatorname{Tr}[A_\mu] + \operatorname{Tr}[\partial_\mu B] \\
&= \operatorname{Tr}[A_\mu] - \partial_\mu (\operatorname{Im} \ln \det U).
\end{aligned} \tag{6.50}$$

In the 1D multiband case, $\Delta\phi$ in Eq. (6.47) is just given by -2π times the winding number of the phase of $\det(U)$ as k_μ runs across the BZ. We can call the gauge change "progressive" if this winding number is zero, or "radical" if it is not, per the discussion of Fig. 3.5 in Chapter 3. For the progressive case, the matrix B can be chosen to be continuous on the BZ regarded as a loop, and the gauge transformation can be continuously deformed to the identity ($U = 1$, $B = 0$). In contrast, for the radical case, there has to be a discontinuity in B (but not in U) somewhere on the loop. More physically, a radical gauge change shifts the choice of Wannier functions belonging to the home unit cell in such a way as to change the branch choice of the electric polarization, while a progressive one does not.

Similar concepts apply to the gauge dependence of the Chern–Simons coupling in 3D insulators. Indeed, we shall see that θ_{CS} can change by an integer multiple of 2π under a general multiband gauge transformation. In view of Eq. (6.27), this is in alignment with the physical arguments of Section 6.2.2 hinting at an ambiguity modulo e^2/h in the surface AHC. A careful derivation of the dependence of Eq. (6.43) on a general multiband gauge transformation in 3D is carried out in Appendix C.1; the result is Eq. (C.19), which we write here as

$$\Delta\theta_{CS} = \frac{1}{12\pi} \int_{BZ} \varepsilon_{\mu\nu\sigma} \operatorname{Tr}\left[U^\dagger(\partial_\mu U)U^\dagger(\partial_\nu U)U^\dagger(\partial_\sigma U)\right] d^3k. \tag{6.51}$$

The resemblance to Eq. (6.47) for the change of Berry phase in 1D is evident. In the following discussion, we lay the groundwork for this derivation and discuss its consequences.

First, recall that a general 3D insulator is characterized by a triplet (C_1, C_2, C_3) of Chern indices for 2D slices of the BZ taken in the three different lattice directions. We have a 3D QAH insulator if one or more of these indices is nonzero, in which case there is a topological obstruction to the existence of a smooth and periodic gauge (i.e., a smooth gauge on the 3-torus). We assume for all the remaining discussion that this is *not* the case, so that the insulator is topologically trivial in the TR-broken QAH context (although it might be a strong or weak TI in the TR-invariant sense).

Next, we could consider gauge changes that are radical in the sense that the phase of $\det(U)$ winds by a multiple of 2π as $\mathbf{k}$ traverses one or more of the three primitive reciprocal lattice vectors; again, these would change the assignment of the Berry-phase polarization $\mathbf{P}$ to a particular one of its lattice of possible values. For simplicity, we shall also exclude this case, insisting that $\det(U)$ is continuously defined on the BZ 3-torus.[5]

Surprisingly, there is another sense in which a multiband gauge transformation can be topologically nontrivial (i.e., "radical") in 3D. Consider an insulator with two occupied bands, so that $U(\mathbf{k})$ is a 2×2 matrix function of $\mathbf{k}$, and let

$$U(\mathbf{k}) = \begin{cases} -e^{-i\mathbf{q}\cdot\boldsymbol{\sigma}}, & q \leq \pi \\ I, & q \geq \pi \end{cases} \tag{6.52}$$

where $\mathbf{q} = \pi\mathbf{k}/k_0$, $q = |\mathbf{q}|$, $\boldsymbol{\sigma}$ is the vector of Pauli matrices, I is the 2×2 identity, and k_0 is a radius of a reciprocal-space sphere centered on the origin and chosen small enough to sit comfortably inside the 3D BZ. In this expression, U is everywhere continuous, including on the surface of the sphere of radius k_0 (i.e., at $q = \pi$). However, it is not possible to choose a representation in the form of Eq. (6.48) without introducing a discontinuity in B on the spherical surface. It is therefore impossible to smoothly connect this $U(\mathbf{k})$ with the identity transformation, so that it is again "radical," but in a new way. Note that $\det(U) = 1$ everywhere, so that this gauge transformation has no effect on Berry phases computed via Eq. (6.44) on any path; in particular, it does not modify the branch choice of the polarization.

The gauge transformation of Eq. (6.52) does, however, shift θ_{CS} by 2π. This remarkable result is demonstrated mathematically in Appendix C.4. It is also shown there that any progressive gauge transformation (i.e., one that can be arrived at by smooth deformations of the gauge choice) leaves θ_{CS} unchanged, so we are free to "smooth out" the nonanalytic behavior of Eq. (6.52) at the sphere surface, or spread the gauge change more evenly over the whole BZ. Subsequent applications of similar radical gauge changes (or their inverses) can generate an overall gauge transformation that changes θ_{CS} by $2\pi m$ for any integer m. It therefore follows that any 2×2 unitary transformation that can be arrived at by any combination of progressive and radical transformations either shifts θ_{CS} by an integer multiple of 2π or not at all.

It can be shown that this process exhausts all possible gauge transformations for a two-band space on the 3-torus. We can regard a given unitary transformation as a smooth mapping from the 3-torus (k_x, k_y, k_z) onto $SU(2)$. A branch of mathematics known as homotopy theory allows us to classify such mappings and extract winding

---

[5] The derivation given in Appendix C does not actually require this condition be met.

numbers. In particular, it is a result of homotopy theory that $\pi_3(SU(2)) = \mathbb{Z}$; that is, the homotopy group of mappings from the 3-sphere (locus of points with $w^2 + x^2 + y^2 + z^2 = 1$) onto the $SU(2)$ (the group of 2×2 unitary matrices with unit determinant) is the group $\mathbb{Z}$ of integers under addition. Essentially this means that such mappings are characterized by an integer winding number. Equation (6.51) (divided by 2π) is said to be a "homotopy invariant," meaning an expression whose evaluation determines the winding number m of the mapping. Technically our mapping onto $SU(2)$ is from $\mathbf{k}$ on the 3-torus T^3, not on the 3-sphere S^3, but the same results follow since our setup of Eq. (6.52) implicitly introduced a secondary mapping from T^3 onto S^3 in which the prefactors of I, σ_x, σ_y, and σ_z in Eq. (C.27) play the roles of (w, x, y, z), respectively.

In short, the essential picture for the two-band case is that any given gauge transformation on the 3-torus is characterized by a winding number m; its application to Eq. (6.51) shifts θ_{CS} by $2\pi m$; and gauge transformations with different m can never be smoothly deformed into one another.

Radical gauge transformations of this type are impossible in the single-band case, where unitary rotations are pure phases and all objects ("1×1 matrices") commute. Thus, the ambiguity of θ_{CS} modulo 2π arises only when there are two or more occupied bands, $J \geq 2$. When $J > 2$, we can imagine composing a sequence of unitary rotations like that in Eq. (6.52), but acting in different two-band subspaces, together with some additional progressive deformations, to arrive at the most general $J \times J$ gauge transformation. Since θ_{CS} can change only by multiples of 2π at each stage, the overall change is also a multiple of 2π. The corresponding mathematical statement would be that $\pi_3(SU(J)) = \mathbb{Z}$ for $J \geq 2$, with Eq. (6.51) again providing the homotopy invariant in the general multiband case.

Let's summarize what we have learned:

- Neither the integrand in Eq. (6.44) for ϕ, nor the integrand in Eq. (6.43) for θ_{CS}, is gauge-invariant.
- Nevertheless, the integrated expressions are both gauge-invariant, but only modulo 2π.
- The Berry phase ϕ is naturally defined in 1D, where a radical gauge transformation shifts ϕ by a multiple of 2π. In 3D, a triplet of integers expresses how a radical gauge transformation shifts the Berry phases along each of the three lattice directions.
- The axion coupling θ_{CS} is naturally defined in 3D, where there is a new kind of radical gauge transformation, operating only when two or more occupied bands are present, that shifts θ_{CS} by a multiple of 2π while leaving the Berry phases unchanged.

These points confirm that there is a close mathematical similarity between the formulation of the Berry phase ϕ in 1D and the Chern–Simons coupling θ_{CS} in 3D.

This mathematical structure correlates with a strong similarity between the physics of the surface charge of a 1D insulator, which is fixed by ϕ modulo e, and the surface AHC of a 3D insulator, which is fixed by θ_{CS} modulo e^2/h.

The similarity can be pursued further by considering cyclic adiabatic variations of the crystal Hamiltonian. To review, if we have an insulating 1D Hamiltonian that depends on an additional cyclic parameter λ, then the change in ϕ over the cycle is given by 2π times the Chern number of the 2D manifold in the (k, λ) space. This corresponds physically to the adiabatic pumping of charge along the chain during the cycle, as discussed in Sections 1.1.2 and 4.2.3. The same mathematics applies if λ is replaced by a second wavevector component for a 2D insulator, in which case the quantized AHC is given, in units of e^2/h, by the Chern number on the 2-torus in (k_x, k_y) space. Specifically, this Chern number can be written as

$$C^{[1]} = \frac{1}{2\pi} \int_{BZ} \varepsilon_{\mu\nu} \, \text{Tr} \, [\partial_\mu A_\nu] \, d^2k \qquad (6.53)$$

where $\varepsilon_{\mu\nu}$ is the second-rank antisymmetric tensor and the trace is over the occupied bands. In the more general context that we are about to encounter, this kind of Chern number is known as a "first Chern number," as denoted by the superscript on the left side of Eq. (6.53).

Similarly, if we have an insulating 3D crystalline Hamiltonian that depends on an additional cyclic parameter λ, then the change in θ_{CS} over the cycle is given by 2π times a "second Chern number" $C^{[2]}$ defined on the 4-torus in (k_x, k_y, k_z, λ) space. This corresponds physically to the adiabatic pumping of surface AHC from, say, the bottom to the top surface of a slab during the cycle. This kind of scenario has been discussed in some detail by Taherinejad and Vanderbilt (2015) and by Olsen et al. (2017). The same mathematics applies if λ is replaced by a fourth wavevector component, as would be appropriate for describing an imaginary insulator in 4D. In this case the second Chern number can be expressed as an integral over the 4-torus in the form

$$C^{[2]} = \frac{1}{32\pi^2} \int_{BZ} \varepsilon_{\mu\nu\sigma\tau} \, \text{Tr} \, [\check{\Omega}_{\mu\nu} \, \check{\Omega}_{\sigma\tau}] \, d^4k \qquad (6.54)$$

where $\varepsilon_{\mu\nu\sigma\tau}$ is the fourth-rank antisymmetric tensor and $\check{\Omega} = \check{\partial}_\mu A_\nu - \check{\partial}_\nu A_\mu = \partial_\mu A_\nu - \partial_\nu A_\mu - i[A_\mu, A_\nu]$ is the gauge-covariant Berry curvature tensor defined by Eqs. (3.134) and (3.149).

We have unwittingly entered here into the mathematics of objects known as Chern classes that arise in algebraic topology. The first Chern class is associated with the first Chern number, an object naturally defined on 2D manifolds such as the 2-torus, and giving the winding number of the Berry phase ϕ defined on 1-cycles (loops) on the 2-torus as these are carried around a closed cycle. Similarly, the second Chern class describes the second Chern number, naturally defined in 4D and describing the winding of θ_{CS} defined on 3D submanifolds as these are carried around a cycle. This pattern continues: There is one new Chern class defined in each higher even-numbered dimension, associated with the winding of a phase-angle object that is well defined only modulo 2π and is defined in one fewer dimension. The interested reader is referred to the books by Frankel (1997), Nakahara (2003), and Eschrig (2011) for the mathematical development, and to the article by Qi et al. (2008) and the book by Bernevig (2013) for further discussion in the context of the electronic structure of insulators.

6.3.3 Symmetry and Connection to Topological Insulators

To review, an isotropic ME coupling α_{iso} describes the magnetization **M** that appears parallel to an applied electric field $\mathcal{E}$, or the polarization **P** that arises parallel to an applied magnetic field **B**. Now **P** and $\mathcal{E}$ are even under TR but odd under inversion, while **M** and **B** are odd under TR but even under inversion. We can therefore describe α_{iso} as a *pseudoscalar*, by which we mean a scalar quantity that reverses sign under either TR or inversion. A *pseudoscalar-symmetric* crystal is one whose symmetry group includes TR, inversion, or any other operator that would reverse the sign of a pseudoscalar.[6] Clearly, then, an ordinary (i.e., single-valued) pseudoscalar quantity would have to vanish in any pseudoscalar-symmetric crystal. This applies, for example, to the contributions to α_{iso} arising from the Kubo terms in Eqs. (6.40) and (6.41).

By contrast, we have seen that θ_{CS}, which determines the Chern–Simons contribution to α_{iso}, is a lattice-valued pseudoscalar quantity; values separated by 2π are identified as equivalent. In a pseudoscalar-symmetric crystal, θ_{CS} is the only contribution to θ, and the requirement that θ has to map onto itself under sign reversal now has *two* solutions:

$$\theta = \begin{cases} 0 & \text{(topologically trivial)} \\ \pi & \text{(topologically nontrivial)}. \end{cases} \tag{6.55}$$

We have thus arrived at a topological $\mathbb{Z}_2$ classification that applies to any pseudoscalar-symmetric insulating crystal!

Consider first the case that the crystal has TR symmetry. We already know of another $\mathbb{Z}_2$ classification that pertains to this case, namely the distinction between trivial and strong TIs that we explored in Section 5.3. Is there a connection between the two classifications? The answer is yes; they are identical! This remarkable

---

[6] Most generally, a crystal is pseudoscalar-symmetric if and only if its magnetic point group includes an improper rotation not composed with TR or a proper rotation composed with TR.

result, first demonstrated by Qi et al. (2008) and elaborated by Essin et al. (2009) and others, forms the nexus of the connection between topological character and ME response.

To make this connection plausible, and indeed to see that it is required, consider a macroscopic crystallite cut from some strong TI such as Bi_2Se_3, and assume that TR remains unbroken at the surface as well as in the bulk. In this material σ_{AHC}^{surf} must vanish at the surface, and since Eq. (6.26) associates this with α_{iso}, we would be tempted to conclude that α_{iso} must vanish.

This conclusion would be incorrect, however, since Eq. (6.26) only applies to insulating surfaces, and we know that the surface of a strong TI is required to have surface states of topological character crossing the gap. We therefore have to use Eq. (6.28) or (6.29) instead, which expresses σ_{AHC}^{surf} on a given patch of surface as being equal, modulo e^2/h, to a sum of a term associated with the bulk θ and another term associated with any metallic behavior that may be present at the surface. In Section 5.3.3, we argued that the Berry phase around a Dirac-cone Fermi loop is exactly π, and that there must be an odd number of these loops on any surface of strong TI. Thus the total ϕ in Eq. (6.29) is an odd multiple of π. Moreover, since the left-hand side vanishes, it follows that θ must also be an odd multiple of π. This corresponds to $\theta = \pi$ in Eq. (6.55), demonstrating the identity between the $\mathbb{Z}_2$ classification based on θ and the strong TI classification discussed in Chapter 5.

Conversely, if the surfaces are gapped, such as by applying local TR-breaking perturbations to the surface, then only the θ term survives in Eq. (6.29), and it is physically correct for a half-integer surface AHC to appear. That is, any gapped surface of a strong TI necessarily displays a half-integer quantum Hall effect. Whether the surface AHC is $e^2/2h$ or $-e^2/2h$ (or perhaps even $3e^2/2h$) will depend on the nature of the surface and of the TR symmetry breaking. Moreover, the sign of the TR symmetry breaking need not be the same everywhere on the surface of a given sample – although if is not, chiral edge channels will appear at the boundaries between surface patches whose anomalous Hall conductivities differ by e^2/h. We shall discuss a case like this in Section 6.4.2 in connection with Fig. 6.8.

Note that TR symmetry is not the only symmetry that can protect a topological state with $\theta = \pi$. In fact, we have seen that any pseudoscalar-symmetric crystal has θ restricted to take values of 0 or π. As an example, if a centrosymmetric TI such as Bi_2Se_3 is placed in a uniform magnetic field, TR is broken but inversion is not. As long as the applied field is weak enough not to close the gap, we know that a quantized invariant such as θ cannot change. Since it started as being equal to π, it will remain so in the presence of the field.

A TI of this type is sometimes called an *axion insulator*, suggesting that the axion coupling of π is constrained not by TR, but rather by something else. Usually this is inversion; the influential paper of Wan et al. (2011) suggested that some pyrochlore iridates might display an inversion-protected axion phase, for example.[7] In general, however, the term can refer to any pseudoscalar-symmetric insulator with $\theta = \pi$. For the inversion-symmetric case, the $\mathbb{Z}_2$ index can be obtained from a parity-counting algorithm: $\theta = 0$ or π, depending on whether the sum of parities of occupied states at the eight time-reversal invariant momenta (TRIM) is of the form $4n$ or $4n + 2$, respectively, where n is an integer (see, for example, Turner et al., 2012). Axion insulators can be regarded as a subclass of the topological crystalline insulators discussed in Section 5.5.1, with the property that insulating surfaces exhibit a half-quantized surface AHC.

6.3.4 Computational Practicalities

Considerable attention was given to the Chern–Simons axion coupling θ_{CS} in the preceding discussion, because of its close connections to polarization, magnetization, AHC, TIs, and other central topics of this book. Nevertheless, its practical importance is not as clear. Calculation of θ_{CS} is not recommended as a way to determine whether a system is a strong TI, or whether it is an axion insulator; this is best done with parity counting algorithms when inversion symmetry is present, and by computing the $\mathbb{Z}_2$ invariants on two TR-invariant planes as in Section 5.3 when TR symmetry is present.

For magnetoelectric materials whose symmetry allows for a nonzero α_{iso}, θ_{CS} should be present, but is typically very small. The work of Coh et al. (2011) provides some benchmarks. For Cr_2O_3, a well-known magnetoelectric, these authors found $\theta_{CS} \sim 1.3 \times 10^{-3}$, which is a very small fraction of 2π; this result perhaps is not surprising since the SOC is rather weak in this material.[8] These authors also did a computational experiment by artificially applying a staggered Zeeman field to the Bi sites of Bi_2Se_3 in such a way as to break both TR and inversion symmetry. They found that θ_{CS} could reach values on the order of 0.5 for fields large enough to induce a magnetic moment of about $0.2\ \mu_B$, albeit in a rather artificial situation. Later attempts to "computationally design" materials with large θ_{CS} by Coh and Vanderbilt (2013) resulted in the suggestion that certain chemically ordered pyrochlore materials such as $Cd_2Ru_2O_{6.5}Te_{0.5}$ could have θ_{CS} as large as 0.21, but such a chemical ordering is not known to occur experimentally.

---

[7] Unfortunately this does not seem to have been borne out by subsequent calculations (see, for example, Zhang et al., 2017).

[8] Subsequent work by Malashevich et al. (2012) gave an even smaller value, with the difference ascribed to a different choice of exchange-correlation potential. In that work, all other contributions to the ME tensor were also calculated, and the Chern–Simons contribution was found to be about an order of magnitude smaller than the Kubo terms in Eqs. (6.40) and (6.41) and about two orders of magnitude smaller than the spin contributions.

In summary, the experience to date has been that θ_{CS} is very small (or zero by symmetry) in all nontopological materials where it has been studied, and is exactly π in some classes of topological materials. However, there is no reason in principle why these should be the only possibilities. Can we find magnetic materials with strong SOC that have a large θ_{CS}, of order unity, but not constrained to equal π? This remains an interesting avenue for future materials research.

From the point of view of computational methodology, the calculation of θ_{CS} is somewhat problematic. The methods used in the works cited earlier relied on a two-step procedure in which a smooth gauge is first constructed on a chosen **k**-space mesh, and then the Berry connections $\mathbf{A}(\mathbf{k})$ and its derivatives are evaluated by finite differences and θ_{CS} is calculated by numerical integration on the mesh. The problem of finding a smooth gauge in **k** is essentially the same as that of finding well-localized Wannier functions, so the maximal localization procedure of Marzari and Vanderbilt (1997; see also Marzari et al., 2012) can be used for this purpose. However, there are situations where it is necessary to break symmetry in the gauge to satisfy the smoothness constraint. For example, this is the case for weak and strong TIs, where TR must be broken in the gauge as discussed at the end of Section 5.3.1. As a result, if one uses this method for computing θ_{CS} for a strong TI, the TR breaking in the gauge propagates into the results such that θ_{CS} is only approximately equal to π. The result must ultimately converge to π in the limit of a fine **k**-point mesh, but the convergence can be rather slow in practice (Coh and Vanderbilt, 2013).

In the case of the Berry phase in 1D, we found a procedure for computing ϕ on a discrete k mesh that involves taking the phase of the determinant of a product of $J \times J$ matrices as in Eq. (4.67). As discussed in the context of that equation and confirmed in Ex. 4.6, this procedure is manifestly gauge-invariant, in the sense that the answer is identical (to numerical precision) no matter how the occupied states are scrambled by unitary mixing matrices $U(\mathbf{k})$ chosen independently at each mesh point **k**. It also has the property that, in situations where ϕ is actually quantized to be 0 or π by some symmetry, the value computed on the discrete mesh will also be precisely 0 or π.

At the time of this writing, there is no known procedure for computing θ_{CS} on a 3D mesh in a way that is manifestly gauge-invariant in the same sense. Such a procedure would be highly desirable, as it would obviate the need for the smooth-gauge construction step and, presumably, preserve symmetries in a similar way as for the Berry phase in 1D. The formulation of such a procedure remains, in my opinion, an important unsolved mathematical problem.

Some other approaches to the computation of θ_{CS} have been formulated, at least partially in an attempt to get around these difficulties. A modification to the previously mentioned strategy is to compute θ_{CS} directly in the Wannier representation, using the formulation in Section III.B of Coh et al. (2011). Because of the localization properties of the Wannier functions in real space, this appears to accelerate the convergence with respect to the **k**-point mesh. A more radical reformulation is to compute θ_{CS} in the hybrid Wannier representation introduced in Section 4.5.4 and used several times in Chapter 5, as described in Taherinejad and Vanderbilt (2015)[9] and Olsen et al. (2017). This approach also has the advantages of being free of gauge-obstruction difficulties, since a global 3D gauge construction is never needed, and of respecting symmetries. As a bonus, it is useful for inspecting surface properties and following the pumping of charge or AHC to a surface over the course of an adiabatic cycle, as discussed in conjunction with Eq. (6.53). At the same time, it requires some additional programming, and it is not yet clear whether it accelerates convergence with respect to **k**-point sampling.

Finally, a related approach, proposed by Liu and Vanderbilt (2015), involves constructing a smooth gauge that is periodic in only two of the three lattice directions. In this approach, one computes θ_{CS} by first integrating Eq. (6.30) over the interior of the 3D BZ using this relaxed gauge, and then adds correction terms to account for the gauge discontinuity at the boundary in the nonperiodic direction.

Overall, it appears that there is still considerable room for improvement in the development of algorithms for the practical and efficient calculation of the Chern–Simons axion coupling.

Exercises

6.8 For the case of a single occupied band, Eq. (6.43) becomes just

$$\theta_{CS} = -\frac{1}{4\pi} \int_{BZ} \varepsilon_{\mu\nu\sigma} A_\mu \partial_\nu A_\sigma.$$

(Why?) Show that this expression is invariant with respect to a single-band gauge transformation as in Eq. (3.15). Note that $\varepsilon_{\mu\nu\sigma} \partial_\mu \partial_\nu \beta = 0$, and be aware that you may need to use an integration by parts.

6.9 Following the example of Appendix C.4, show that the two-band gauge transformation of Eq. (C.32) results in a change of θ by $2\pi m$.

6.4 Axion Electrodynamics

In this section, we return to the physical consequences of an isotropic ME coupling. Such a term gives rise to physical effects that can be described in the framework of

---

[9] There is a sign error in Eq. (12) of this paper; see Olsen et al. (2017) for a corrected version.

axion electrodynamics as described by the Lagrangian of Eq. (6.31), which can be regarded either as the fundamental Lagrangian of the vacuum or as an effective Lagrangian of a material system (Wilczek, 1987). We focus now on the latter case. Some discussion of axion electrodynamics in this context can be found in Qi et al. (2008), Essin et al. (2009), Nomura and Nagaosa (2011), Chen and Lee (2011), and Wu et al. (2016), as well as in Section V.A of the review by Hasan and Kane (2010).

In general the ME coupling is not isotropic, such that the following analysis would have to be supplemented by additional terms to represent the anisotropic responses. This is beyond our present scope, so we restrict ourselves to high-symmetry crystals in which such anisotropies are absent or we neglect their effects if present. Henceforth we suppress the 'iso' subscript on α_{iso} for the sake of conciseness, writing it simply as α.

We do, however, include all relevant physical contributions to α, including those from θ_{Kubo} and θ_{CS} and, if we are interested in responses at sub-phonon (e.g., terahertz) frequencies, those from the lattice-mediated mechanism discussed on p. 286. We also allow for the fact that α, or equivalently θ, may be slowly varying in space or time. As a reminder, α and θ are related by

$$\alpha(\mathbf{r}, t) = \frac{e^2}{2\pi \hbar c} \theta(\mathbf{r}, t) = \frac{\mathfrak{a}}{4\pi^2} \theta(\mathbf{r}, t) \tag{6.56}$$

where $\mathfrak{a}$ is the fine structure constant. The spatial dependence of α will be of most use to us, such as when we are modeling a surface or interface as a discontinuity in α, but it is also interesting to see what a time dependence would do.[10]

6.4.1 Derivation from Maxwell's Equations

We now derive the axion electrodynamics just by applying the standard Maxwell's equations to the case of an isotropic $\alpha(\mathbf{r}, t)$. In Gaussian units, these read

$$\nabla \cdot \boldsymbol{\mathcal{E}} = 4\pi\rho, \tag{6.57a}$$

$$\nabla \cdot \mathbf{B} = 0, \tag{6.57b}$$

$$\nabla \times \boldsymbol{\mathcal{E}} = -\frac{1}{c} \frac{\partial \mathbf{B}}{\partial t}, \tag{6.57c}$$

$$\nabla \times \mathbf{B} = \frac{4\pi}{c} \mathbf{J} + \frac{1}{c} \frac{\partial \boldsymbol{\mathcal{E}}}{\partial t}. \tag{6.57d}$$

---

[10] We could equally well write all the following equations in terms of θ, but the formulation in terms of α has been chosen to keep the discussion closer to the spirit of the classical Maxwell's equations.

It is standard practice to define bound charges, bound currents, and polarization currents as

$$\rho_b = -\nabla \cdot \mathbf{P}, \tag{6.58a}$$

$$\mathbf{J}_b = \nabla \times \mathbf{M}, \tag{6.58b}$$

$$\mathbf{J}_p = \frac{\partial \mathbf{P}}{\partial t} \tag{6.58c}$$

with the free charge and current as the remainders,

$$\rho_f = \rho - \rho_b, \tag{6.59a}$$

$$\mathbf{J}_f = \mathbf{J} - \mathbf{J}_b - \mathbf{J}_p. \tag{6.59b}$$

Here, we take the polarization and magnetization to be

$$\mathbf{P} = \mathbf{P}_0 + \alpha \, \mathbf{B}, \tag{6.60a}$$

$$\mathbf{M} = \mathbf{M}_0 + \alpha \, \mathcal{E}, \tag{6.60b}$$

where $\mathbf{P}_0$ is the polarization in the absence of the magnetic field and $\mathbf{M}_0$ is the magnetization in the absence of the electric field. We also define $\tilde{\rho}_b = -\nabla \cdot \mathbf{P}_0$, $\tilde{\mathbf{J}}_b = \nabla \times \mathbf{M}_0$, and $\tilde{\mathbf{J}}_p = \partial \mathbf{P}_0/\partial t$ as the parts of the objects in Eq. (6.58) that would remain in the absence of the axion coupling. Straightforward algebra then leads to

$$\rho_b = \tilde{\rho}_b - (\nabla \alpha) \cdot \mathbf{B}, \tag{6.61a}$$

$$\mathbf{J}_b = \tilde{\mathbf{J}}_b + c \, (\nabla \alpha) \times \mathcal{E} + c\alpha \, \nabla \times \mathcal{E}, \tag{6.61b}$$

$$\mathbf{J}_p = \tilde{\mathbf{J}}_p + \frac{\partial \alpha}{\partial t} \mathbf{B} + \alpha \, \frac{\partial \mathbf{B}}{\partial t}, \tag{6.61c}$$

where Eq. (6.57b) was used in arriving at Eq. (6.61a). When adding $\mathbf{J}_b + \mathbf{J}_p$, the last terms above cancel by virtue of Eq. (6.57c). Thus, we arrive at modified versions of Eq. (6.57a) and Eq. (6.57d):

$$\nabla \cdot \mathcal{E} = 4\pi \left(\rho_f + \tilde{\rho}_b - (\nabla \alpha) \cdot \mathbf{B} \right), \tag{6.62a}$$

$$\nabla \times \mathbf{B} = \frac{4\pi}{c} \left(\mathbf{J}_f + \tilde{\mathbf{J}}_b + \tilde{\mathbf{J}}_p + c \, (\nabla \alpha) \times \mathcal{E} + \frac{\partial \alpha}{\partial t} \mathbf{B} \right) + \frac{1}{c} \frac{\partial \mathcal{E}}{\partial t}. \tag{6.62b}$$

Together with Eqs. (6.57b) and (6.57c), which remain unchanged, these relations express Maxwell's equations in the presence of an axion coupling.

The only change in these equations from the standard case is the presence of the $\nabla \alpha$ terms in the first and last of Maxwell's equations, and of the $\partial \alpha / \partial t$ term in the last Maxwell equation. In other words, the equations of motion depend only on spatial or temporal derivatives of $\alpha(\mathbf{r}, t)$, and not on α itself.

As mentioned earlier, the equations of axion electrodynamics have been introduced and discussed in two very different contexts. Notably, they have been explored in elementary particle physics and cosmology, where the term in Eq. (6.31) has been proposed as an addition to the fundamental Lagrangian. Through Eq. (6.56) this leads to a new set of Maxwell's equations in the form of Eq. (6.62), but without the $\tilde{\rho}$ and $\tilde{\mathbf{J}}$ terms. The addition of the axion terms represents "new physics" and, as discussed on p. 292, could have consequences for magnetic monopoles or produce dark-matter candidates.

In our case, we are assuming that the usual Maxwell's equations continue to hold as in Eq. (6.57); all we have done is to rewrite ρ and $\mathbf{J}$ as sums of several contributions in an effort to clarify the physical responses of a material system. In this context, all terms on the right-hand side of Eqs. (6.61), including the ones involving α, can represent perfectly ordinary bound charges, bound currents, and polarization currents whose presence arises from the linear ME coupling of the material. However, because of the 2π ambiguity of θ, we shall see that they can also have a kind of topological character.

We saw on p. 290 that $\theta = 2\pi$ in Eq. (6.56) amounts to a ME coupling of $(a/2\pi)$ g.u.', or about 10^{-3} g.u.', while values for known ME materials are even smaller, typically closer to 10^{-5} g.u.' Thus, α is very small under ordinary circumstances. At the other extreme, a value of α on the order of unity sets a scale of ME strength at which magnetic and electric responses would become strongly mixed. The prospects for reaching such a colossal ME coupling seem remote using ordinary materials, but as we shall see in Section 6.4.2, this goal may not be entirely out of reach using topological materials.

In the case of material bodies at rest, we can drop the time dependence of $\alpha(\mathbf{r}, t)$ and take it to be a function of space only. We note in passing, however, that the chiral magnetic effect discussed in Section 5.4.3 can be described in terms of axion electrodynamics if one assigns a linear time dependence to α for each Weyl point surrounded by a Fermi pocket. In this interpretation, the current appearing parallel to an applied $\mathbf{B}$ as given by Eq. (5.46) is identified with the $(\partial \alpha/\partial t)\mathbf{B}$ term in Eq. (6.61c) or (6.62b). This connection is nicely summarized in Section (II.C.4) of Armitage et al. (2018). Here we instead focus on insulators, and consequently drop the time dependence of α in the remainder of our discussion.

6.4.2 Magnetoelectric Effects Revisited

The $\nabla\alpha$ terms in Eqs. (6.62a) and (6.62b) have a fairly straightforward interpretation. To focus on the effects associated with the ME coupling, we write Maxwell's equations one more time, this time assuming (1) static field configurations,

(2) the absence of free charge, and (3) the absence of ordinary polarization $\mathbf{P}_0$ and magnetization $\mathbf{M}_0$. In this case Maxwell's equations simplify to

$$\nabla \cdot \boldsymbol{\mathcal{E}} = -4\pi (\nabla \alpha) \cdot \mathbf{B}, \tag{6.63a}$$

$$\nabla \cdot \mathbf{B} = 0, \tag{6.63b}$$

$$\nabla \times \boldsymbol{\mathcal{E}} = 0, \tag{6.63c}$$

$$\nabla \times \mathbf{B} = 4\pi (\nabla \alpha) \times \boldsymbol{\mathcal{E}}. \tag{6.63d}$$

To see the meaning of the $\nabla \alpha$ contributions, consider a cylindrical sample with isotropic ME coupling α_0 filling $z_1 < z < z_2$ in the z direction and $s < R$ in the radial direction in (s, φ, z) cylindrical coordinates. In this case $\nabla \alpha$ takes the form of delta-function concentrations on the surfaces: $-\alpha_0 \delta(s - R)\, \hat{s}$ on the side surface, $-\alpha_0 \delta(z - z_2)\, \hat{z}$ on the top cap, and $\alpha_0 \delta(z - z_1)\, \hat{z}$ on the bottom cap. In the presence of an electric field $\boldsymbol{\mathcal{E}} = \mathcal{E}\,\hat{z}$, Eq. (6.63d) predicts a surface current flow along $+\hat{\varphi}$ as shown in Fig. 6.6(a). For an ordinary ME material, this is easily understood as arising from the surface bound current $\mathbf{K}_b = \mathbf{M} \times \hat{n}$, where $\mathbf{M}$ is the bulk magnetization induced via α and $\hat{n}$ is the outward-directed unit normal.

In Section 6.2.2, however, we found that if we start with some isotropic ME material and surround it on all sides with a layer of Chern insulator with $C = 1$, as shown in Fig. 6.4, this shifts the effective ME coupling of the material by $-e^2/hc$. As an extreme version of this argument, we can assume that the starting "material" is actually just the vacuum (or some magnetoelectrically inert material such as glass), and wrap it with a Chern insulator layer. We do this now in the cylindrical geometry we have been discussing, covering both the cylindrical side surface and the two circular end caps with a Chern insulator layer having a consistent index $C = -1$ on all surfaces as defined relative to the outward-directed surface normal $\hat{n}$.

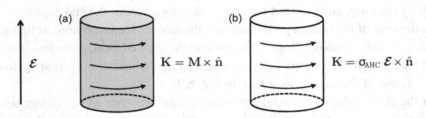

Figure 6.6 Current response of a cylindrical sample to an applied electric field $\boldsymbol{\mathcal{E}}$ along $\hat{z}$. (a) The sample is composed of an ordinary ME material whose induced $\mathbf{M} = \alpha_0 \boldsymbol{\mathcal{E}}$ generates a surface current in the $\hat{\varphi}$ direction. (b) The sample is a Chern wrapper surrounding a vacuum (see text for explanation). An identical current flows, but is now attributed to the AHC on the lateral surface.

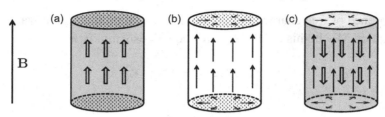

Figure 6.7 Response of a cylindrical sample as a magnetic field $\mathbf{B} = B(t)\hat{\mathbf{z}}$ is turned on in the vertical direction. (a) Naive view of an ordinary ME material. The outline-formatted arrows indicate a bulk polarization current $\mathbf{J}_p = d\mathbf{P}/dt = \alpha_0\, d\mathbf{B}/dt$ in the interior, producing surface charges on the end caps (hashing). (b) Faraday-induced response of the Chern wrapper; the line arrows indicate surface sheet currents whose divergence gives rise to surface charges. (c) Additional current flows associated with Faraday induction that should be added to those in (a) to arrive at a picture of current flow consistent with (b).

We shall refer to this configuration as a "Chern wrapper."[11] Then we again have a configuration with $\alpha = \alpha_0$ inside and 0 outside, where $\alpha_0 = e^2/hc$.

This configuration is sketched in Fig. 6.6(b). In this case the Chern wrapper carries a 2D AHC of $-c\alpha_0 = -e^2/h$, which, in the presence of the electric field along $\hat{\mathbf{z}}$, generates exactly the same current distribution as in Fig. 6.6(a), consisting of a sheet current on the lateral surface in the $+\hat{\varphi}$ direction. It may appear remarkable that the current distributions generated by the two different mechanisms in the two cases – simple bound current in the "ordinary" case and anomalous Hall response in the "topological" case – should be identical, but we should have anticipated this outcome from the discussion in Section 6.2.2.

It is potentially more confusing to consider the case of a uniform magnetic field $\mathbf{B} = B\hat{\mathbf{z}}$ applied along the axis. A naive inspection of Eq. (6.63) suggests that only Eq. (6.63a) contributes to the response, giving rise to a surface charge density $\sigma_{surf} = \pm\alpha_0 B$ on the top and bottom surfaces, respectively. But what is the nature of this charge density, and how did the charge density arrive at the end caps?

For the case of the ordinary ME material, the answer seems obvious at first sight: A polarization $\mathbf{P} = \alpha_0\mathbf{B}$ develops in response to the applied field, providing a surface bound charge $\mathbf{P}\cdot\hat{\mathbf{n}}$ that is apparently delivered by the polarization current $\mathbf{J}_p$ flowing in the interior of the sample as shown in Fig. 6.7(a).

For the case of the Chern wrapper, however, the presence of the charge density on the caps is associated with the physics of the Středa formula, Eq. (1.34). This formula basically describes a change in the density of quantum states, and thus in

---

[11] Inclusion of the end caps has the effect of eliminating all edge channels, so that the Chern wrapper is uniformly insulating everywhere. This becomes more relevant when considering an applied $\mathbf{B}$ field, as we shall do shortly.

the charge density, when a Berry curvature and magnetic field are simultaneously present. Using $\sigma_{\text{surf}} = -en_{\text{surf}}$ and $\sigma_{\text{AHC}} = -c\alpha_0$, Eq. (1.34) implies that

$$\sigma_{\text{surf}} = \frac{-\sigma_{\text{AHC}}}{c} B_\perp = \alpha_0 B_\perp . \tag{6.64}$$

But in this case, how did the charge arrive there? Certainly not through a volume current in the interior. To solve this riddle, it is necessary to imagine slowly turning on the magnetic field from zero, and to take into account the effects of Faraday induction during this process. By Eq. (6.57c), the dB_z/dt generates an azimuthal electric field in the $-\hat{\varphi}$ direction that, acting on the QAH nature of the Chern wrapper, generates a current pattern as shown in Fig. 6.7(b): upward along the lateral surface, inward on the top cap, and outward on the bottom cap. It is then easy to see that the resulting charge distribution is consistent with this current flow pattern, and to verify that the total charge pumped is insensitive to the rate at which the magnetic field is turned on.

Again, the results are the same but the mechanisms are rather different: ordinary bound charge versus Středa charge; ordinary polarization current versus Faraday-induced anomalous Hall conductivity. There does appear to be a distinction this time, however, since the current appears to flow through the bulk in one case, and only on the skin in the other. Is this physically correct, and does it allow us to distinguish the two kinds of response experimentally?

Alas, the answer is no; the current flow patterns are actually identical. To see this, we must also include the effects of Faraday induction in the case of the ordinary material. The resulting electric field in the $-\hat{\varphi}$ direction generates a transient magnetization $M_\varphi = \alpha_0 \mathcal{E}_\varphi$, proportional to $-s\hat{\varphi}$, which in turn gives rise to current densities as shown in Fig. 6.7(c). There is a volume current $\nabla \times \mathbf{M}$, which is in the $-\hat{z}$ direction, and there are surface currents $\mathbf{M} \times \hat{\mathbf{n}}$ flowing upward on the lateral surface, inward on the top cap, and outward on the bottom cap. This flow pattern of the bound currents, shown in Fig. 6.7(c), is divergence-free (arising as it does from a curl), and so does not affect the charge densities deposited on the end caps. However, when combined with the current flows shown in Fig. 6.7(a), the volume currents exactly cancel, leaving a net pattern of surface-only currents identical to that of Fig. 6.7(b). This cancellation is the same as the one that eliminated the last terms in Eqs. (6.61b) and (6.61c) when writing Eq. (6.62b). The conclusion is that the conventional ME material and the Chern wrapper really do show the same pattern of current responses implied by Eqs. (6.63). It is worth emphasizing that all the effects discussed here involve dissipationless currents.

What happens in the case of a Chern wrapper *without* end caps? In this scenario, circular chiral edge channels are present at the top and bottom of the open cylinder. Equations (6.63) are not capable of treating cases like this without some modification to deal with the edge channels explicitly, since one cannot treat α_0 (proportional to θ) as single-valued in the presence of an edge channel. However, it is not hard to see what should happen on physical grounds, using arguments similar to those in Chapter 1.

Under an electric field applied along $\hat{z}$, the open Chern wrapper behaves much the same as in Fig. 6.6(b), with an azimuthal current flowing on the cylindrical surface such that the sample displays a magnetic dipole moment along $\hat{z}$. However, in the presence of the field, the Fermi levels on the top and bottom channels are shifted and pushed out of equilibrium. If charge is allowed to flow so as to restore a common Fermi level, the change of occupations of the states in the edge channels will generate edge currents as described in Section 1.3.1, and a simple calculation shows that these will add up so as to cancel the surface sheet current (see Ex. 6.12). In other words, the overall magnetic dipole moment of the cylinder will vanish. In the case of an applied B_z, the vertical current flow on the side surface shown in Fig. 6.7(b) will pump charge from one of these edge channels to the other as the magnetic field is applied. As a result, the pumped charge will accumulate in these "wires," giving the sample a net electric dipole moment. Again, however, if the two circular wires are equilibrated by connecting them, this pumped charge will relax away and the object as a whole will no longer have any electric dipole moment. In either case, then, the edge channels should *not* be equilibrated if one wants to use this open-cylinder geometry as a ME coupler.

The Chern wrapper discussed here requires us to have at hand a QAH insulator, and to date the only known realizations have been at sub-Kelvin temperatures. If we want to build a ME coupler that works on the principle of the Chern wrapper, it is more promising in the short term to use the surface properties of strong TIs. Recall that we argued in Section 6.3.3 that any surface of a strong TI should exhibit an AHC of $\pm e^2/2h$ if the surface is fully gapped. This requires breaking of TR at the surface, which can be accomplished, for example, by proximity to a ferromagnetic surrounding material. Because we have a half quantum instead of a full quantum of AHC, the ME effect is only half as strong as that of a Chern wrapper, but it would still be quite large compared to most known bulk ME materials. Such a phenomenon has been labeled the *topological magnetoelectric effect* and has received considerable attention because it offers an attractive route to the exploration of phenomena like those we are discussing here in thin-film heterostructures.[12]

One has to be careful to distinguish between two geometries, however. Figure 6.8(a) shows a sketch of a strong TI thin film surrounded by a cladding made of an

---

[12] See, for example, Section V.A of the review by Hasan and Kane (2010) as well as the articles by Nomura and Nagaosa (2011), Morimoto et al. (2015), and Wu et al. (2016).

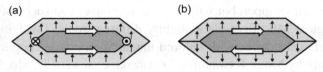

Figure 6.8 Two configurations of a TI film whose surfaces are gapped by a magnetic proximity effect. (a) Strong TI (dark shading) in contact with a surrounding insulating ferromagnet (light shading) with uniform magnetization direction (arrows). Open block arrows indicate the direction of current flow at the interface when an electric field is applied into the plane of the paper. Circles indicate the location of chiral edge channels; a circled cross or dot indicates the channel direction into or out of the paper, respectively. (b) Same but for a two-domain ferromagnet arranged so that the magnetization direction is outward on all surfaces of the TI.

insulating ferromagnet with its magnetization pointing uniformly upward. In this case the sign of the TR-symmetry breaking imparted by the interface proximity effect is the same for the top and bottom interfaces in an absolute sense; thus, for example, the surface AHC is positive when defined relative to the $+\hat{z}$ direction at both the top and bottom interfaces. That is, each interface carries an AHC of $+e^2/2h$, and the thin-film structure as a whole behaves like a QAH layer with Chern number of $+1$. But if so, it should have a chiral edge channel running around its perimeter. How does this come about?

The picture becomes clearer if we view the situation from the perspective of the TI material and take note of whether any changes occur in the sign of the half-quantized surface AHC as defined relative to an *outward-directed* unit normal. If so, these are the locations where chiral edge channels must be located. In fact, we do find two such locations at the left and right edges of the configuration in Fig. 6.8(a), corresponding to edge channels into and out of the paper.

In other words, we have built a QAH system out of a TI cladded with an insulating ferromagnet.[13] But if our goal is still to create a ME coupler, we can do so with a slightly different configuration of ferromagnetic cladding, as shown in Fig. 6.8(b). This time we arrange matters so that the magnetization, and thus the sign of the surface AHC, is consistent when defined relative to an outward-directed unit normal. To do this we need two magnetic domains with a tail-to-tail domain wall that meets the edge of the TI material as shown. Now there are no edge channels present, and the interface between the TI and the ferromagnet acts

---

[13] For a TI with inversion symmetry, a simpler way to arrive at this kind of configuration is just to apply a strong external magnetic field, since inversion symmetry enforces that θ remains quantized to π. Observations of Faraday and Kerr rotation at terahertz frequencies made by Wu et al. (2016) on high-quality Bi_2Se_3 thin films has confirmed a quantized state of this kind, with the observed optical rotation agreeing with the quantized prediction within 1%.

just like our Chern wrapper, but with half the response. The successful realization of this topological ME effect via a slightly different route, involving the growth of thin-film heterostructures with tailored magnetic interface couplings, has been reported recently by Mogi, Kawamura, Tsukazaki, et al. (2017); Mogi, Kawamura, Yoshimi, et al. (2017); Wang et al. (2015); and Xiao et al. (2018).

As more practical 2D and 3D QAH materials become available, however, we may begin to think about structures that could attain enormous ME couplings. For example, consider what would happen if we arranged for 100 Chern wrappers like those in Figs. 6.6 and 6.7 to be concentrically wrapped around a cylindrical glass rod. The result would be a rod that acts as if it has a ME coupling of $\alpha = 100e^2/hc$, which is competitive with or superior to the best known ME materials today. A still more impressive feat would be building a ME coupler out of a 3D QAH material, if one is available, as shown in Fig. 6.9. Our coupler is constructed from six panels of some 3D Chern insulator arranged around an inert core, with the Chern vector (C_x, C_y, C_z) (see Section 5.1.6) of each panel oriented to point inward [i.e., $(1, 0, 0)$ on the left panel, and so on]. A cross-sectional view showing the Chern vectors appears in Fig. 6.9(b). The joints between the panels are beveled and smoothly joined so that no edge channels occur; technically speaking, this requires the curl of the Chern-vector field to vanish, so we design these joints to occur at 45° angles.

If each of these panels has a thickness of N unit cells, it acts like a QAH system with a Chern number of N, and the structure as a whole behaves like a material with a ME coupling of $\alpha = Ne^2/hc$. As a reminder, even for small N, this performance is comparable to that of the best known ME materials. Since $e^2/hc = \mathfrak{a}/2\pi$ in our Gaussian units (where $\mathfrak{a} \simeq 1/137$ is the fine structure constant), a thickness of just $N = 10^3$ unit cells would be enough to reach the regime of $\alpha \simeq 1$ where Maxwell's electric and magnetic equations begin to become strongly coupled. There appears

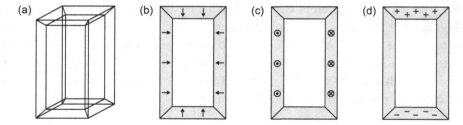

Figure 6.9 (a) Wire-frame view of an electromagnetic coupler assembled from six beveled panels of a 3D Chern insulator surrounding an empty core. (b) Cross-section view; the arrows indicate the direction of the Chern vector in each panel (shaded). (c) Current response to an upward-directed electric field. (d) Charge response to an upward-directed magnetic field.

to be no barrier, in principle, to the fabrication of a system with $N = 10^6$ or more, which would correspond to a truly enormous ME coupling of about 10^3 in natural units.

The practical physical behavior of a ME coupler of this kind is not immediately obvious. Because of the strong cross-coupling, any theory would certainly have to be solved self-consistently for both electric and magnetic fields from the outset. Insofar as the solution involves magnetic fields parallel to the Chern vector in one of the panels, an enormous volume charge density of Středa origin would develop in its interior, probably leading to dielectric breakdown. Enormous local currents would also be flowing, but since these would be dissipationless, they might be less problematic. At the same time, piezoelectric and piezomagnetic effects might become severe, leading to structural failure. All of these are interesting questions that have received little attention to date, but they will become more pressing if and when 3D QAH insulators become an experimental reality.

Exercises

6.10 Imagine we have a solid spherical shell extending from inner radius a to outer radius b, with $\alpha = \alpha_a$ at $r = a$, $\alpha = \alpha_b$ at $r = b$, and varying linearly as a function of radius in between. (Of course, $\alpha = 0$ for $r < a$ and $r > b$.) Describe the configuration of volume and surface charges, or volume and surface currents, that result if a uniform electric field is applied, or if a uniform magnetic field is applied.

6.11 A material with uniform axion coupling α fills the half-space $z < 0$. By definition, a magnetic monopole of strength Q_m generates a magnetic field $\mu_0 Q_m / 4\pi r^2$.

    (a) A point charge Q_e is placed at a height h above the surface. Show that the magnetic field configuration above the surface is that of an image magnetic monopole a distance h below the surface.

    (b) A point magnetic monopole Q_m is placed at a height h above the surface. Show that the configuration of electric fields above the surface is that of an image electric charge a distance h below the surface.

6.12 Justify the statement made on p. 312 that for the open Chern-wrapped cylinder in applied $\mathcal{E}_z$ discussed there, the net magnetic moment will vanish if the edge channels are equilibrated.

Appendix A

Fourier Transform Conventions

Some basic formulas for carrying out Fourier transforms between real and reciprocal space are recorded here. These are given in three dimensions (3D), but the generalizations to 2D and 1D are straightforward.

A.1 3D Continuous Fourier Transforms

$(\mathbf{r} \in \mathcal{R}^3, \mathbf{q} \in \mathcal{R}^3)$

Fourier transform pair:

$$f(\mathbf{r}) = \int e^{i\mathbf{q}\cdot\mathbf{r}} f(\mathbf{q})\, d^3q \qquad (A.1)$$

$\updownarrow$ FT

$$f(\mathbf{q}) = \frac{1}{(2\pi)^3} \int e^{-i\mathbf{q}\cdot\mathbf{r}} f(\mathbf{r})\, d^3r . \qquad (A.2)$$

Orthogonality relations:

$$\frac{1}{(2\pi)^3} \int e^{i\mathbf{q}\cdot\mathbf{r}}\, d^3q = \delta^3(\mathbf{r}) , \qquad (A.3)$$

$$\frac{1}{(2\pi)^3} \int e^{i\mathbf{q}\cdot\mathbf{r}}\, d^3r = \delta^3(\mathbf{q}) . \qquad (A.4)$$

Note that the conventions used in Eqs. (A.1) and (A.2) are not unique. Alternatives can be arrived at by (1) flipping the signs in both exponents and/or (2) modifying the prefactors of these two equations, keeping the product equal to $1/(2\pi)^3$. In any case, the orthogonality relations remain unchanged.

A.2 3D Continuous Real Space and Discrete Reciprocal Space

$(\mathbf{r} \in V_{\text{cell}}, \mathbf{G} = m_1 \mathbf{b}_1 + m_2 \mathbf{b}_2 + m_3 \mathbf{b}_3)$

Fourier transform pair:

$$f(\mathbf{r}) = \sum_{\mathbf{G}} e^{i\mathbf{G}\cdot\mathbf{r}} f(\mathbf{G}) \tag{A.5}$$

$$\Updownarrow \text{ FT}$$

$$f(\mathbf{G}) = \frac{1}{V_{\text{cell}}} \int_{V_{\text{cell}}} e^{-i\mathbf{G}\cdot\mathbf{r}} f(\mathbf{r}) \, d^3r. \tag{A.6}$$

Orthogonality relations:

$$\frac{1}{V_{\text{cell}}} \int_{V_{\text{cell}}} e^{i\mathbf{G}\cdot\mathbf{r}} \, d^3r = \delta_{\mathbf{G},0}, \tag{A.7}$$

$$\frac{1}{V_{\text{cell}}} \sum_{\mathbf{G}} e^{i\mathbf{G}\cdot\mathbf{r}} = \sum_{\mathbf{R}} \delta^3(\mathbf{r} - \mathbf{R}). \tag{A.8}$$

If $\mathbf{r}$ is known to lie in the home unit cell, the right-hand side of Eq. (A.8) can be written as just $\delta^3(\mathbf{r})$. Once again, alternative versions of Eqs. (A.5) and (A.6) can be arrived at by flipping signs and/or by modifying the prefactors so as to keep their product equal to $1/V_{\text{cell}}$.

A.3 3D Continuous Reciprocal Space and Discrete Real Space

$(\mathbf{q} \in \text{BZ}, \mathbf{R} = n_1 \mathbf{a}_1 + n_2 \mathbf{a}_2 + n_3 \mathbf{a}_3)$

Fourier transform pair:

$$f(\mathbf{q}) = \sum_{\mathbf{R}} e^{i\mathbf{q}\cdot\mathbf{R}} f(\mathbf{R}) \tag{A.9}$$

$$\Updownarrow \text{ FT}$$

$$f(\mathbf{R}) = \frac{V_{\text{cell}}}{(2\pi)^3} \int_{\text{BZ}} e^{-i\mathbf{q}\cdot\mathbf{R}} f(\mathbf{q}) \, d^3q. \tag{A.10}$$

Orthogonality relations:

$$\frac{V_{\text{cell}}}{(2\pi)^3} \int_{\text{BZ}} e^{i\mathbf{q}\cdot\mathbf{R}} \, d^3q = \delta_{\mathbf{R},0}, \tag{A.11}$$

$$\frac{V_{\text{cell}}}{(2\pi)^3} \sum_{\mathbf{R}} e^{i\mathbf{q}\cdot\mathbf{R}} = \sum_{\mathbf{G}} \delta^3(\mathbf{q} - \mathbf{G}). \tag{A.12}$$

If **q** is known to lie in the first Brillouin zone, the right-hand side of Eq. (A.12) can be written as just $\delta^3(\mathbf{q})$. Once again, alternative versions of Eqs. (A.5) and (A.6) can be arrived at by flipping signs and/or by modifying the prefactors so as to keep their product equal to $1/V_{BZ} = V_{cell}/(2\pi)^3$.

A.4 Orthogonality Relation for Bloch Functions

A proof of Eq. (3.85) is given in this section. We consider any two functions of Bloch form, $|\psi_\mathbf{k}\rangle$ and $|\chi_{\mathbf{k}'}\rangle$, normalized according to Eq. (3.82), whose respective periodic parts $|u_\mathbf{k}\rangle$ and $|v_{\mathbf{k}'}\rangle$ are normalized according to Eq. (3.81). From Eq. (3.83), we have that

$$
\begin{aligned}
\langle \psi_\mathbf{k} | \chi_{\mathbf{k}'} \rangle &= \int_{\text{all space}} e^{-i(\mathbf{k}-\mathbf{k}')\cdot\mathbf{r}} u_\mathbf{k}^*(\mathbf{r}) v_{\mathbf{k}'}(\mathbf{r})\, d^3r \\
&= \sum_\mathbf{R} \int_{\text{cell}} e^{-i(\mathbf{k}-\mathbf{k}')\cdot(\mathbf{R}+\mathbf{r})} u_\mathbf{k}^*(\mathbf{r}+\mathbf{R}) v_{\mathbf{k}'}(\mathbf{r}+\mathbf{R})\, d^3r \\
&= \sum_\mathbf{R} \int_{\text{cell}} e^{-i(\mathbf{k}-\mathbf{k}')\cdot(\mathbf{R}+\mathbf{r})} u_\mathbf{k}^*(\mathbf{r}) v_{\mathbf{k}'}(\mathbf{r})\, d^3r \\
&= \left(\sum_\mathbf{R} e^{-i(\mathbf{k}-\mathbf{k}')\cdot\mathbf{R}} \right) \int_{\text{cell}} e^{-i(\mathbf{k}-\mathbf{k}')\cdot\mathbf{r}} u_\mathbf{k}^*(\mathbf{r}) v_{\mathbf{k}'}(\mathbf{r})\, d^3r \\
&= \frac{(2\pi)^3}{V_{cell}} \delta^3(\mathbf{k}-\mathbf{k}') \int_{\text{cell}} u_\mathbf{k}^*(\mathbf{r}) v_\mathbf{k}(\mathbf{r})\, d^3r \\
&= \frac{(2\pi)^3}{V_{cell}} \delta^3(\mathbf{k}-\mathbf{k}')\, \langle u_\mathbf{k} | v_\mathbf{k} \rangle
\end{aligned}
\tag{A.13}
$$

as claimed.

Appendix B

Optimal Alignment and the Singular Value Decomposition

It sometimes happens that we are given two orthonormal sets of J ket vectors $|u_n\rangle$ and $|v_n\rangle$ $(n = 1,\ldots,J)$, and we would like to rotate one set of states to be in optimal alignment with the other. We consider a case where the subspaces $\mathcal{P}_u = \sum |u_n\rangle\langle u_n|$ and $\mathcal{P}_v = \sum |v_n\rangle\langle v_n|$ spanned by these two sets of vectors are similar but not identical. Then we want to find a $J\times J$ unitary matrix U_{mn} that can be applied to the $|v_n\rangle$ such that the rotated states

$$|\tilde{v}_n\rangle = \sum_m^J U_{mn}|v_m\rangle \tag{B.1}$$

are as similar to the $|u_n\rangle$ as possible, in both phase and character. In the case that $J = 1$, the solution is to rotate the phase via $|\tilde{v}\rangle = e^{-i\varphi}|v\rangle$, where $\varphi = \mathrm{Im}\ln\langle u|v\rangle$; then the inner product $\langle u|\tilde{v}\rangle$ will be real and positive, and not much less than unity if the two states are not very different. In the multistate case, if the spanned subspaces are identical, $\mathcal{P}_u = \mathcal{P}_v$, the solution is obviously just to set $|v_n\rangle = |u_n\rangle$. We seek a general solution that applies beyond these special cases.

The optimal alignment will occur when the matrix

$$\tilde{M}_{mn} = \langle u_m|\tilde{v}_n\rangle \tag{B.2}$$

is as close to the unit matrix as possible, in a sense to be clarified shortly. To evaluate the original degree of alignment, we compute the overlap matrix

$$M_{mn} = \langle u_m|v_n\rangle. \tag{B.3}$$

We then subject M to a standard mathematical operation known as the *singular value decomposition* (SVD):

$$M = V\Sigma W^\dagger \tag{B.4}$$

where V and W are $J \times J$ unitary matrices and $\Sigma_{mn} = s_n\delta_{mn}$ is a diagonal matrix whose elements, the "singular values" s_n, are all real and positive. The

319

SVD is a standard operation that is implemented in virtually all linear algebra software packages.

One way to think about the meaning of the SVD is to imagine rotating each set of states separately according to the following equations:

$$|u'_n\rangle = \sum_m V_{mn}|u_m\rangle, \tag{B.5}$$

$$|v'_n\rangle = \sum_m W_{mn}|v_m\rangle. \tag{B.6}$$

This yields new states satisfying $\langle u'_m|v'_n\rangle = s_n\delta_{mn}$, which are therefore optimally aligned. Essentially, the extent to which the singular values s_n appearing in Σ are less than unity is a measure of the difference between the subspaces $\mathcal{P}_u$ and $\mathcal{P}_v$. Now if we were to apply a common unitary rotation to both sets of states, they will still be optimally aligned with each other. Since the $|u_n\rangle$ were given and should not have been rotated, the trick is to apply $V^\dagger$ to both sets of states given previously. This operation returns the $|u'_n\rangle$ back to the original $|u_n\rangle$, while the $|v'_n\rangle$ now become

$$|\tilde{v}_n\rangle = \sum_m (WV^\dagger)_{mn}|v_m\rangle. \tag{B.7}$$

These are the desired states that are in optimal alignment with the $|u_n\rangle$.

Another way to think about this procedure is to start from Eq. (B.4) and construct the unitary matrix

$$\mathcal{M} = VW^\dagger, \tag{B.8}$$

which can be regarded as the best unitary approximation to M – in other words, the unitary matrix that tells us approximately how the states $|u_n\rangle$ got rotated in going to the $|v_n\rangle$. To undo this rotation, we apply the unitary matrix $\mathcal{M}^\dagger$ to the $|v_n\rangle$, just as in Eq. (B.7), to obtain the optimally aligned states.

A few lines of algebra starting from Eq. (B.2) show that the matrix $\tilde{M}$ associated with the optimally rotated states is

$$\tilde{M}_{mn} = V\Sigma V^\dagger, \tag{B.9}$$

which is a Hermitian matrix whose eigenvalues are again the same singular values s_n. Insofar as all of these values are close to unity, the matrix $\tilde{M}$ is close to the unit matrix. We can thus say that the unitary rotation in Eq. (B.1) that aligns the states as closely as possible is also the one that makes the overlap matrix of Eq. (B.2) be Hermitian and positive definite (that is, with all positive eigenvalues).

Note that this optimal alignment procedure fails if the overlap matrix M_{mn} is singular, such that one or more of the singular values s_n are zero. In this case there is a subspace of vectors in $\mathcal{P}_v$ that have no projection at all in $\mathcal{P}_u$, so it is hopeless to try to align them.

Appendix C

Gauge Transformation of the Chern–Simons Axion Coupling

In Section 6.3, we derived Eq. (6.43) for the Chern–Simons axion coupling in the form

$$\theta_{CS} = -\frac{1}{4\pi} \int_{BZ} \varepsilon_{\mu\nu\sigma} \, \text{Tr} \left[A_\mu \partial_\nu A_\sigma - \frac{2i}{3} A_\mu A_\nu A_\sigma \right] d^3k \qquad \text{(C.1)}$$

where $\varepsilon_{\mu\nu\sigma}$ is the third-rank antisymmetric tensor, A_μ is the $J \times J$ Berry connection matrix in Cartesian direction μ as in Eq. (3.133) for an insulator with J occupied bands, and there is an implied sum over Cartesian indices. In the remainder of this appendix, we drop the 'CS' subscript from θ_{CS} for conciseness of notation, but it is understood that θ always refers to the Chern–Simons contribution.

C.1 General Gauge Transformation

Here we derive the dependence of θ on a general multiband gauge transformation as given by Eq. (3.107) or Eq. (3.140), which we write here as

$$|u_{nk}\rangle \longrightarrow \sum_m U_{mn} |u_{mk}\rangle \qquad \text{(C.2)}$$

such that, according to Eq. (3.143),

$$A_\mu \longrightarrow U^\dagger A_\mu U + iU^\dagger \partial_\mu U . \qquad \text{(C.3)}$$

This will take

$$\theta \longrightarrow \theta + \Delta\theta. \qquad \text{(C.4)}$$

Our job is to compute $\Delta\theta$.

Before proceeding, it is useful to take note of a set of rules that apply to any three Cartesian matrices A_μ, B_μ, and C_μ that might appear in an expression like

321

Eq. (C.1). Using the antisymmetry of the indices together with the cyclic properties of the trace, we have that

$$\varepsilon_{\mu\nu\sigma}\mathrm{Tr}\,[A_\mu B_\nu C_\sigma] = \varepsilon_{\mu\nu\sigma}\mathrm{Tr}\,[B_\mu C_\nu A_\sigma] = \varepsilon_{\mu\nu\sigma}\mathrm{Tr}\,[C_\mu A_\nu B_\sigma]. \qquad (C.5)$$

We adopt a notation in which this can be rewritten as

$$A_\mu B_\nu C_\sigma \doteq B_\mu C_\nu A_\sigma \doteq C_\mu A_\nu B_\sigma \qquad (C.6)$$

where the symbol '$\doteq$' indicates that the two sides are not necessarily equal per se, but rather make equal contributions when inserted into the brackets in Eq. (C.1). The same notation can be used when two terms make equal contributions after an integration by parts, so that, for example,

$$A_\mu \partial_\nu B_\sigma \doteq (\partial_\mu A_\nu)B_\sigma. \qquad (C.7)$$

There is no overall minus sign in this equation because the sign change from the integration by parts is canceled by another sign change from the reordering of indices.[1]

It is convenient to define

$$W_\mu = i(\partial_\mu U)U^\dagger \qquad (C.8)$$

so that

$$A_\mu \longrightarrow U^\dagger(A_\mu + W_\mu)U. \qquad (C.9)$$

Since $\partial_\mu(UU^\dagger) = 0$, it also follows that $W_\mu = -iU(\partial_\mu U^\dagger)$, so that W_μ is Hermitian and

$$\partial_\mu U = -iW_\mu U, \qquad (C.10a)$$
$$\partial_\mu U^\dagger = iU^\dagger W_\mu. \qquad (C.10b)$$

We will also need $\partial_\mu W_\nu = i(\partial_\mu \partial_\nu U)U^\dagger + i(\partial_\nu U)(\partial_\mu U^\dagger)$. When inserted in Eq. (C.1) the first term vanishes under the $\varepsilon_{\mu\nu\sigma}$ by the equality of mixed partials, and it follows that

$$\partial_\mu W_\nu \doteq -iW_\mu W_\nu. \qquad (C.11)$$

Returning to Eq. (C.1), the first term transforms as

$$\mathrm{Tr}\,[A_\mu \partial_\nu A_\sigma] \longrightarrow \mathrm{Tr}\left[U^\dagger(A_\mu + W_\mu)U\partial_\nu\left(U^\dagger(A_\sigma + W_\sigma)U\right)\right]. \qquad (C.12)$$

---

[1] In eliminating the surface term in the integration by parts, we have implicitly assumed a smooth gauge, such that A_μ is smooth and periodic in the 3D Brillouin zone. For this reason the present analysis does not apply to 3D Chern insulators (see Section 5.1.6), since there is a topological obstruction to the construction of a smooth gauge in that case.

Using the cyclic property of the trace, the initial $U^\dagger$ can be moved to the end, and the quantity following $(A_\mu + W_\mu)$ becomes

$$U\partial_\nu(...)U^\dagger \doteq iW_\nu(A_\sigma + W_\sigma) + \partial_\nu(A_\sigma + W_\sigma) - i(A_\sigma + W_\sigma)W_\nu$$

$$\doteq \partial_\nu A_\sigma + iA_\nu W_\sigma + iW_\nu A_\sigma + iW_\nu W_\sigma \tag{C.13}$$

with the help of Eqs. (C.10) and (C.11).

Defining

$$\mathcal{F}_{\mu\nu\sigma}[\dots] = -\frac{1}{4\pi}\int_{BZ}\varepsilon_{\mu\nu\sigma}\,\text{Tr}\,[\dots]\,d^3k, \tag{C.14}$$

the change in θ arising from the first term in Eq. (C.1) is

$$\Delta\theta_1 = \mathcal{F}_{\mu\nu\sigma}[W_\mu\partial_\nu A_\sigma + (A_\mu + W_\mu)(iA_\nu W_\sigma + iW_\nu A_\sigma + iW_\nu W_\sigma)]. \tag{C.15}$$

Equations (C.7) and (C.11), when applied to the first term, yield $W_\mu\partial_\nu A_\sigma \doteq (\partial_\mu W_\nu)A_\sigma \doteq -iW_\mu W_\nu A_\sigma$. We then use Eq. (C.6) to move the A's to the front and combine terms to get

$$\Delta\theta_1 = \mathcal{F}_{\mu\nu\sigma}[2iA_\mu A_\nu W_\sigma + 2iA_\mu W_\nu W_\sigma + iW_\mu W_\nu W_\sigma]. \tag{C.16}$$

The same strategy applied to the second term in Eq. (C.1) gives

$$\Delta\theta_2 = \frac{-2i}{3}\mathcal{F}_{\mu\nu\sigma}[3A_\mu A_\nu W_\sigma + 3A_\mu W_\nu W_\sigma + W_\mu W_\nu W_\sigma]. \tag{C.17}$$

Combining these, we find a miraculous cancellation of terms leading to

$$\Delta\theta = \frac{i}{3}\mathcal{F}_{\mu\nu\sigma}[W_\mu W_\nu W_\sigma]$$

$$= \frac{-i}{12}\int_{BZ}\varepsilon_{\mu\nu\sigma}\,\text{Tr}\,[W_\mu W_\nu W_\sigma]\,d^3k. \tag{C.18}$$

This result implies that the change in axion coupling θ under a gauge transformation is completely independent of the underlying Berry connection or curvature. Instead, it depends only on the unitary transformation itself.

Restoring the explicit notation in terms of the gauge transformation field $U(\mathbf{k})$, the result is

$$\Delta\theta = \frac{1}{12\pi}\int_{BZ}\varepsilon_{\mu\nu\sigma}\,\text{Tr}\,[(\partial_\mu U)U^\dagger(\partial_\nu U)U^\dagger(\partial_\sigma U)U^\dagger]\,d^3k. \tag{C.19}$$

Some discussion of this result in the field-theory context can be found in Section III.2 of Coleman (1979) and on pp. 429–430 of Deser et al. (1982). Equation (C.19) can be compared with the corresponding formula for the Berry phase in 1D, Eq. (6.44), which can be written in the same language as

$$\Delta\phi = i \int_{\text{BZ}} \text{Tr}\left[(\partial_\mu U)U^\dagger\right] dk_\mu, \tag{C.20}$$

which is the same as Eq. (6.47).

C.2 Single-Band Case

For the single-band case ($J = 1$), all Hermitian matrices (A_μ, U, and so on) become real scalars. Then A_μ, A_ν, and A_σ obviously commute, so that the second term in Eq. (C.1) vanishes and

$$\theta = -\frac{1}{4\pi} \int_{\text{BZ}} \varepsilon_{\mu\nu\sigma} A_\mu (\partial_\nu A_\sigma) \, d^3k. \tag{C.21}$$

The unitary gauge transformation $U(\mathbf{k})$ is just a phase $e^{-i\beta(\mathbf{k})}$, and Eq. (C.19) reduces to

$$\Delta\theta = \frac{i}{12\pi} \int_{\text{BZ}} \varepsilon_{\mu\nu\sigma} (\partial_\mu\beta)(\partial_\nu\beta)(\partial_\sigma\beta) \, d^3k$$

$$= 0, \tag{C.22}$$

using once more the fact that scalars commute. In the single-band case, then, θ is uniquely defined with no gauge ambiguity whatsoever.

Before returning to the general multiband case, it is important to realize that, unlike the Berry phase, the θ coupling is *not* band-additive. That is, if the occupied subspace is composed of J isolated bands (i.e., each being free of degeneracies with the next lower or higher band), then one could imagine computing θ_n separately for each band n using Eq. (C.21) and summing them:

$$\theta_{\text{band sum}} = -\frac{1}{4\pi} \sum_n \int_{\text{BZ}} \varepsilon_{\mu\nu\sigma} A_{nn,\mu}(\partial_\nu A_{nn,\sigma}) \, d^3k. \tag{C.23}$$

Nevertheless, this would *not* yield the correct overall θ as given by Eq. (C.1). The band trace in Eq. (C.1) includes terms like $A_{mn,x}(\partial_y A_{nm,z})$ for $m \neq n$, and $A_{lm,x}A_{mn,y}A_{nl,z}$ for $l \neq m \neq n$, that are not included in the sum of single-band quantities. Thus, the multiband gauge invariance does not simply follow from the the results of the previous section, even in the favorable case of isolated bands.

C.3 Infinitesimal and Progressive Gauge Transformations

We now return to the general multiband case. Near the end of Section 3.1.2 we introduced a distinction between "progressive" and "radical" gauge transformations, where a progressive gauge transformation is one that can be reached by a smooth deformation of $U(\mathbf{k})$ along some path in the space of all transformations. We now wish to show that such a smooth deformation cannot result in any change of θ.

Consider first an infinitesimal gauge transformation

$$U(\mathbf{k}) = e^{-i\epsilon B(\mathbf{k})} \tag{C.24}$$

for some infinitesimal ϵ and Hermitian $B(\mathbf{k})$ (U and B are $J \times J$ matrices). At first order in ϵ, this is $I - i\epsilon B$ and Eq. (C.8) gives $W_\mu = \epsilon \partial_\mu B$. Using Eq. (C.18), it follows that $\Delta\theta$ is proportional to ϵ^3, so that

$$\frac{d\theta}{d\epsilon} = \lim_{\Delta\epsilon \to 0} \frac{\Delta\theta}{\Delta\epsilon} = 0. \tag{C.25}$$

Thus, a first-order change in the gauge makes no first-order change in θ.

This immediately implies that θ remains invariant under any progressive gauge transformation, since we should be able to represent the total change in θ resulting from the deformation of the gauge $U_\lambda(\mathbf{k})$ along some path parametrized by λ as

$$\Delta\theta = \int_{\lambda_i}^{\lambda_f} \frac{d\theta}{d\lambda} d\lambda. \tag{C.26}$$

But Eq. (C.25) states that the integrand vanishes, demonstrating the claim.

C.4 Radical Gauge Transformations

Can this result be generalized to cover all gauge transformations, including radical ones? To see that it cannot, it is enough to demonstrate the existence of a gauge transformation that does change θ. For this purpose it is sufficient to work in the two-band case ($J = 2$) and consider the gauge transformation specified by Eq. (6.52), which we now write in polar coordinates, $\mathbf{q} \to (q, \vartheta, \varphi)$, as

$$U(q, \vartheta, \varphi) = \begin{cases} -\cos q\, I + i \sin q\, \sigma_q, & q \leq \pi \\ I, & q \geq \pi \end{cases} \tag{C.27}$$

where the Pauli matrices are defined as $\sigma_q = \hat{q} \cdot \boldsymbol{\sigma}$, $\sigma_\vartheta = \hat{\vartheta} \cdot \boldsymbol{\sigma}$, and $\sigma_\varphi = \hat{\varphi} \cdot \boldsymbol{\sigma}$. Using $\partial_q \sigma_q = 0$, $\partial_\vartheta \sigma_q = \sigma_\vartheta$, and $\partial_\varphi \sigma_q = \sin\vartheta\, \sigma_\varphi$ together with the algebra of the Pauli matrices ($\sigma_q \sigma_\vartheta = i\sigma_\varphi$, and so on), it is straightforward to derive that

$$W_q = \sigma_q,$$
$$W_\vartheta = \sin q \cos q \, \sigma_\vartheta + \sin^2 q \, \sigma_\varphi,$$
$$W_\varphi = \sin \vartheta \, (- \sin^2 q \, \sigma_\vartheta + \sin q \cos q \, \sigma_\varphi). \tag{C.28}$$

Now we want to evaluate Eq. (C.18), which takes the form of an integral over Cartesian **k**-space, but we can integrate over our (q, ϑ, φ) coordinates instead. There is no need to introduce the determinant of the Jacobean matrix relating the two coordinate systems because the derivatives with respect to the coordinates, as seen explicitly in Eq. (C.19), provide a canceling factor. Then Eq. (C.18) becomes

$$\Delta\theta = \frac{-i}{4\pi} \int_0^\pi dq \int_0^\pi d\vartheta \int_0^{2\pi} d\varphi \, \mathrm{Tr} \, [W_q W_\vartheta W_\varphi - W_q W_\varphi W_\vartheta]$$
$$= \int_0^\pi dq \int_0^\pi d\vartheta \, \mathrm{Im} \, \mathrm{Tr} \, [W_q W_\vartheta W_\varphi]. \tag{C.29}$$

But $\mathrm{Im} \, \mathrm{Tr} \, [\sigma_q \sigma_\vartheta \sigma_\varphi] = 2$, $\mathrm{Im} \, \mathrm{Tr} \, [\sigma_q \sigma_\varphi \sigma_\vartheta] = -2$, and other combinations give zero, so that

$$\mathrm{Im} \, \mathrm{Tr} \, [W_q W_\vartheta W_\varphi] = 2 \sin \vartheta \sin^2 q. \tag{C.30}$$

The final result is

$$\Delta\theta = 2 \left(\int_0^\pi \sin \vartheta \, d\vartheta \right) \left(\int_0^\pi \sin^2 q \, dq \right)$$
$$= 2\pi. \tag{C.31}$$

This demonstrates that the axion coupling θ is shifted by 2π by this gauge transformation.

It is straightforward to show that a generalization of Eq. (C.27) in the form

$$U(q, \vartheta, \varphi) = \begin{cases} (-)^m \, e^{-imq\sigma_q}, & q \leq \pi \\ I, & q \geq \pi \end{cases} \tag{C.32}$$

for integer m leads to a change of θ by $2\pi m$. This gives us an example of each member of the topological $\mathbb{Z}$ classification of gauge transformations, and subsequent progressive gauge transformations allow us access to other members of any given class.

We have not attempted a general proof that any gauge transformation can change θ only by a multiple of 2π, but the result should now be plausible. References to a more general approach using the methods of algebraic topology are given at the end of Section 6.3.2.

Appendix D
The PYTHTB Package

Numerous examples and exercises in this book make use of the PYTHTB software package. This package, which is written in the PYTHON programming language, is designed to allow the user to construct and solve tight-binding (TB) models of the electronic structure of finite clusters and of systems that display periodicity in one or more dimensions, such as polymer chains, ribbons, slabs, and 3D crystals. It is also designed to provide convenient features for computing geometric or topological properties such as Berry phases, Berry curvatures, and Chern numbers. The code was developed beginning in 2010 and is maintained principally by Sinisa Coh and David Vanderbilt, although significant contributions have been provided by others (see http://www.physics.rutgers.edu/pythtb/about.html#History).

The examples and exercises used in this book assume the use of version 1.7.2 of PYTHTB, which is the current version at the time of this writing. PYTHTB version 1.7.2 is designed for compatibility with both PYTHON2.7 and PYTHON3.x (PYTHON2.6 and below are not recommended). If you already have one of these PYTHON versions and PIP installed on a Linux system, you should be able to install the latest version of PYTHTB just by issuing the command

```
pip install pythtb --upgrade
```

at a command prompt if you have root permission, or

```
pip install pythtb --upgrade --user
```

to install it in your home folder if you do not. If you wish to install the package without PIP or otherwise need further help with installation, see the package web pages at http://www.physics.rutgers.edu/pythtb. You will also need access to the standard NUMPY (numerical) and MATPLOTLIB (plotting) PYTHON packages; install these too if they are not already present.

The example programs, together with their printed and plotted outputs, are available at `http://www.cambridge.org/9781107157651#resources`. This is referred to as "the website" in the remainder of this appendix. All efforts will be made to keep future upgrades of PYTHTB backwardcompatible with the example programs provided here, but in case of doubt you can obtain version 1.7.2 specifically by issuing the command

```
pip install pythtb==1.7.2
```

or download and install PYTHTB 1.7.2 from the "Installation" link on the website.

Each program starts with the header lines

```
#!/usr/bin/env python
from __future__ import print_function # python3 style print
```

The first identifies the script as a PYTHON program. The second is ignored when running under PYTHON3, but is needed for compatibility with PYTHON2 so that `print` is interpreted as a function, as is standard in PYTHON3. These two lines are typically followed by some comment lines (starting with '#') that identify the program.

The next few lines import the PYTHTB module and any other modules needed for execution of the script. Typical entries include

```
from pythtb import *
import numpy as np
import matplotlib.pyplot as plt
```

The first line imports the PYTHTB module. All of the example programs also make use of the NUMPY mathematical subroutine library, which is imported on the second line in such a way that NUMPY 'func' will be called as `np.func`. This line is actually unnecessary because the import is done automatically when PYTHTB is imported, but it may be included to make the code more readable; this has been done in the first example in Appendix D.1, but not in subsequent ones. Finally, the import of `matplotlib.pyplot` on the third line provides a MATLIB-like plotting environment for those example programs that produce plotted output.[1]

PYTHTB defines two *classes*: `tb_model` and `wf_array`.[2] A command such as `my_model=tb_model`(*arguments*) defines an *instance* of the `tb_model` class, an object that carries all the relevant information about the TB model. The (*arguments*) given initially define some general characteristics of the model (dimensionality, whether it is a spinor model, and so on), but then parameters such as on-site energies and hopping amplitudes are set subsequently by making

---

[1] The precise appearance of a plot may depend on the installed version of MATPLOTLIB (version 1.5.3 was used here) and on any `matplotlibrc` configuration files that may be present.

[2] It also defines a third, w90, which provides an interface to the WANNIER90 code package (Mostofi et al., 2008, 2014). We shall not make use of this class here.

use of the set_onsite and add_hop methods. For periodic models, the next step is usually to set up a mesh of **k**-points on which the model is to be solved, with the solution typically done by calling the solve_all method. Alternatively, an array object can be defined as an instance of the wf_array class and solved by the solve_on_grid method of this class; this is especially useful when computing Berry phases and curvatures, since the wf_array class provides methods specifically for such purposes. Finally, the results are printed or plotted for inspection.

All this is best illustrated by browsing the example programs and their outputs in the sections that follow.

D.1 Water Molecule

Program h2o.py computes the eigenstates of a TB model for the water molecule as described in Section 2.2.2. The first few lines define the molecular geometry (bond length and angle) and the TB parameters (*s* and *p* site energies and *s-s* and *p-p* hoppings). The lat= line specifies three basis vectors (here Cartesian), in terms of which the atom coordinates will be given, and the orb line specifies those coordinates. The next few lines define the TB model my_model, while my_model.display prints a summary of it and my_model.solve_all solves for the eigenvalues and eigenvectors. Since the model is described by a real Hamiltonian, the imaginary parts of the eigenvectors are zero to numerical accuracy and are discarded. The last few lines print the result; np.set_printoptions is a NUMPY function that sets the format for subsequent print statement, and the eigenvalues and eigenvectors are printed at the end.

h2o.py

```
#!/usr/bin/env python
from __future__ import print_function # python3 style print

# ------------------------------------------------------------
# Tight-binding model for H2O molecule
# ------------------------------------------------------------

# import the pythtb module
from pythtb import *
import numpy as np

# geometry: bond length and half bond-angle
b=1.0; angle=54.0*np.pi/180

# site energies [O(s), O(p), H(s)]
eos=-1.5; eop=-1.2; eh=-1.0

# hoppings [O(s)-H(s), O(p)-H(s)]
ts=-0.4; tp=-0.3
```

```
# define frame for defining vectors: 3D Cartesian
lat=[[1.0,0.0,0.0],[0.0,1.0,0.0],[0.0,0.0,1.0]]

# define coordinates of orbitals: O(s,px,py,pz) ; H(s) ; H(s)
orb=[ [0.,0.,0.], [0.,0.,0.], [0.,0.,0.], [0.,0.,0.],
      [b*np.cos(angle), b*np.sin(angle),0.],
      [b*np.cos(angle),-b*np.sin(angle),0.] ]

# define model
my_model=tbmodel(0,3,lat,orb)
my_model.set_onsite([eos,eop,eop,eop,eh,eh])
my_model.set_hop(ts,0,4)
my_model.set_hop(ts,0,5)
my_model.set_hop(tp*np.cos(angle),1,4)
my_model.set_hop(tp*np.cos(angle),1,5)
my_model.set_hop(tp*np.sin(angle),2,4)
my_model.set_hop(-tp*np.sin(angle),2,5)

# print model
my_model.display()

# solve model
(eval,evec)=my_model.solve_all(eig_vectors=True)

# the model is real, so OK to discard imaginary parts of eigenvectors
evec=evec.real

# optional: choose overall sign of evec according to some specified rule
# (here, we make the average oxygen p component positive)
for i in range(len(eval)):
  if sum(evec[i,1:4]) < 0:
    evec[i,:]=-evec[i,:]

# print results, setting numpy to format floats as xx.xxx
np.set_printoptions(formatter={'float': '{: 6.3f}'.format})
# print eigenvalues and real parts of eigenvectors, one to a line
print(" n    eigval    eigvec")
for n in range(6):
    print(" %2i  %7.3f  " % (n,eval[n]), evec[n,:])
```

D.2 Benzene Molecule

Program benzene.py, discussed in Section 2.2.2, computes the eigenstates for a TB model of the p_π manifold (i.e., p orbitals oriented normal to the plane of the molecule) for benzene (CH_6). The structure is much like that of h2o.py in the preceding section. See also Ex. 2.5.

benzene.py

```
#!/usr/bin/env python
from __future__ import print_function # python3 style print

# ------------------------------------------------------------
# Tight-binding model for p_z states of benzene molecule
# ------------------------------------------------------------

from pythtb import *

# set up molecular geometry
lat=[[1.0,0.0],[0.0,1.0]]              # define coordinate frame: 2D Cartesian
```

```
r=1.2                              # distance of atoms from center
orb=np.zeros((6,2),dtype=float)    # initialize array for orbital positions
for i in range(6):                 # define coordinates of orbitals
    angle=i*np.pi/3.0
    orb[i,:]= [r*np.cos(angle), r*np.sin(angle)]

# set site energy and hopping amplitude, respectively
ep=-0.4
t=-0.25

# define model
my_model=tbmodel(0,2,lat,orb)
my_model.set_onsite([ep,ep,ep,ep,ep,ep])
my_model.set_hop(t,0,1)
my_model.set_hop(t,1,2)
my_model.set_hop(t,2,3)
my_model.set_hop(t,3,4)
my_model.set_hop(t,4,5)
my_model.set_hop(t,5,0)

# print model
my_model.display()

# solve model and print results
(eval,evec)=my_model.solve_all(eig_vectors=True)

# print results, setting numpy to format floats as xx.xxx
np.set_printoptions(formatter={'float': '{: 6.3f}'.format})
# print eigenvalues and real parts of eigenvectors, one to a line
print("  n    eigval   eigvec")
for n in range(6):
    print(" %2i   %7.3f   " % (n,eval[n]), evec[n,:].real)
```

D.3 bcc Li Crystal

Program li.py computes the band structure of bcc Li based on a TB model with one *s* orbital per atomic site, as introduced in Section 2.2.4. Now the vectors lat have to be a set of primitive real-space lattice vectors that define the periodicity of the crystal, and subsequent coordinates must be given in terms of these. In particular, the neighboring orbitals specified by the calls to set_hop are given in these lattice coordinates, with the translation to Cartesian coordinates given in comments at the end of each line. Orbital locations also need to be defined in lattice coordinates, but there is only one atom in the unit cell and we place it at the origin. (The value of the lattice constant has no effect on the results to be computed here, so it is set to unity for convenience.)

The next few lines after my_model.display() specify a list of three special **k** points, or "nodes," that define the path along which the band structure will be computed and plotted. The path and label variables are lists of the node coordinates and their labels, respectively, and the k_path method takes path as input and constructs a list of **k** points tracing a path along straight-line segments between these nodes. The returned variables k_vec and k_dist are arrays containing the coordinates of the **k** points and corresponding accumulated distance

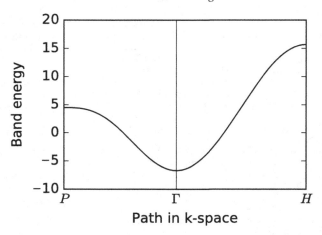

Figure D.1 Band structure of bcc Li as generated by the li.py program.

along the path for each point, while k_node gives the accumulated distance just for the nodes. These are used in the calls to the PYPLOT plotting routines to generate the band-structure plot which is output as li_band.pdf, shown in Fig. D.1. This reproduces the results shown in Fig. 2.4 in Section 2.2.3.

Note that the node positions in path must be specified in *reciprocal lattice coordinates*, as linear combinations of the primitive reciprocal lattice vectors that are dual to the real-space ones. (The same convention is used for the output array k_vec.) It is the responsibility of the user to specify path in this way, but the "k_path report" generated by the k_path function is useful for checking that this has been done correctly. For this program, for example, see the lines between k_path report begin and k_path report end in the printed output on the website.

li.py

```
#!/usr/bin/env python
from __future__ import print_function # python3 style print

# 3D model of Li on bcc lattice, with s orbitals only

from pythtb import * # import TB model class
import matplotlib.pyplot as plt

# define lattice vectors
lat=[[-0.5, 0.5, 0.5],[ 0.5,-0.5, 0.5],[ 0.5, 0.5,-0.5]]
# define coordinates of orbitals
orb=[[0.0,0.0,0.0]]

# make 3D model
my_model=tb_model(3,3,lat,orb)

# set model parameters
# lattice parameter implicitly set to a=1
Es= 4.5    # site energy
t =-1.4    # hopping parameter
```

```
# set on-site energy
my_model.set_onsite([Es])
# set hoppings along four unique bonds
# note that neighboring cell must be specified in lattice coordinates
# (the corresponding Cartesian coords are given for reference)
my_model.set_hop(t, 0, 0, [1,0,0])      # [-0.5, 0.5, 0.5] cartesian
my_model.set_hop(t, 0, 0, [0,1,0])      # [ 0.5,-0.5, 0.5] cartesian
my_model.set_hop(t, 0, 0, [0,0,1])      # [ 0.5, 0.5,-0.5] cartesian
my_model.set_hop(t, 0, 0, [1,1,1])      # [ 0.5, 0.5, 0.5] cartesian

# print tight-binding model
my_model.display()


# generate k-point path and labels
# again, specified in reciprocal lattice coordinates
k_P     = [0.25,0.25,0.25]              # [ 0.5, 0.5, 0.5] cartesian
k_Gamma = [ 0.0, 0.0, 0.0]              # [ 0.0, 0.0, 0.0] cartesian
k_H     = [-0.5, 0.5, 0.5]              # [ 1.0, 0.0, 0.0] cartesian
path=[k_P,k_Gamma,k_H]
label=(r'$P$',r'$\Gamma $',r'$H$')
(k_vec,k_dist,k_node)=my_model.k_path(path,101)

print('----------------------------------------')
print('starting calculation')
print('----------------------------------------')
print('Calculating bands...')

# solve for eigenenergies of Hamiltonian on
# the set of k-points from above
evals=my_model.solve_all(k_vec)

# plotting of band structure
print('Plotting band structure...')

# first make a figure object
fig, ax = plt.subplots(figsize=(4.,3.))

# specify horizontal axis details
ax.set_xlim([0,k_node[-1]])
ax.set_xticks(k_node)
ax.set_xticklabels(label)
for n in range(len(k_node)):
  ax.axvline(x=k_node[n], linewidth=0.5, color='k')

# plot bands
ax.plot(k_dist,evals[0],color='k')
# put title
ax.set_xlabel("Path in k-space")
ax.set_ylabel("Band energy")
# make a PDF figure of a plot
fig.tight_layout()
fig.savefig("li_bsr.pdf")

print('Done.\n')
```

D.4 Alternating Site Model

Program chain_alt.py computes the band structure of the 1D alternating site chain model of Fig. 2.6 in Section 2.2.4. The lattice constant is set to unity for convenience. The orbitals at $x = 0$ and $x = 1/2$ are assigned site energies Δ and

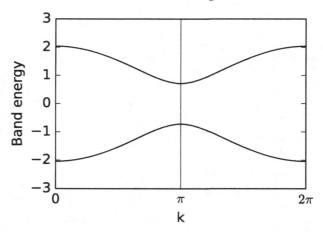

Figure D.2 Band structure resulting from the 1D alternating chain model of Fig. 2.6 as computed by `chain_alt.py`.

$-\Delta$, respectively, and the alternating hopping strengths are $t + \delta t$ and $t - \delta t$. This time a subroutine `set_model` is defined to specify the model; this will be convenient later when looping over parameters of the model. The resulting plot appears in Fig. D.2.

chain_alt.py

```
#!/usr/bin/env python
from __future__ import print_function # python3 style print

# Chain with alternating site energies and hoppings

from pythtb import *
import matplotlib.pyplot as plt

# define function to set up model for a given paramter set
def set_model(t,del_t,Delta):
  # 1D model with two orbitals per cell
  lat=[[1.0]]
  orb=[[0.0],[0.5]]
  my_model=tbmodel(1,1,lat,orb)
  # alternating site energies (let average be zero)
  my_model.set_onsite([Delta,-Delta])
  # alternating hopping strengths
  my_model.add_hop(t+del_t, 0, 1, [0])
  my_model.add_hop(t-del_t, 1, 0, [1])
  return my_model

# set reference hopping strength to unity to set energy scale
t=-1.0
# set alternation strengths
del_t=-0.3    # bond strength alternation
Delta= 0.4    # site energy alternation

# set up the model
my_model=set_model(t,del_t,Delta)
```

```
# construct the k-path
(k_vec,k_dist,k_node)=my_model.k_path('full',121)
k_lab=(r'0',r'$\pi$',r'$2\pi$')

# solve for eigenvalues at each point on the path
evals=my_model.solve_all(k_vec)

# set up the figure and specify details
fig, ax = plt.subplots(figsize=(4.,3.))
ax.set_xlim([0,k_node[-1]])
ax.set_xticks(k_node)
ax.set_xticklabels(k_lab)
ax.axvline(x=k_node[1],linewidth=0.5, color='k')
ax.set_xlabel("k")
ax.set_ylabel("Band energy")

# plot first and second bands
ax.plot(k_dist,evals[0],color='k')
ax.plot(k_dist,evals[1],color='k')

# save figure as a PDF
fig.tight_layout()
fig.savefig("chain_alt.pdf")
```

D.5 Graphene

Program graphene.py computes the band structure for the model of the π orbitals of graphene discussed in Section 2.2.4 and illustrated in Fig. 2.7(a). The conventional Wigner–Seitz Brillouin zone (BZ) is shown in Fig. 2.7(b) with high-symmetry points labeled; the my_model.k_path line of the program specifies a path from Γ to K to M and back to Γ (see the "k_path report" in the printed output available on the website). The resulting band structure plot appears as Fig. 2.8 in Section 2.2.4.

graphene.py

```
#!/usr/bin/env python
from __future__ import print_function # python3 style print

# Simple model of pi manifold of graphene

from pythtb import * # import TB model class
import matplotlib.pyplot as plt

# define lattice vectors
lat=[[1.0,0.0],[0.5,np.sqrt(3.0)/2.0]]
# define coordinates of orbitals
orb=[[1./3.,1./3.],[2./3.,2./3.]]

# make 2D tight-binding graphene model
my_model=tb_model(2,2,lat,orb)

# set model parameters
delta=0.0
t=-1.0
```

```
my_model.set_onsite([-delta,delta])
my_model.set_hop(t, 0, 1, [ 0, 0])
my_model.set_hop(t, 1, 0, [ 1, 0])
my_model.set_hop(t, 1, 0, [ 0, 1])

# print out model details
my_model.display()

# list of k-point nodes and their labels defining the path for the
#    band structure plot
path=[[0.,0.],[2./3.,1./3.],[.5,.5],[0.,0.]]
label=(r'$\Gamma $',r'$K$', r'$M$', r'$\Gamma $')

# construct the k-path
nk=121
(k_vec,k_dist,k_node)=my_model.k_path(path,nk)

# solve for eigenvalues at each point on the path
evals=my_model.solve_all(k_vec)

# generate band structure plot

fig, ax = plt.subplots(figsize=(4.,3.))
# specify horizontal axis details
ax.set_xlim([0,k_node[-1]])
ax.set_ylim([-3.4,3.4])
ax.set_xticks(k_node)
ax.set_xticklabels(label)
# add vertical lines at node positions
for n in range(len(k_node)):
    ax.axvline(x=k_node[n],linewidth=0.5, color='k')
# put titles
ax.set_xlabel("Path in k-space")
ax.set_ylabel("Band energy")

# plot first and second bands
ax.plot(k_dist,evals[0],color='k')
ax.plot(k_dist,evals[1],color='k')

# save figure as a PDF
fig.tight_layout()
fig.savefig("graphene.pdf")
```

D.6 Trimer Molecule

Program `trimer.py` computes the eigenfunctions of the trimer molecule discussed in Section 3.2.4. In addition, it produces plots of Berry phases and Berry curvatures of the ground state resulting from variation of the parameters.

A `set_model` subroutine has been defined so as to return an instance of the model for a given set of the parameters defined in Eq. (3.43), and another subroutine `get_evecs` calls `set_model` and returns the computed eigenvectors. Predefining these functions simplifies the process of looping over the values of φ and α as needed later. This time one of the hoppings is generally complex, so TR symmetry is broken and the eigenvectors are generally complex as well.

The plotted output of this program appears as Fig. 3.11 in the main text. For pedagogical purposes the Berry phase is computed in several ways. First, for a single given value of α, the program computes the discrete Berry phase explicitly according to Eq. (3.1) as φ runs from 0 to 2π on a dense mesh (increments of $\pi/30$) at a fixed value of α. The product of the inner products appearing in Eq. (3.1) is accumulated in the variable prod and the Berry phase is computed explicitly by taking the imaginary part of the log using the NUMPY function angle.

Second, on a coarse mesh of α values (in increments of $\pi/6$), the eigenvectors on a dense mesh of φ values are stored into a 1D PYTHTB wave-function array object evec_array that is initialized as an instance of the PYTHTB wf_array class. The Berry phase is computed by the berry_phase method associated with this class, and the results are plotted as the black dots in Fig. 3.11(a). The same procedure is repeated to compute the Berry phases in the α direction on a coarse mesh of φ values and plot them as the dots in Fig. 3.11(b).

Next, a 2D wf_array object (also named evec_array) is filled with eigenvectors on a fine mesh in both the φ and α directions. The berry_phase method is then used to compute the Berry phases first in the φ direction and then in the α direction, and these (together with values shifted by integer multiples of 2π) are plotted as the continuous lines in Fig. 3.11(a–b). By default, the berry_phase procedure enforces a smooth evolution of the Berry phase, so we do not see the same kind of discontinuities that appeared when using the previous method.

Finally, the berry_flux method is applied to the same 2D evec_array array to compute the Berry flux through each of the 60×60 plaquettes, and the results are divided by parametric area to obtain the Berry curvature and used to generate the contour plot in Fig. 3.11(c). The PYPLOT subplots method is used to combine the three panels into a single PDF output file.

trimer.py

```
#!/usr/bin/env python
from __future__ import print_function # python3 style print

# Tight-binding model for trimer with magnetic flux

from pythtb import *
import matplotlib.pyplot as plt

# ------------------------------------------------------------------
# define function to set up model for given (t0,s,phi,alpha)
# ------------------------------------------------------------------
def set_model(t0,s,phi,alpha):

    # coordinate space is 2D
    lat=[[1.0,0.0],[0.0,1.0]]
    # finite model with three orbitals forming a triangle at unit
    # distance from the origin
    sqr32=np.sqrt(3.)/2.
```

```
orb=np.zeros((3,2),dtype=float)
orb[0,:]=[0.,1.]              # orbital at top vertex
orb[1,:]=[-sqr32,-0.5]        # orbital at lower left
orb[2,:]=[ sqr32,-0.5]        # orbital at lower right

# compute hoppings [t_01, t_12, t_20]
# s is distortion amplitude; phi is "pseudorotation angle"
tpio3=2.0*np.pi/3.0
t=[ t0+s*np.cos(phi), t0+s*np.cos(phi-tpio3), t0+s*np.cos(phi-2.0*tpio3) ]

# alpha is fraction of flux quantum passing through the triangle
# magnetic flux correction, attached to third bond
t[2]=t[2]*np.exp((1.j)*alpha)

# set up model (leave site energies at zero)
my_model=tbmodel(0,2,lat,orb)
my_model.set_hop(t[0],0,1)
my_model.set_hop(t[1],1,2)
my_model.set_hop(t[2],2,0)
return(my_model)

# ----------------------------------------------------------------
# define function to return eigenvectors for given (t0,s,phi,alpha)
# ----------------------------------------------------------------
def get_evecs(t0,s,phi,alpha):
  my_model=set_model(t0,s,phi,alpha)
  (eval,evec)=my_model.solve_all(eig_vectors=True)
  return(evec)           # evec[bands,orbitals]

# ----------------------------------------------------------------
# begin regular execution
# ----------------------------------------------------------------
# for the purposes of this problem we keep t0 and s fixed
t0 =-1.0
s  =-0.4
ref_model=set_model(t0,s,0.,1.)  # reference with phi=alpha=0
ref_model.display()

# define two pi
twopi=2.*np.pi

# ----------------------------------------------------------------
# compute Berry phase for phi loop explicitly at alpha=pi/3
# ----------------------------------------------------------------
alpha=np.pi/3.
n_phi=60
psi=np.zeros((n_phi,3),dtype=complex)      # initialize wavefunction array
for i in range(n_phi):
  phi=float(i)*twopi/float(n_phi)          # 60 equal intervals
  psi[i]=get_evecs(t0,s,phi,alpha)[0]      # psi[i] is short for psi[i,:]
prod=1.+0.j                                # final [0] picks out band 0
for i in range(1,n_phi):
  prod=prod*np.vdot(psi[i-1],psi[i]) # <psi_0|psi_1>...<psi_58|psi_59>
prod=prod*np.vdot(psi[-1],psi[0])    # include <psi_59|psi_0>
berry=-np.angle(prod)                # compute Berry phase
print("Explicitly computed phi Berry phase at alpha=pi/3 is %6.3f"% berry)

# ----------------------------------------------------------------
# compute Berry phases for phi loops for several alpha values
# using pythtb wf_array() method
# ----------------------------------------------------------------
```

```
alphas=np.linspace(0.,twopi,13)      # 0 to 2pi in 12 increments
berry_phi=np.zeros_like(alphas)         # same shape and type array (empty)
print("\nBerry phases for phi loops versus alpha")
for j,alpha in enumerate(alphas):

    # let phi range from 0 to 2pi in equally spaced steps
    n_phi=61
    phit=np.linspace(0.,twopi,n_phi)

    # set up empty wavefunction array object using pythtb wf_array()
    # creates 1D array of length [n_phi], with hidden [nbands,norbs]
    evec_array=wf_array(ref_model,[n_phi])

    # run over values of phi and fill the array
    for k,phi in enumerate(phit[0:-1]):     # skip last point of loop
      evec_array[k]=get_evecs(t0,s,phi,alpha)
    evec_array[-1]=evec_array[0]    # copy first point to last point of loop

    # now compute and store the Berry phase
    berry_phi[j]=evec_array.berry_phase([0])    # [0] specifices lowest band
    print("%3d %7.3f %7.3f"% (j, alpha, berry_phi[j]))

# -------------------------------------------------------------
# compute Berry phases for alpha loops for several phi values
# using pythtb wf_array() method
# -------------------------------------------------------------
phis=np.linspace(0.,twopi,13)    # 0 to 2pi in 12 increments
berry_alpha=np.zeros_like(phis)
print("\nBerry phases for alpha loops versus phi")
for j,phi in enumerate(phis):
  n_alpha=61
  alphat=np.linspace(0.,twopi,n_alpha)
  evec_array=wf_array(ref_model,[n_alpha])
  for k,alpha in enumerate(alphat[0:-1]):
    evec_array[k]=get_evecs(t0,s,phi,alpha)
  evec_array[-1]=evec_array[0]
  berry_alpha[j]=evec_array.berry_phase([0])
  print("%3d %7.3f %7.3f"% (j, phi, berry_alpha[j]))

# -------------------------------------------------------------
# now illustrate use of wf_array() to set up 2D array
# recompute Berry phases and compute Berry curvature
# -------------------------------------------------------------
n_phi=61
n_alp=61
n_cells=(n_phi-1)*(n_alp-1)
phi=np.linspace(0.,twopi,n_phi)
alp=np.linspace(0.,twopi,n_alp)
evec_array=wf_array(ref_model,[n_phi,n_alp])  # empty 2d wavefunction array
for i in range(n_phi):
  for j in range(n_alp):
    evec_array[i,j]=get_evecs(t0,s,phi[i],alp[j])
evec_array.impose_loop(0)    # copy first to last points in each dimension
evec_array.impose_loop(1)

bp_of_alp=evec_array.berry_phase([0],0)  # compute phi Berry phases vs. alpha
bp_of_phi=evec_array.berry_phase([0],1)  # compute alpha Berry phases vs. phi

# compute 2D array of Berry fluxes for band 0
flux=evec_array.berry_flux([0])
print("\nFlux = %7.3f = 2pi * %7.3f"% (flux, flux/twopi))
```

```
curvature=evec_array.berry_flux([0],individual_phases=True)*float(n_cells)

# ----------------------------------------------------------------
# plots
# ----------------------------------------------------------------
fig,ax=plt.subplots(1,3,figsize=(10,4),gridspec_kw={'width_ratios':[1,1,2]})
(ax0,ax1,ax2)=ax

ax0.set_xlim(0.,1.)
ax0.set_ylim(-6.5,6.5)
ax0.set_xlabel(r"$\alpha/2\pi$")
ax0.set_ylabel(r"Berry phase $\phi(\alpha)$ for $\varphi$ loops")
ax0.set_title("Berry phase")
for shift in (-twopi,0.,twopi):
  ax0.plot(alp/twopi,bp_of_alp+shift,color='k')
ax0.scatter(alphas/twopi,berry_phi,color='k')

ax1.set_xlim(0.,1.)
ax1.set_ylim(-6.5,6.5)
ax1.set_xlabel(r"$\varphi/2\pi$")
ax1.set_ylabel(r"Berry phase $\phi(\varphi)$ for $\alpha$ loops")
ax1.set_title("Berry phase")
for shift in (-twopi,0.,twopi):
  ax1.plot(phi/twopi,bp_of_phi+shift,color='k')
ax1.scatter(phis/twopi,berry_alpha,color='k')

X=alp[0:-1]/twopi + 0.5/float(n_alp-1)
Y=phi[0:-1]/twopi + 0.5/float(n_phi-1)
cs=ax2.contour(X,Y,curvature,colors='k')
ax2.clabel(cs, inline=1, fontsize=10)
ax2.set_title("Berry curvature")
ax2.set_xlabel(r"$\alpha/2\pi$")
ax2.set_xlim(0.,1.)
ax2.set_ylim(0.,1.)
ax2.set_ylabel(r"$\varphi/2\pi$")

fig.tight_layout()
fig.savefig("trimer.pdf")
```

D.7 Berry Phase of Alternating Site Chain

Program `chain_alt_bp.py` returns to the same alternating chain model presented in Appendix D.4 and computes the Berry phase of the lower band as the Bloch wavevector k is cycled around the BZ, as presented in Section 3.4. The Berry phase is again calculated first explicitly and then again using the `wf_array` method. In the explicit calculation, one has to be sure to account for the extra phase factor in Eq. (3.74) as discussed on p. 108. The `wf_array` method, by contrast, takes care of this automatically.

chain_alt_bp.py

```
#!/usr/bin/env python
from __future__ import print_function # python3 style print

# Chain with alternating site energies and hoppings

from pythtb import *
import matplotlib.pyplot as plt
```

```
# define function to set up model for a given parameter set
def set_model(t,del_t,Delta):
  lat=[[1.0]]
  orb=[[0.0],[0.5]]
  my_model=tbmodel(1,1,lat,orb)
  my_model.set_onsite([Delta,-Delta])
  my_model.add_hop(t+del_t, 0, 1, [0])
  my_model.add_hop(t-del_t, 1, 0, [1])
  return my_model


# set parameters of model
t=-1.0        # average hopping
del_t=-0.3    # bond strength alternation
Delta= 0.4    # site energy alternation
my_model=set_model(t,del_t,Delta)
my_model.display()

# -----------------------------------
# explicit calculation of Berry phase
# -----------------------------------


# set up and solve the model on a discretized k mesh
nk=61              # 60 equal intervals around the unit circle
(k_vec,k_dist,k_node)=my_model.k_path('full',nk,report=False)
(eval,evec)=my_model.solve_all(k_vec,eig_vectors=True)
evec=evec[0]    # pick band=0 from evec[band,kpoint,orbital]
                # now just evec[kpoint,orbital]

# k-points 0 and 60 refer to the same point on the unit circle
# so we will work only with evec[0],...,evec[59]

# compute Berry phase of lowest band
prod=1.+0.j
for i in range(1,nk-1):                    # <evec_0|evec_1>...<evec_58|evec_59>
  prod*=np.vdot(evec[i-1],evec[i])  # a*=b means a=a*b

# now compute the phase factors needed for last inner product
orb=np.array([0.0,0.5])              # relative coordinates of orbitals
phase=np.exp((-2.j)*np.pi*orb)      # construct phase factors
evec_last=phase*evec[0]             # evec[60] constructed from evec[0]
prod*=np.vdot(evec[-2],evec_last)  # include <evec_59|evec_last>

print("Berry phase is %7.3f"% (-np.angle(prod)))

# -----------------------------------
# Berry phase via the wf_array method
# -----------------------------------

evec_array=wf_array(my_model,[61])            # set array dimension
evec_array.solve_on_grid([0.])                # fill with eigensolutions
berry_phase=evec_array.berry_phase([0])      # Berry phase of bottom band

print("Berry phase is %7.3f"% berry_phase)
```

D.8 Infinite and Finite Three-Site Chain

Program `chain_3_site.py` solves a 1D TB model with three sites per unit cell and computes the Wannier function centers in the multiband context as discussed in Section 3.6.3. The nearest-neighbor hoppings are all equal, but the site energies are modulated according to Eq. (3.132) in such a way that the evolution of parameter λ from 0 to 2π represents a sliding charge-density wave, as illustrated in Fig. 1.4 in Chapter 1.

The Wannier centers are first computed for the periodic bulk via calculations of multiband Berry phases, first for the lowest-energy band, then for the next lowest, then by computing the centers of the maximally localized Wannier functions (MLWFs) from the multiband Berry phases for the two lowest bands taken as a group (this requires specifying `berry_evals=True` in the call to `berry_phase`).

Next, the finite-system MLWF centers are computed for a finite chain of 10 unit cells cut from the infinite bulk model using the `cut_piece` method of the PYTHTB `tb_model` class. Again this is done for the lowest band alone (more precisely, the 10 lowest eigenstates), the second band (next 10 eigenstates), and the joint group including both (20 lowest states). The MLWF centers are obtained by diagonalizing the 10×10 or 20×20 X position matrix of Eq. (3.129) using the `position_hwf` method of the PYTHTB package.

The resulting printout is reproduced on p. 134. The format is designed to facilitate comparison between the two sets of results. In each case it is evident that while the Wannier centers deviate from the bulk positions near the ends of the chain, they converge rapidly to those bulk values as one goes deeper into bulk-like region of the chain.

chain_3_site.py

```python
#!/usr/bin/env python
from __future__ import print_function # python3 style print

# Chain with three sites per cell

from pythtb import *
import matplotlib.pyplot as plt

# define function to construct model
def set_model(t,delta,lmbd):
    lat=[[1.0]]
    orb=[[0.0],[1.0/3.0],[2.0/3.0]]
    model=tb_model(1,1,lat,orb)
    model.set_hop(t, 0, 1, [0])
    model.set_hop(t, 1, 2, [0])
    model.set_hop(t, 2, 0, [1])
    onsite_0=delta*(-1.0)*np.cos(2.0*np.pi*(lmbd-0.0/3.0))
    onsite_1=delta*(-1.0)*np.cos(2.0*np.pi*(lmbd-1.0/3.0))
    onsite_2=delta*(-1.0)*np.cos(2.0*np.pi*(lmbd-2.0/3.0))
    model.set_onsite([onsite_0,onsite_1,onsite_2])
    return(model)

# construct the model
t=-1.3
delta=2.0
lmbd=0.3
my_model=set_model(t,delta,lmbd)
```

```
# compute the results on a uniform k-point grid
evec_array=wf_array(my_model,[21])          # set array dimension
evec_array.solve_on_grid([0.])              # fill with eigensolutions

# obtain Berry phases and convert to Wannier center positions
#    constrained to the interval [0.,1.]
wfc0=evec_array.berry_phase([0])/(2.*np.pi)%1.
wfc1=evec_array.berry_phase([1])/(2.*np.pi)%1.
x=evec_array.berry_phase([0,1],berry_evals=True)/(2.*np.pi)%1.
gwfc0=x[0]
gwfc1=x[1]

print ("Wannier centers of bands 0 and 1:")
print(("  Individual"+" Wannier centers: "+2*"%7.4f") % (wfc0,wfc1))
print(("  Multiband "+" Wannier centers: "+2*"%7.4f") % (gwfc1,gwfc0))
print()

# construct and solve finite model by cutting 10 cells from infinite chain
finite_model=my_model.cut_piece(10,0)
(feval,fevec)=finite_model.solve_all(eig_vectors=True)

print ("Finite-chain eigenenergies associated with")
print(("Band 0:"+10*"%6.2f")% tuple(feval[0:10]))
print(("Band 1:"+10*"%6.2f")% tuple(feval[10:20]))

# find maxloc Wannier centers in each band subspace
xbar0=finite_model.position_hwf(fevec[0:10,],0)
xbar1=finite_model.position_hwf(fevec[10:20,],0)
xbarb=finite_model.position_hwf(fevec[0:20,],0)

print ("\nFinite-chain Wannier centers associated with band 0:")
print((10*"%7.4f")% tuple(xbar0))
x=10*(wfc0,)
print(("Compare with bulk:\n"+10*"%7.4f")% x)
print ("\nFinite-chain Wannier centers associated with band 1:")
print((10*"%7.4f")% tuple(xbar1))
x=10*(wfc1,)
print(("Compare with bulk:\n"+10*"%7.4f")% x)
print ("\nFirst 10 finite-chain Wannier centers associated with bands"+
    "0 and 1:")
print((10*"%7.4f")% tuple(xbarb[0:10]))
x=5*(gwfc0,gwfc1)
print(("Compare with bulk:\n"+10*"%7.4f")% x)
```

D.9 Adiabatic Cycle for Infinite Three-Site Chain

Program chain_3_cycle.py treats the same model as in Appendix D.8, but this time treating λ as a cyclic parameter that runs from 0 to 2π and focusing on tracking the bulk Wannier center position of the lowest band through this cycle. The approach is almost the same as is used in the "one-dimensional cycle of 1D tight-binding model" example program in the standard PYTHTB distribution, except for the usage of the wf_array method in two-dimensional (k, λ) space there, which has the advantage of providing automatic enforcement of continuity with respect to λ. By contrast, continuity is enforced explicitly in the version here in the lines following the comment line #enforce smooth evolution of xbar. The plotted output of this program appears as Fig. 4.5 in Section 4.2.3.

chain_3_cycle.py

```python
#!/usr/bin/env python
from __future__ import print_function # python3 style print

# Chain with three sites per cell - cyclic variation

from pythtb import *
import matplotlib.pyplot as plt

# define function to construct model
def set_model(t,delta,lmbd):
    lat=[[1.0]]
    orb=[[0.0],[1.0/3.0],[2.0/3.0]]
    model=tb_model(1,1,lat,orb)
    model.set_hop(t, 0, 1, [0])
    model.set_hop(t, 1, 2, [0])
    model.set_hop(t, 2, 0, [1])
    onsite_0=delta*(-1.0)*np.cos(lmbd)
    onsite_1=delta*(-1.0)*np.cos(lmbd-2.0*np.pi/3.0)
    onsite_2=delta*(-1.0)*np.cos(lmbd-4.0*np.pi/3.0)
    model.set_onsite([onsite_0,onsite_1,onsite_2])
    return(model)

def get_xbar(band,model):
    evec_array=wf_array(model,[21])          # set array dimension
    evec_array.solve_on_grid([0.])           # fill with eigensolutions
    wfc=evec_array.berry_phase([band])/(2.*np.pi)  # Wannier centers
    return(wfc)

# set fixed parameters
t=-1.3
delta=2.0

# obtain results for an array of lambda values
lmbd=np.linspace(0.,2.*np.pi,61)
xbar=np.zeros_like(lmbd)
for j,lam in enumerate(lmbd):
    my_model=set_model(t,delta,lam)
    xbar[j]=get_xbar(0,my_model)      # Wannier center of bottom band

# enforce smooth evolution of xbar
for j in range(1,61):
    delt=xbar[j]-xbar[j-1]
    delt=-0.5+(delt+0.5)%1.  # add integer to enforce |delt| < 0.5
    xbar[j]=xbar[j-1]+delt

# set up the figure
fig, ax = plt.subplots(figsize=(5.,3.))
ax.set_xlim([0.,2.*np.pi])
ax.set_ylim([-0.6,1.1])
ax.set_xlabel(r"Parameter $\lambda$")
ax.set_ylabel(r"Wannier center position")
xlab=[r"0",r"$\pi/3$",r"$2\pi/3$",r"$\pi$",r"$4\pi/3$",r"$5\pi/3$",r"$2\pi$"]
ax.set_xticks(np.linspace(0.,2.*np.pi,num=7))
ax.set_xticklabels(xlab)
ax.plot(lmbd,xbar,'k')     # plot Wannier center and some periodic images
ax.plot(lmbd,xbar-1.,'k')
ax.plot(lmbd,xbar+1.,'k')
ax.axhline(y=1.,color='k',linestyle='dashed') # horizontal reference lines
ax.axhline(y=0.,color='k',linestyle='dashed')
fig.tight_layout()
fig.savefig("chain_3_cycle.pdf")
```

D.10 Surface Properties of Alternating Site Model

Program `chain_alt_surf.py` returns to the alternating site model of Appendices D.4 and D.7, but now focuses on computing the surface properties of a finite chain. Two types of variation are considered: modification of the site energy of the last orbital on the chain, and modification of the bulk Hamiltonian along an adiabatic cycle corresponding to a charge pump. The calculations are discussed in Section 4.5.3, where the resulting plot is presented as Fig. 4.16. For each kind of parameter variation, three plots are presented, showing the variation of the energy eigenvalues, the Wannier center positions (computed along the same lines as in Appendix D.8), and the surface charge computed using the ramp-function method of Eqs. (4.78) and (4.79).

chain_alt_surf.py

```
#!/usr/bin/env python
from __future__ import print_function # python3 style print

# Chain with alternating site energies and hoppings
# Study surface properties of finite chain

from pythtb import *
import matplotlib as mpl
import matplotlib.pyplot as plt

# to set up model for given surface energy shift and lambda
def set_model(n_cell,en_shift,lmbd):

  # set parameters of model
  t=-1.0          # average hopping
  Delta=-0.4*np.cos(lmbd)    # site energy alternation
  del_t=-0.3*np.sin(lmbd)    # bond strength alternation

  # construct bulk model
  lat=[[1.0]]
  orb=[[0.0],[0.5]]
  bulk_model=tbmodel(1,1,lat,orb)
  bulk_model.set_onsite([Delta,-Delta])
  bulk_model.add_hop(t+del_t, 0, 1, [0])
  bulk_model.add_hop(t-del_t, 1, 0, [1])

  # cut chain of length n_cell and shift energy on last site
  finite_model=bulk_model.cut_piece(n_cell,0)
  finite_model.set_onsite(en_shift,ind_i=2*n_cell-1,mode='add')

  return finite_model

# set Fermi energy and number of cells
Ef=0.18
n_cell=20
n_orb=2*n_cell

# set number of parameter values to run over
n_param=101

# initialize arrays
params=np.linspace(0.,1.,n_param)
eig_sav=np.zeros((n_orb,n_param),dtype=float)
```

```
xbar_sav=np.zeros((n_orb,n_param),dtype=float)
nocc_sav=np.zeros((n_param),dtype=int)
surf_sav=np.zeros((n_param),dtype=float)
count=np.zeros((n_orb),dtype=float)

# initialize plots
mpl.rc('font',size=10)  # set global font size
fig,ax=plt.subplots(3,2,figsize=(7.,6.),
    gridspec_kw={'height_ratios':[2,1,1]},sharex="col")

# loop over two cases: vary surface site energy, or vary lambda
for mycase in ['surface energy','lambda']:

  if mycase == 'surface energy':
     (ax0,ax1,ax2)=ax[:,0]    # axes for plots in left panels
     ax0.text(-0.30,0.90,'(a)',size=22.,transform=ax0.transAxes)
     lmbd=0.15*np.pi*np.ones((n_param),dtype=float)
     en_shift=-3.0+6.0*params
     abscissa=en_shift
  elif mycase == 'lambda':
     (ax0,ax1,ax2)=ax[:,1]    # axes for plots in right panels
     ax0.text(-0.30,0.90,'(b)',size=22.,transform=ax0.transAxes)
     lmbd=params*2.*np.pi
     en_shift=0.2*np.ones((n_param),dtype=float)
     abscissa=params

  # loop over parameter values
  for j in range(n_param):

     # set up and solve model; store eigenvalues
     my_model=set_model(n_cell,en_shift[j],lmbd[j])
     (eval,evec)=my_model.solve_all(eig_vectors=True)

     # find occupied states
     nocc=(eval<Ef).sum()
     ovec=evec[0:nocc,:]

     # get Wannier centers
     xbar_sav[0:nocc,j]=my_model.position_hwf(ovec,0)

     # get electron count on each site
     # convert to charge (2 for spin; unit nuclear charge per site)
     # compute surface charge down to depth of 1/3 of chain
     for i in range(n_orb):
        count[i]=np.real(np.vdot(evec[:nocc,i],evec[:nocc,i]))
     charge=-2.*count+1.
     n_cut=int(0.67*n_orb)
     surf_sav[j]=0.5*charge[n_cut-1]+charge[n_cut:].sum()

     # save information for plots
     nocc_sav[j]=nocc
     eig_sav[:,j]=eval

  ax0.set_xlim(0.,1.)
  ax0.set_ylim(-2.8,2.8)
  ax0.set_ylabel(r"Band energy")
  ax0.axhline(y=Ef,color='k',linewidth=0.5)
  for n in range(n_orb):
     ax0.plot(abscissa,eig_sav[n,:],color='k')

  ax1.set_xlim(0.,1.)
```

```
ax1.set_ylim(n_cell-4.6,n_cell+0.4)
ax1.set_yticks(np.linspace(n_cell-4,n_cell,5))
#ax1.set_ylabel(r"$\bar{x}$")
ax1.set_ylabel(r"Wannier centers")
for j in range(n_param):
  nocc=nocc_sav[j]
  ax1.scatter([abscissa[j]]*nocc,xbar_sav[:nocc,j],color='k',
    s=3.,marker='o',edgecolors='none')

ax2.set_ylim(-2.2,2.2)
ax2.set_yticks([-2.,-1.,0.,1.,2.])
ax2.set_ylabel(r"Surface charge")
if mycase == 'surface energy':
  ax2.set_xlabel(r"Surface site energy")
elif mycase == 'lambda':
  ax2.set_xlabel(r"$\lambda/2\pi$")
ax2.set_xlim(abscissa[0],abscissa[-1])
ax2.scatter(abscissa,surf_sav,color='k',s=3.,marker='o',edgecolors='none')

# vertical lines denote surface state at right end crossing the
# Fermi energy
for j in range(1,n_param):
  if nocc_sav[j] != nocc_sav[j-1]:
    n=min(nocc_sav[j],nocc_sav[j-1])
    frac=(Ef-eig_sav[n,j-1])/(eig_sav[n,j]-eig_sav[n,j-1])
    a_jump=(1-frac)*abscissa[j-1]+frac*abscissa[j]
    if mycase == 'surface energy' or nocc_sav[j] < nocc_sav[j-1]:
      ax0.axvline(x=a_jump,color='k',linewidth=0.5)
      ax1.axvline(x=a_jump,color='k',linewidth=0.5)
      ax2.axvline(x=a_jump,color='k',linewidth=0.5)

fig.tight_layout()
plt.subplots_adjust(left=0.12,wspace=0.4)
fig.savefig("chain_alt_surf.pdf")
```

D.11 Band Structure of the Haldane Model

Program `haldane_bsr.py` computes the band structure of the Haldane model
of Eq. (5.1) for four different parameter sets, with the plotted output appearing as
Fig. 5.2 in Section 5.1.1. In addition to plotting the band structure, the program
places filled or open circles on the bands at the high-symmetry K and K' points
according to which site dominates the character of the lower-energy band, thereby
highlighting the band inversion.

haldane_bsr.py

```
#!/usr/bin/env python
from __future__ import print_function # python3 style print

# Band structure of Haldane model

from pythtb import * # import TB model class
import matplotlib.pyplot as plt

# set model parameters
```

```
delta=0.7    # site energy shift
t=-1.0       # real first-neighbor hopping
t2=0.15      # imaginary second-neighbor hopping

def set_model(delta,t,t2):
  lat=[[1.0,0.0],[0.5,np.sqrt(3.0)/2.0]]
  orb=[[1./3.,1./3.],[2./3.,2./3.]]
  model=tb_model(2,2,lat,orb)
  model.set_onsite([-delta,delta])
  for lvec in ([ 0, 0], [-1, 0], [ 0,-1]):
    model.set_hop(t, 0, 1, lvec)
  for lvec in ([ 1, 0], [-1, 1], [ 0,-1]):
    model.set_hop(t2*1.j, 0, 0, lvec)
  for lvec in ([-1, 0], [ 1,-1], [ 0, 1]):
    model.set_hop(t2*1.j, 1, 1, lvec)
  return model

# construct path in k-space and solve model
path=[[0.,0.],[2./3.,1./3.],[.5,.5],[1./3.,2./3.], [0.,0.]]
label=(r'$\Gamma $',r'$K$', r'$M$', r'$K^\prime$', r'$\Gamma $')
(k_vec,k_dist,k_node)=set_model(delta,t,t2).k_path(path,101)

# set up band structure plots
fig, ax = plt.subplots(2,2,figsize=(8.,6.),sharex=True,sharey=True)
ax=ax.flatten()

t2_values=[0.,-0.06,-0.1347,-0.24]
labs=['(a)','(b)','(c)','(d)']
for j in range(4):

  my_model=set_model(delta,t,t2_values[j])
  evals=my_model.solve_all(k_vec)

  ax[j].set_xlim([0,k_node[-1]])
  ax[j].set_xticks(k_node)
  ax[j].set_xticklabels(label)
  for n in range(len(k_node)):
    ax[j].axvline(x=k_node[n],linewidth=0.5, color='k')
  ax[j].set_ylabel("Energy")
  ax[j].set_ylim(-3.8,3.8)
  for n in range(2):
    ax[j].plot(k_dist,evals[n],color='k')

  # filled or open dots at K and K' following band inversion
  for m in [1,3]:
    kk=k_node[m]
    (en,ev)=my_model.solve_one(path[m],eig_vectors=True)
    if np.abs(ev[0,0]) > np.abs(ev[0,1]):   #ev[band,orb]
      en=[en[1],en[0]]
    ax[j].scatter(kk,en[0],s=40.,marker='o',edgecolors='k',
      facecolors='w',zorder=4)
    ax[j].scatter(kk,en[1],s=40.,marker='o',color='k',zorder=6)

  ax[j].text(0.20,3.1,labs[j],size=18.)

# save figure as a PDF
fig.tight_layout()
fig.savefig("haldane_bsr.pdf")
```

D.12 Berry Curvature of the Haldane Model

Program `haldane_bcurv.py` generates contour plots of the Berry curvature of the Haldane model for three parameter sets. The model is solved on a 2D mesh of **k**-points using the `solve_on_grid` method of `wf_array`. Then the `berry_flux` method is used in two different ways: once with `individual_phases=True` to obtain the flux through each plaquette of the mesh, and once with the default `False` value to obtain the total flux through the entire 2D BZ. When normalized, the first gives the Berry curvature, which is plotted using the `contour` method of PYPLOT, and the second gives the Chern number, which is printed out. The plot appears as Fig. 5.3 in Section 5.1.1.

haldane_bcurv.py

```
#!/usr/bin/env python
from __future__ import print_function # python3 style print

# Berry curvature of Haldane model

from pythtb import * # import TB model class
import matplotlib.pyplot as plt

# define setup of Haldane model
def set_model(delta,t,t2):
    lat=[[1.0,0.0],[0.5,np.sqrt(3.0)/2.0]]
    orb=[[1./3.,1./3.],[2./3.,2./3.]]
    model=tb_model(2,2,lat,orb)
    model.set_onsite([-delta,delta])
    for lvec in ([ 0, 0], [-1, 0], [ 0,-1]):
        model.set_hop(t, 0, 1, lvec)
    for lvec in ([ 1, 0], [-1, 1], [ 0,-1]):
        model.set_hop(t2*1.j, 0, 0, lvec)
    for lvec in ([-1, 0], [ 1,-1], [ 0, 1]):
        model.set_hop(t2*1.j, 1, 1, lvec)
    return model

# miscellaneous setup
delta=0.7    # site energy shift
t=-1.0       # real first-neighbor hopping

nk=61
dk=2.*np.pi/(nk-1)
k0=(np.arange(nk-1)+0.5)/(nk-1)
kx=np.zeros((nk-1,nk-1),dtype=float)
ky=np.zeros((nk-1,nk-1),dtype=float)
sq3o2=np.sqrt(3.)/2.
for i in range(nk-1):
    for j in range(nk-1):
        kx[i,j]=sq3o2*k0[i]
        ky[i,j]= -0.5*k0[i]+k0[j]

fig,ax=plt.subplots(1,3,figsize=(11,4))
labs=['(a)','(b)','(c)']
```

```
# compute Berry curvature and Chern number for three values of t2
for j,t2 in enumerate([0.,-0.06,-0.24]):
  my_model=set_model(delta,t,t2)
  my_array=wf_array(my_model,[nk,nk])
  my_array.solve_on_grid([0.,0.])
  bcurv=my_array.berry_flux([0],individual_phases=True)/(dk*dk)
  chern=my_array.berry_flux([0])/(2.*np.pi)
  print('Chern number =',"%8.5f"%chern)

  # make contour plot of Berry curvature
  pos_lvls= 0.02*np.power(2.,np.linspace(0,8,9))
  neg_lvls=-0.02*np.power(2.,np.linspace(8,0,9))
  ax[j].contour(kx,ky,bcurv,levels=pos_lvls,colors='k')
  ax[j].contour(kx,ky,bcurv,levels=neg_lvls,colors='k',linewidths=1.4)

  # remove rectangular box and draw parallelogram, etc.
  ax[j].xaxis.set_visible(False)
  ax[j].yaxis.set_visible(False)
  for loc in ["top","bottom","left","right"]:
    ax[j].spines[loc].set_visible(False)
  ax[j].set(aspect=1.)
  ax[j].plot([0,sq3o2,sq3o2,0,0],[0,-0.5,0.5,1,0],color='k',linewidth=1.4)
  ax[j].set_xlim(-0.05,sq3o2+0.05)
  ax[j].text(-.35,0.88,labs[j],size=24.)

fig.savefig("haldane_bcurv.pdf")
```

D.13 Hybrid Wannier Centers and Edge States of the Haldane Model

Program haldane_topo.py produces plots of the hybrid Wannier center flow in
the bulk, and of the edge band structure of a finite-width ribbon, for the Haldane
model with two different parameter sets covering both the trivial and topological
cases. The plotted output appears as Fig. 5.4 in Section 5.1.1. The methods used
to obtain the bulk Wannier centers should be familiar by now. The cut_piece
method of the tb_model class is applied to the 2D bulk model my_model
to obtain the new 1D ribbon_model, extending indefinitely in the horizontal
direction but of finite width vertically, here chosen to be 20 cells high. The band
structure of the ribbon is plotted using the PYPLOT scatter method, instead of as
a continuous line, so that the identity of the states localized at the top and bottom
edges of the ribbon can be distinguished. This is done by assigning each plotted
point a weight that is normally 1, but that is reduced for states at the bottom edge.
These are identified by using the position_expectation method of PYTHTB
to compute the component of the position expectation value in the vertical direction
for each Bloch eigenvector.

haldane_topo.py

```
#!/usr/bin/env python
from __future__ import print_function # python3 style print

# Band structure of Haldane model
```

```
from pythtb import * # import TB model class
import matplotlib.pyplot as plt

# define setup of Haldane model
def set_model(delta,t,t2):
  lat=[[1.0,0.0],[0.5,np.sqrt(3.0)/2.0]]
  orb=[[1./3.,1./3.],[2./3.,2./3.]]
  model=tb_model(2,2,lat,orb)
  model.set_onsite([-delta,delta])
  for lvec in ([ 0, 0], [-1, 0], [ 0,-1]):
    model.set_hop(t, 0, 1, lvec)
  for lvec in ([ 1, 0], [-1, 1], [ 0,-1]):
    model.set_hop(t2*1.j, 0, 0, lvec)
  for lvec in ([-1, 0], [ 1,-1], [ 0, 1]):
    model.set_hop(t2*1.j, 1, 1, lvec)
  return model

# set model parameters and construct bulk model
delta=0.7    # site energy shift
t=-1.0       # real first-neighbor hopping
nk=51

# For the purposes of plot labels:
#    Real space is (r1,r2) in reduced coordinates
#    Reciprocal space is (k1,k2) in reduced coordinates
# Below, following Python, these are (r0,r1) and (k0,k1)

# set up figures
fig,ax=plt.subplots(2,2,figsize=(7,6))

# run over two choices of t2
for j2,t2 in enumerate([-0.06,-0.24]):

  # solve bulk model on grid and get hybrid Wannier centers along r1
  # as a function of k0
  my_model=set_model(delta,t,t2)
  my_array=wf_array(my_model,[nk,nk])
  my_array.solve_on_grid([0.,0.])
  rbar_1 = my_array.berry_phase([0],1,contin=True)/(2.*np.pi)

  # set up and solve ribbon model that is finite along direction 1
  width=20
  nkr=81
  ribbon_model=my_model.cut_piece(width,fin_dir=1,glue_edgs=False)
  (k_vec,k_dist,k_node)=ribbon_model.k_path('full',nkr,report=False)
  (rib_eval,rib_evec)=ribbon_model.solve_all(k_vec,eig_vectors=True)

  nbands=rib_eval.shape[0]
  (ax0,ax1)=ax[j2,:]

  # hybrid Wannier center flow
  k0=np.linspace(0.,1.,nk)
  ax0.set_xlim(0.,1.)
  ax0.set_ylim(-1.3,1.3)
  ax0.set_xlabel(r"$\kappa_1/2\pi$")
  ax0.set_ylabel(r"HWF centers")
  for shift in (-2.,-1.,0.,1.):
    ax0.plot(k0,rbar_1+shift,color='k')

  # edge band structure
  k0=np.linspace(0.,1.,nkr)
```

```
ax1.set_xlim(0.,1.)
ax1.set_ylim(-2.5,2.5)
ax1.set_xlabel(r"$\kappa_1/2\pi$")
ax1.set_ylabel(r"Edge band structure")
for (i,kv) in enumerate(k0):

    # find expectation value <r1> at i'th k-point along direction k0
    pos_exp=ribbon_model.position_expectation(rib_evec[:,i],dir=1)

    # assign weight in [0,1] to be 1 except for edge states near bottom
    weight=3.0*pos_exp/width
    for j in range(nbands):
      weight[j]=min(weight[j],1.)

    # scatterplot with symbol size proportional to assigned weight
    s=ax1.scatter([k_vec[i]]*nbands, rib_eval[:,i],
        s=0.6+2.5*weight, c='k', marker='o', edgecolors='none')

# save figure as a PDF
aa=ax.flatten()
for i,lab in enumerate(['(a)','(b)','(c)','(d)']):
  aa[i].text(-0.45,0.92,lab,size=18.,transform=aa[i].transAxes)
fig.tight_layout()
plt.subplots_adjust(left=0.15,wspace=0.6)
fig.savefig("haldane_topo.pdf")
```

D.14 Entanglement Spectrum of the Haldane Model

Program `haldane_entang.py` produces plots of the entanglement spectrum for the Haldane model for two parameter sets, one in the trivial phase and one in the topological phase. The entanglement spectrum, mentioned briefly at the end of Section 5.1.1, is defined as follows. For an infinite 1D chain, one arbitrarily chooses a dividing surface that partitions the chain into a lower region A and an upper region B. The entanglement spectrum is then given by computing the eigenvalues of the reduced density matrix $\rho_A = \mathcal{P}_A \rho \mathcal{P}_A$, where $\rho_{ij} = \sum_n^{\mathrm{occ}} \psi_{nk}^*(i)\psi_{nk}(j)$ is the one-particle density matrix in the TB basis and $\mathcal{P}_A$ is the projector onto subregion A. For a 2D system, one plots these eigenvalues for the effective 1D system extending along lattice direction $\mathbf{a}_2$ as a function of κ_1, as shown in Fig. D.3. In practice, `haldane_entang.py` obtains the entanglement spectrum from the same finite-width ribbon constructed in `haldane_topo.py`, partitioned in the middle; edge states on the physical boundary of the ribbon have little influence since they are far from the dividing surface.

A comparison of Fig. D.3(b) and Fig. 5.4(c) shows that an upward flow of the hybrid Wannier function (HWF) centers with increasing κ_1 corresponds to a downward flow of the ρ_A eigenvalues. This is only natural, since an upward migration of Wannier functions corresponds to a depletion of occupation in the lower region A.

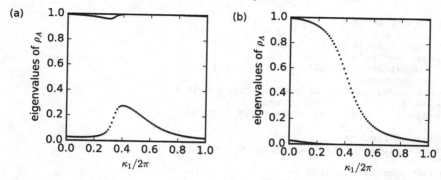

Figure D.3 Entanglement spectrum of the Haldane model with $\Delta = 0.7$ and $t_1 = -1$. A ribbon geometry of width 20 unit cells along the κ_2 direction was subdivided into bottom and top halves A and B, respectively; eigenvalues of the reduced density matrix ρ_A (see text) are plotted vertically. (a) Trivial phase ($t_2 = 0.10$) with $C = 0$. (b) Topological phase ($t_2 = -0.24$) with $C = 1$.

The origin of the name "entanglement spectrum" is associated with the definition of an *entanglement entropy* given by $\mathrm{Tr}\,[-\rho_A \ln \rho_A - (1 - \rho_A) \ln (1 - \rho_A)]$. This object is dominated by eigenvalues of ρ_A lying in the middle of the interval $[0, 1]$, so roughly speaking it counts the number of states that cannot be localized entirely in A or in B, but instead are "entangled" across the dividing surface.

The flow of the single-particle entanglement spectrum, as shown in Fig. D.3, can be used as a tool to extract the Chern index of a 2D magnetic insulator, much like the edge band structures and HWF flow patterns in Fig. 5.4. The three methods are, in a sense, close cousins. However, the HWF flow approach has the advantage of requiring only a simple bulk calculation, with no introduction of artificial bounding surfaces or edge terminations. When generalized to the many-body context, however, the entanglement entropy becomes a useful tool for the study of strongly correlated topological states with long-range entanglement, such as those mentioned in Section 5.5.3.

haldane_entang.py

```
#!/usr/bin/env python
from __future__ import print_function # python3 style print

# Entanglement spectrum of Haldane model

from pythtb import * # import TB model class
import matplotlib.pyplot as plt

# define setup of Haldane model
def set_model(delta,t,t2):
  lat=[[1.0,0.0],[0.5,np.sqrt(3.0)/2.0]]
  orb=[[1./3.,1./3.],[2./3.,2./3.]]
  model=tb_model(2,2,lat,orb)
  model.set_onsite([-delta,delta])
  for lvec in ([ 0, 0], [-1, 0], [ 0,-1]):
```

```
      model.set_hop(t, 0, 1, lvec)
    for lvec in ([ 1, 0], [-1, 1], [ 0,-1]):
      model.set_hop(t2*1.j, 0, 0, lvec)
    for lvec in ([-1, 0], [ 1,-1], [ 0, 1]):
      model.set_hop(t2*1.j, 1, 1, lvec)
    return model
# set model parameters and construct bulk model
delta=0.7    # site energy shift
t=-1.0       # real first-neighbor hopping

# set up figures
fig,ax=plt.subplots(1,2,figsize=(7,3))

# run over two choices of t2
for j2,t2 in enumerate([-0.10,-0.24]):

  my_model=set_model(delta,t,t2)

  # set up and solve ribbon model that is finite along direction 1
  width=20
  nkr=81
  ribbon_model=my_model.cut_piece(width,fin_dir=1,glue_edgs=False)
  (k_vec,k_dist,k_node)=ribbon_model.k_path('full',nkr,report=False)
  (rib_eval,rib_evec)=ribbon_model.solve_all(k_vec,eig_vectors=True)

  nbands=rib_eval.shape[0]
  ax1=ax[j2]

  # entanglement spectrum
  k0=np.linspace(0.,1.,nkr)
  ax1.set_xlim(0.,1.)
  ax1.set_ylim(0.,1)
  ax1.set_xlabel(r"$\kappa_1/2\pi$")
  ax1.set_ylabel(r"eigenvalues of $\rho_A$")

  (nband,nk,norb)=rib_evec.shape
  ncut=norb/2
  nocc=nband/2

  for (i,kv) in enumerate(k0):

    # construct reduced density matrix for half of the chain
    dens_mat=np.zeros((ncut,ncut),dtype=complex)
    for nb in range(nocc):
      for j1 in range(ncut):
        for j2 in range(ncut):
          dens_mat[j1,j2] += np.conj(rib_evec[nb,i,j1])*rib_evec[nb,i,j2]

    # diagonalize
    spect=np.real(np.linalg.eigvals(dens_mat))
    # scatterplot
    s=ax1.scatter([k_vec[i]]*nocc, spect,
        s=4, c='k', marker='o', edgecolors='none')

# save figure as a PDF
aa=ax.flatten()
for i,lab in enumerate(['(a)','(b)']):
  aa[i].text(-0.45,0.92,lab,size=18.,transform=aa[i].transAxes)
fig.tight_layout()
plt.subplots_adjust(left=0.15,wspace=0.6)
fig.savefig("haldane_entang.pdf")
```

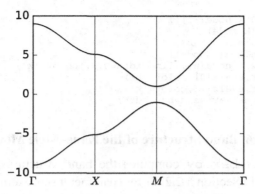

Figure D.4 Band structure of the 2D checkerboard model.

D.15 Band Structure of the Checkerboard Model

Program `checkerboard.py` computes the band structure for the 2D checkerboard model introduced in Ex. 5.3. The output appears here as Fig. D.4.

checkerboard.py

```python
#!/usr/bin/env python
from __future__ import print_function # python3 style print

from pythtb import * # import TB model class
import matplotlib.pyplot as plt

# set geometry
lat=[[1.0,0.0],[0.0,1.0]]
orb=[[0.0,0.0],[0.5,0.5]]
my_model=tbmodel(2,2,lat,orb)

# set model
Delta  = 5.0
t_0    = 1.0
tprime = 0.4
my_model.set_sites([-Delta,Delta])
my_model.add_hop(-t_0, 0, 0, [ 1, 0])
my_model.add_hop(-t_0, 0, 0, [ 0, 1])
my_model.add_hop( t_0, 1, 1, [ 1, 0])
my_model.add_hop( t_0, 1, 1, [ 0, 1])
my_model.add_hop( tprime   , 1, 0, [ 1, 1])
my_model.add_hop( tprime*1j, 1, 0, [ 0, 1])
my_model.add_hop(-tprime   , 1, 0, [ 0, 0])
my_model.add_hop(-tprime*1j, 1, 0, [ 1, 0])
my_model.display()

# generate k-point path and labels and solve Hamiltonian
path=[[0.0,0.0],[0.0,0.5],[0.5,0.5],[0.0,0.0]]
k_lab=(r'$\Gamma $',r'$X$', r'$M$', r'$\Gamma $')
(k_vec,k_dist,k_node)=my_model.k_path(path,121)
evals=my_model.solve_all(k_vec)

# plot band structure
fig, ax = plt.subplots(figsize=(4.,3.))
```

```
ax.set_xlim([0,k_node[-1]])
ax.set_xticks(k_node)
ax.set_xticklabels(k_lab)
for n in range(len(k_node)):
  ax.axvline(x=k_node[n], linewidth=0.5, color='k')
ax.plot(k_dist,evals[0],color='k')
ax.plot(k_dist,evals[1],color='k')
fig.savefig("checkerboard_bsr.pdf")
```

D.16 Band Structure of the Kane–Mele Model

Program `kanemele_bsr.py` computes the band structure of the Kane–Mele model of Eq. (5.17) in Section 5.2.2 for two parameter sets exemplifying the normal and topological regimes. This is a spinor model, so we set `nspin=2` when creating the model. To simplify the specification of the spin-dependent hoppings in the model, we have predefined Pauli matrices σ_x, σ_y, and σ_z in Cartesian directions, and then σ_a, σ_b, and σ_c in three relevant in-plane directions related by 120° rotations. Note that `mode="add"` is needed when we specify a second component of the first-neighbor hoppings.

The results are plotted as Fig. 5.12. Filled and open circles are added to denote the orbital character for the highest occupied and lowest unoccupied states at K and K' to highlight the band inversion. Here we have cheated, with the filled and open circles simply exchanged in the second plot; a proper implementation should follow the example of `haldane_bsr.py` in Appendix D.11.

kanemele_bsr.py

```python
#!/usr/bin/env python
from __future__ import print_function # python3 style print

# Tight-binding 2D Kane-Mele model
# C.L. Kane and E.J. Mele, PRL 95, 146802 (2005)

from pythtb import * # import TB model class
import matplotlib.pyplot as plt

# set model parameters
delta=0.7      # site energy
t=-1.0         # spin-independent first-neighbor hop
rashba=0.05    # spin-flip first-neighbor hop
soc_list=[-0.06,-0.24] # spin-dependent second-neighbor hop

def set_model(t,soc,rashba,delta):

  # set up Kane-Mele model
  lat=[[1.0,0.0],[0.5,np.sqrt(3.0)/2.0]]
  orb=[[1./3.,1./3.],[2./3.,2./3.]]
  model=tb_model(2,2,lat,orb,nspin=2)
  model.set_onsite([delta,-delta])

  # definitions of Pauli matrices
  sigma_x=np.array([0.,1.,0.,0])
  sigma_y=np.array([0.,0.,1.,0])
```

```
sigma_z=np.array([0.,0.,0.,1])
r3h =np.sqrt(3.0)/2.0
sigma_a= 0.5*sigma_x-r3h*sigma_y
sigma_b= 0.5*sigma_x+r3h*sigma_y
sigma_c=-1.0*sigma_x

# spin-independent first-neighbor hops
for lvec in ([ 0, 0], [-1, 0], [ 0,-1]):
  model.set_hop(t, 0, 1, lvec)
# spin-dependent second-neighbor hops
for lvec in ([ 1, 0], [-1, 1], [ 0,-1]):
  model.set_hop(soc*1.j*sigma_z, 0, 0, lvec)
for lvec in ([-1, 0], [ 1,-1], [ 0, 1]):
  model.set_hop(soc*1.j*sigma_z, 1, 1, lvec)
# spin-flip first-neighbor hops
model.set_hop(1.j*rashba*sigma_a, 0, 1, [ 0, 0], mode="add")
model.set_hop(1.j*rashba*sigma_b, 0, 1, [-1, 0], mode="add")
model.set_hop(1.j*rashba*sigma_c, 0, 1, [ 0,-1], mode="add")

return model

# construct path in k-space and solve model
path=[[0.,0.],[2./3.,1./3.],[.5,.5],[1./3.,2./3.], [0.,0.]]
label=(r'$\Gamma $',r'$K$', r'$M$', r'$K^\prime$', r'$\Gamma $')
(k_vec,k_dist,k_node)=set_model(t,0.,rashba,delta).k_path(path,101)

# set up band structure plots
fig, ax = plt.subplots(1,2,figsize=(8.,3.))

labs=['(a)','(b)']
for j in range(2):

  my_model=set_model(t,soc_list[j],rashba,delta)
  evals=my_model.solve_all(k_vec)

  ax[j].set_xlim([0,k_node[-1]])
  ax[j].set_xticks(k_node)
  ax[j].set_xticklabels(label)
  for n in range(len(k_node)):
    ax[j].axvline(x=k_node[n],linewidth=0.5, color='k')
  ax[j].set_ylabel("Energy")
  ax[j].set_ylim(-3.8,3.8)
  for n in range(4):
    ax[j].plot(k_dist,evals[n],color='k')

  for m in [1,3]:
    kk=k_node[m]
    en=my_model.solve_one(path[m])
    en=en[1:3]  # pick out second and third bands
    if j==1:    # exchange them in second plot
      en=[en[1],en[0]]
    ax[j].scatter(kk,en[0],s=40.,marker='o',color='k',zorder=6)
    ax[j].scatter(kk,en[1],s=40.,marker='o',edgecolors='k',
      facecolors='w',zorder=4)

  ax[j].text(-0.45,3.5,labs[j],size=18.)

# save figure as a PDF
fig.tight_layout()
plt.subplots_adjust(wspace=0.35)
fig.savefig("kanemele_bsr.pdf")
```

D.17 Hybrid Wannier Centers and Edge States of the Kane–Mele Model

Program kanemele_topo.py produces plots of the hybrid Wannier center flow in the bulk, and of the edge band structure of a finite-width ribbon, for the Kane–Mele model with two different parameter sets covering both the trivial and topological cases, following the example of haldane_topo.py in Appendix D.13. The plotted output appears as Fig. 5.13 in Section 5.2.2.

kanemele_topo.py

```
#!/usr/bin/env python
from __future__ import print_function # python3 style print

# Tight-binding 2D Kane-Mele model
# C.L. Kane and E.J. Mele, PRL 95, 146802 (2005)

from pythtb import * # import TB model class
import matplotlib.pyplot as plt

# set model parameters
delta=0.7      # site energy
t=-1.0         # spin-independent first-neighbor hop
soc=0.06       # spin-dependent second-neighbor hop
rashba=0.05    # spin-flip first-neighbor hop
soc_list=[-0.06,-0.24] # spin-dependent second-neighbor hop

def set_model(t,soc,rashba,delta):

    # set up Kane-Mele model
    lat=[[1.0,0.0],[0.5,np.sqrt(3.0)/2.0]]
    orb=[[1./3.,1./3.],[2./3.,2./3.]]
    model=tb_model(2,2,lat,orb,nspin=2)
    model.set_onsite([delta,-delta])

    # definitions of Pauli matrices
    sigma_x=np.array([0.,1.,0.,0])
    sigma_y=np.array([0.,0.,1.,0])
    sigma_z=np.array([0.,0.,0.,1])
    r3h =np.sqrt(3.0)/2.0
    sigma_a= 0.5*sigma_x-r3h*sigma_y
    sigma_b= 0.5*sigma_x+r3h*sigma_y
    sigma_c=-1.0*sigma_x

    # spin-independent first-neighbor hops
    for lvec in ([ 0, 0], [-1, 0], [ 0,-1]):
      model.set_hop(t, 0, 1, lvec)
    # spin-dependent second-neighbor hops
    for lvec in ([ 1, 0], [-1, 1], [ 0,-1]):
      model.set_hop(soc*1.j*sigma_z, 0, 0, lvec)
    for lvec in ([-1, 0], [ 1,-1], [ 0, 1]):
      model.set_hop(soc*1.j*sigma_z, 1, 1, lvec)
    # spin-flip first-neighbor hops
    model.set_hop(1.j*rashba*sigma_a, 0, 1, [ 0, 0], mode="add")
    model.set_hop(1.j*rashba*sigma_b, 0, 1, [-1, 0], mode="add")
    model.set_hop(1.j*rashba*sigma_c, 0, 1, [ 0,-1], mode="add")

    return model

# For the purposes of plot labels:
#    Real space is (r1,r2) in reduced coordinates
```

```
#    Reciprocal space is (k1,k2) in reduced coordinates
# Below, following Python, these are (r0,r1) and (k0,k1)

# set up figures
fig,ax=plt.subplots(2,2,figsize=(7,6))

nk=51
# run over two choices of t2
for je,soc in enumerate(soc_list):

    # solve bulk model on grid and get hybrid Wannier centers along r1
    # as a function of k0
    my_model=set_model(t,soc,rashba,delta)
    my_array=wf_array(my_model,[nk,nk])
    my_array.solve_on_grid([0.,0.])
    rbar = my_array.berry_phase([0,1],1,berry_evals=True,contin=True)/ \
        (2.*np.pi)

    # set up and solve ribbon model that is finite along direction 1
    width=20
    nkr=81
    ribbon_model=my_model.cut_piece(width,fin_dir=1,glue_edgs=False)
    (k_vec,k_dist,k_node)=ribbon_model.k_path('full',nkr,report=False)
    (rib_eval,rib_evec)=ribbon_model.solve_all(k_vec,eig_vectors=True)

    nbands=rib_eval.shape[0]
    (ax0,ax1)=ax[je,:]

    # hybrid Wannier center flow
    k0=np.linspace(0.,1.,nk)
    ax0.set_xlim(0.,1.)
    ax0.set_ylim(-1.3,1.3)
    ax0.set_xlabel(r"$\kappa_1/2\pi$")
    ax0.set_ylabel(r"HWF centers")
    ax0.axvline(x=0.5,linewidth=0.5, color='k')
    for shift in (-1.,0.,1.,2.):
      ax0.plot(k0,rbar[:,0]+shift,color='k')
      ax0.plot(k0,rbar[:,1]+shift,color='k')

    # edge band structure
    k0=np.linspace(0.,1.,nkr)
    ax1.set_xlim(0.,1.)
    ax1.set_ylim(-2.5,2.5)
    ax1.set_xlabel(r"$\kappa_1/2\pi$")
    ax1.set_ylabel(r"Edge band structure")
    for (i,kv) in enumerate(k0):

        # find expectation value <r1> at i'th k-point along direction k0
        pos_exp=ribbon_model.position_expectation(rib_evec[:,i],dir=1)

        # assign weight in [0,1] to be 1 except for edge states near bottom
        weight=3.0*pos_exp/width
        for j in range(nbands):
          weight[j]=min(weight[j],1.)

        # scatterplot with symbol size proportional to assigned weight
        s=ax1.scatter([k_vec[i]]*nbands, rib_eval[:,i],
            s=0.6+2.5*weight, c='k', marker='o', edgecolors='none')

    #    ax0.text(-0.45,0.92,'(a)',size=18.,transform=ax0.transAxes)
    #    ax1.text(-0.45,0.92,'(b)',size=18.,transform=ax1.transAxes)
```

```
# save figure as a PDF
aa=ax.flatten()
for i,lab in enumerate(['(a)','(b)','(c)','(d)']):
    aa[i].text(-0.45,0.92,lab,size=18.,transform=aa[i].transAxes)
fig.tight_layout()
plt.subplots_adjust(left=0.15,wspace=0.6)
fig.savefig("kanemele_topo.pdf")
```

D.18 Fu–Kane–Mele Model

Program `fkm.py` solves the Fu–Kane–Mele model of Eq. (5.28) in Section 5.3.4. This is again a spinor model, and the spin-dependent hoppings are now specified using the convention that `spin` is a NUMPY array containing the coefficients of $\mathbb{1}$, σ_x, σ_y, and σ_z respectively, with the list of directions `dir_list` specifying the last three of those coefficients (the first is set to zero).

The program generates two plot files, `fkm_bsr.pdf` and `fkm_topo.pdf`. The first is the band-structure plot that appears as Fig. D.5. The second is a plot of the hybrid Wannier center flow in two different κ_1–κ_2 planes, at $\kappa_3 = 0$ and at $\kappa_3 = \pi$, which appears as Fig. 5.19 in Section 5.3.4.

fkm.py

```
#!/usr/bin/env python
from __future__ import print_function # python3 style print

# Three-dimensional Fu-Kane-Mele model
# Fu, Kane and Mele, PRL 98, 106803 (2007)

from pythtb import * # import TB model class
import matplotlib.pyplot as plt

# set model parameters
t=1.0        # spin-independent first-neighbor hop
```

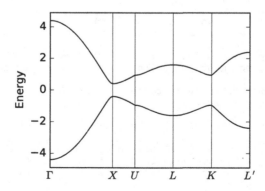

Figure D.5 Band structure of the Fu–Kane-Mele model of Eq. (5.28). Parameter values are $t_0 = 1$, $\Delta t = 0.4$, and $\lambda_{SO} = 0.125$.

```
dt=0.4      # modification to t for (111) bond
soc=0.125   # spin-dependent second-neighbor hop

def set_model(t,dt,soc):

    # set up Fu-Kane-Mele model
    lat=[[.0,.5,.5],[.5,.0,.5],[.5,.5,.0]]
    orb=[[0.,0.,0.],[.25,.25,.25]]
    model=tb_model(3,3,lat,orb,nspin=2)

    # spin-independent first-neighbor hops
    for lvec in ([0,0,0],[-1,0,0],[0,-1,0],[0,0,-1]):
      model.set_hop(t,0,1,lvec)
    model.set_hop(dt,0,1,[0,0,0],mode="add")

    # spin-dependent second-neighbor hops
    lvec_list=([1,0,0],[0,1,0],[0,0,1],[-1,1,0],[0,-1,1],[1,0,-1])
    dir_list=([0,1,-1],[-1,0,1],[1,-1,0],[1,1,0],[0,1,1],[1,0,1])
    for j in range(6):
      spin=np.array([0.]+dir_list[j])
      model.set_hop( 1.j*soc*spin,0,0,lvec_list[j])
      model.set_hop(-1.j*soc*spin,1,1,lvec_list[j])

    return model

my_model=set_model(t,dt,soc)
my_model.display()

# first plot: compute band structure
# --------------------------------

# construct path in k-space and solve model
path=[[0.,0.,0.],[0.,.5,.5],[0.25,.625,.625],
      [.5,.5,.5],[.75,.375,.375],[.5,0.,0.]]
label=(r'$\Gamma$',r'$X$',r'$U$',r'$L$',r'$K$',r'$L^\prime$')
(k_vec,k_dist,k_node)=my_model.k_path(path,101)

evals=my_model.solve_all(k_vec)

# band structure plot
fig, ax = plt.subplots(1,1,figsize=(4.,3.))
ax.set_xlim([0,k_node[-1]])
ax.set_xticks(k_node)
ax.set_xticklabels(label)
for n in range(len(k_node)):
  ax.axvline(x=k_node[n],linewidth=0.5, color='k')
ax.set_ylabel("Energy")
ax.set_ylim(-4.9,4.9)
for n in range(4):
  ax.plot(k_dist,evals[n],color='k')
fig.tight_layout()
fig.savefig("fkm_bsr.pdf")

# second plot: compute Wannier flow
# --------------------------------

# initialize plot
fig, ax = plt.subplots(1,2,figsize=(5.4,2.6),sharey=True)

# Obtain eigenvectors on 2D grid on slices at fixed kappa_3
# Note physical (kappa_1,kappa_2,kappa_3) have python indices (0,1,2)
```

```
kappa2_values=[0.,0.5]
labs=[r'$\kappa_3$=0',r'$\kappa_3$=$\pi$']
nk=41
dk=1./float(nk-1)
wf=wf_array(my_model,[nk,nk])

#loop over slices
for j in range(2):
  for k0 in range(nk):
    for k1 in range(nk):
      kvec=[k0*dk,k1*dk,kappa2_values[j]]
      (eval,evec)=my_model.solve_one(kvec,eig_vectors=True)
      wf[k0,k1]=evec
  wf.impose_pbc(mesh_dir=0,k_dir=0)
  wf.impose_pbc(mesh_dir=1,k_dir=1)
  hwfc=wf.berry_phase([0,1],dir=1,contin=True,berry_evals=True)/\
    (2.*np.pi)

  ax[j].set_xlim([0.,1.])
  ax[j].set_xticks([0.,0.5,1.])
  ax[j].set_xlabel(r"$\kappa_1/2\pi$")
  ax[j].set_ylim(-0.5,1.5)
  for n in range(2):
    for shift in [-1.,0.,1.]:
      ax[j].plot(np.linspace(0.,1.,nk),hwfc[:,n]+shift,color='k')
  ax[j].text(0.08,1.20,labs[j],size=12.,bbox=dict(facecolor='w',
    edgecolor='k'))

ax[0].set_ylabel(r"HWF center $\bar{s}_2$")
fig.tight_layout()
plt.subplots_adjust(left=0.15,wspace=0.2)
fig.savefig("fkm_topo.pdf")
```

References

Alexandradinata, A., Dai, X., and Bernevig, B. A. 2014. Wilson-loop characterization of inversion-symmetric topological insulators. *Phys. Rev. B*, **89**, 155114.

Armitage, N. P., Mele, E. J., and Vishwanath, A. Weyl and Dirac semimetals in three dimensional solids. *Rev. Mod. Phys.*, **90**, 015001.

Bansil, A., Lin, H., and Das, T. 2016. *Colloquium*: Topological band theory. *Rev. Mod. Phys.*, **88**, 021004.

Baroni, S., De Gironcoli, S., Dal Corso, A., and Giannozzi, P. 2001. Phonons and related crystal properties from density-functional perturbation theory. *Rev. Mod. Phys.*, **73**, 515.

Baroni, S., Giannozzi, P., and Testa, A. 1987. Green's-function approach to linear response in solids. *Phys. Rev. Lett.*, **58**, 1861–1864.

Baroni, S., and Resta, R. 1986. *Ab initio* calculation of the low-frequency Raman cross section in silicon. *Phys. Rev. B*, **33**, 5969–5971.

Beenakker, C. W. J. 1990. Edge channels for the fractional quantum Hall effect. *Phys. Rev. Lett.*, **64**, 216.

Berger, L. 1964. Influence of spin-orbit interaction on the transport processes in ferromagnetic nickel alloys, in the presence of a degeneracy of the 3D band. *Physica*, **30**, 1141–1159.

Berger, L. 1970. Side-jump mechanism for the Hall effect of ferromagnets. *Phys. Rev. B*, **2**, 4559–4566.

Bernevig, B. A. 2013. *Topological Insulators and Topological Superconductors*. Princeton University Press.

Bernevig, B. A., Hughes, T. L., and Zhang, S. C. 2006. Quantum spin Hall effect and topological phase transition in HgTe quantum wells. *Science*, **314**, 1757–1761.

Berry, M. V. 1984. Quantal phase factors accompanying adiabatic changes. *Proc. R. Soc. Lond. A*, **392**, 45.

Berry, M. V. 1985. Aspects of degeneracy. Pages 123–140 of: Casati, Giulio (ed.), *Chaotic Behavior in Quantum Systems: Theory and Applications*. Springer.

Blount, E. I. 1962. Formalisms of band theory. *Solid State Phys.*, **13**, 305.

Bohm, A., Kendrick, B., and Loewe, Mark E. 1992. The Berry phase in molecular physics. *Int. J. Quantum Chem.*, **41**, 53–75.

Boys, S. F. 1960. Construction of some molecular orbitals to be approximately invariant for changes from one molecule to another. *Rev. Mod. Phys.*, **32**, 296.

Boys, S. F. 1966. Localized orbitals and localized adjustment functions. Page 253 of: Löwdin, P.-O. (ed.), *Quantum Theory of Atoms, Molecules, and the Solid State*. Academic Press.

Bradlyn, B., Elcoro, L., Cano, J., Wang, Z., Vergniory, M. G., Felser, C., Aroyo, M. I., and Bernevig, B. A. 2017. Topological quantum chemistry. *Nature*, **547**, 298–305.

Brouder, C., Panati, G., Calandra, M., Mourougane, C., and Marzari, N. 2007. Exponential localization of Wannier functions in insulators. *Phys. Rev. Lett.*, **98**, 046402.

Bukov, M., D'Alessio, L., and Polkovnikov, A. 2015. Universal high-frequency behavior of periodically driven systems: from dynamical stabilization to Floquet engineering. *Adv. Phys.*, **64**, 139–226.

Cayssol, J., Dóra, B., Simon, F., and Moessner, R. 2013. Floquet topological insulators. *Phys. Status Solidi-R*, **7**, 101–8.

Ceresoli, D., Gerstmann, U., Seitsonen, A. P., and Mauri, F. 2010. First-principles theory of orbital magnetization. *Phys. Rev. B*, **81**, 060409.

Ceresoli, D., Thonhauser, T., Vanderbilt, D., and Resta, R. 2006. Orbital magnetization in crystalline solids: multi-band insulators, Chern insulators, and metals. *Phys. Rev. B*, **74**, 024408.

Chang, C.-Z., Zhang, J., Feng, X., Shen, J., Zhang, Z., Guo, M., Li, K., Ou, Y., Wei, P., Wang, L.-L., et al. 2013. Experimental observation of the quantum anomalous Hall effect in a magnetic topological insulator. *Science*, **340**, 167.

Chang, C.-Z., Zhao, W., Kim, D. Y., Zhang, H., Assaf, B. A., Heiman, D., Zhang, S.-C., Liu, C., Chan, M. H. W., and Moodera, J. S. 2015. High-precision realization of robust quantum anomalous Hall state in a hard ferromagnetic topological insulator. *Nat. Mater.*, **14**, 473–477.

Chang, M.-C., and Niu, Q. 1996. Berry phase, hyperorbits, and the Hofstadter spectrum: semiclassical dynamics in magnetic Bloch bands. *Phys. Rev. B*, **53**, 7010–7023.

Chang, M.-C., and Niu, Q. 2008. Berry curvature, orbital moment, and effective quantum theory of electrons in electromagnetic fields. *J. Phys. Condens. Matter*, **20**, 193202.

Chen, K.-T., and Lee, P. A. 2011. Topological insulator and the θ vacuum in a system without boundaries. *Phys. Rev. B*, **83**, 125119.

Coh, S., and Vanderbilt, D. 2013. Canonical magnetic insulators with isotropic magnetoelectric coupling. *Phys. Rev. B*, **88**, 121106.

Coh, S., Vanderbilt, D., Malashevich, A., and Souza, I. 2011. Chern–Simons orbital magnetoelectric coupling in generic insulators. *Phys. Rev. B*, **83**, 085108.

Coleman, S. 1979. The uses of instantons. Pages 805–941 of: *The Whys of Subnuclear Physics*. Springer.

Deser, S., Jackiw, R., and Templeton, S. 1982. Topologically massive gauge theories. *Ann. Phys.*, **140**, 372–411.

Dziawa, P., Kowalski, B. J., Dybko, K., Buczko, R., Szczerbakow, A., Szot, M., Lusakowska, E., Balasubramanian, T., Wojek, Bastian M., Berntsen, M. H., et al. 2012. Topological crystalline insulator states in $Pb_{1-x}Sn_xTe$. *Nat. Mater.*, **11**, 1023–1027.

Edmiston, C., and Ruedenberg, K. 1963. Localized atomic and molecular orbitals. *Rev. Mod. Phys.*, **35**, 457.

Eschrig, H. 2011. *Topology and Geometry for Physics*. Vol. 822. Springer Science & Business Media.

Essin, A. M., Moore, J. E., and Vanderbilt, D. 2009. Magnetoelectric polarizability and axion electrodynamics in crystalline insulators. *Phys. Rev. Lett.*, **102**, 146805. See also Erratum: *Phys. Rev. Lett.*, **103**, 259902.

Essin, A. M., Turner, A. M., Moore, J. E., and Vanderbilt, D. 2010. Orbital magnetoelectric coupling in band insulators. *Phys. Rev. B*, **81**, 205104.

Fang, C., Weng, H., Dai, X., and Fang, Z. 2016. Topological nodal line semimetals. *Chinese Physics B*, **25**, 117106.

Fang, Z., Nagaosa, N., Takahashi, K. S., Asamitsu, A., Mathieu, R., Ogasawara, T., Yamada, H., Kawasaki, M., Tokura, Y., and Terakura, K. 2003. The anomalous Hall effect and magnetic monopoles in momentum space. *Science*, **302**, 92–95.

Fiebig, M. 2005. Revival of the magnetoelectric effect. *J. Phys. D Appl. Phys.*, **38**, R123–R152.

Foster, J. M., and Boys, S. F. 1960a. A quantum variational calculation for HCHO. *Rev. Mod. Phys.*, **32**, 303.

Foster, J. M., and Boys, S. F. 1960b. Canonical configurational interaction procedure. *Rev. Mod. Phys.*, **32**, 300.

Frankel, T. 1997. *The Geometry of Physics*. Cambridge University Press.

Freed, D. S. and Moore, G. W. 2013. Twisted equivariant matter. *Annales Henri Poincaré*, **14**, 1927–2023.

Fruchart, M. and Carpentier, D. 2013. An introduction to topological insulators. *Comptes Rendus Physique*, **14**, 779–815.

Fu, L., and Kane, C. L. 2007. Topological insulators with inversion symmetry. *Phys. Rev. B*, **76**, 045302.

Fu, L., Kane, C. L., and Mele, E. J. 2007. Topological insulators in three dimensions. *Phys. Rev. Lett.*, **98**, 106803.

Fu, L. 2011. Topological crystalline insulators. *Phys. Rev. Lett.*, **106**, 106802.

Fu, L., and Kane, C. L. 2006. Time reversal polarization and a Z_2 adiabatic spin pump. *Phys. Rev. B*, **74**, 195312.

Fuh, H.-R., and Guo, G.-Y. 2011. Intrinsic anomalous Hall effect in nickel: a GGA+U study. *Phys. Rev. B*, **84**, 144427.

Fukui, T., and Hatsugai, Y. 2007. Quantum spin Hall effect in three dimensional materials: lattice computation of Z_2 topological invariants and its application to Bi and Sb. *J. Phys. Soc. Japan*, **76**, 053702.

Gao, Y., Yang, S. A., and Niu, Q. 2014. Field induced positional shift of Bloch electrons and its dynamical implications. *Phys. Rev. Lett.*, **112**, 166601.

Gmitra, M., Konschuh, S., Ertler, C., Ambrosch-Draxl, C., and Fabian, J. 2009. Band-structure topologies of graphene: spin-orbit coupling effects from first principles. *Phys. Rev. B*, **80**, 235431.

Goldman, N., Budich, J. C., and Zoller, P. 2016. Topological quantum matter with ultracold gases in optical lattices. *Nat. Phys.*, **12**, 639–645.

Gonze, X. 1997. First-principles responses of solids to atomic displacements and homogeneous electric fields: implementation of a conjugate-gradient algorithm. *Phys. Rev. B*, **55**, 10337.

Gonze, X., and Lee, C. 1997. Dynamical matrices, Born effective charges, dielectric permittivity tensors, and interatomic force constants from density-functional perturbation theory. *Phys. Rev. B*, **55**, 10355.

Gonze, X., and Zwanziger, J. W. 2011. Density-operator theory of orbital magnetic susceptibility in periodic insulators. *Phys. Rev. B*, **84**, 064445.

Gosálbez-Martínez, D., Souza, I., and Vanderbilt, D. 2015. Chiral degeneracies and Fermi-surface Chern numbers in bcc Fe. *Phys. Rev. B*, **92**, 085138.

Gradhand, M., Fedorov, D. V., Pientka, Falko, Zahn, P., Mertig, I., and Györffy, B. L. 2012. First-principle calculations of the Berry curvature of Bloch states for charge and spin transport of electrons. *J. Phys. Condens. Matter*, **24**, 213202.

Gresch, D., Autès, G., Yazyev, O. V., Troyer, M., Vanderbilt, D., Bernevig, B. A., and Soluyanov, A. A. 2017. Z2Pack: numerical implementation of hybrid Wannier centers for identifying topological materials. *Phys. Rev. B*, **95**, 075146.

Haldane, F. D. M. 1983. Nonlinear field theory of large-spin Heisenberg antiferromagnets: semiclassically quantized solitons of the one-dimensional easy-axis Néel state. *Phys. Rev. Lett.*, **50**, 1153–1156.

Haldane, F. D. M. 1988. Model for a quantum Hall effect without Landau levels: condensed-matter realization of the "parity anomaly"? *Phys. Rev. Lett.*, **61**, 2015–18.

Haldane, F. D. M. 2004. Berry curvature on the Fermi surface: anomalous Hall effect as a topological Fermi-liquid property. *Phys. Rev. Lett.*, **93**, 206602.

Haldane, F. D. M. 2014. Attachment of surface "Fermi arcs" to the bulk Fermi surface: "Fermi-level plumbing" in topological metals. *ArXiv e-prints*.

Hall, E. H. 1879. On a new action of the magnet on electric currents. *Am. J. Math*, **2**, 287–92.

Hall, E. H. 1881. On the possibility of transverse currents in ferromagnets. *Philos. Mag.*, **12**, 157–60.

Hamamoto, K., Ezawa, M., and Nagaosa, N. 2015. Quantized topological Hall effect in skyrmion crystal. *Phys. Rev. B*, **92**, 115417.

Hanke, J.-P., Freimuth, F., Nandy, A. K., Zhang, H., Blügel, S., and Mokrousov, Y. 2016. Role of Berry phase theory for describing orbital magnetism: from magnetic heterostructures to topological orbital ferromagnets. *Phys. Rev. B*, **94**, 121114.

Harrison, W. A. 1989. *Electronic Structure and the Properties of Solids*. Dover.

Hasan, M. Z., and Kane, C. L. 2010. Colloquium: topological insulators. *Rev. Mod. Phys.*, **82**, 3045.

Hasan, Z., Xu, S.-Y., Belopolski, I., and Huang, S.-M. 2017. Discovery of Weyl fermion semimetals and topological Fermi arc states. *Annu. Rev. Condens. Matt. Phys.*, **8**, 289–309.

Hirst, L. L. 1997. The microscopic magnetization: Concept and application. *Rev. Mod. Phys.*, **69**, 607–28.

Hohenberg, P., and Kohn, W. 1964. Inhomogeneous electron gas. *Phys. Rev.*, **136**, B864.

Hsieh, D., Qian, D., Wray, L., Xia, Y., Hor, Y. S., Cava, R. J., and Hasan, M. Z. 2008. A topological Dirac insulator in a quantum spin Hall phase. *Nature*, **452**, 970–U5.

Hsieh, T. H., Lin, H., Liu, J., Duan, W., Bansil, A., and Fu, L. 2012. Topological crystalline insulators in the SnTe material class. *Nat. Commun.*, **3**, 982.

Huang, S.-M., Xu, S.-Y., Belopolski, I., Lee, C.-C., Chang, G., Wang, B., Alidoust, N., Bian, G., Neupane, M., Zhang, C., et al. 2015. A Weyl fermion semimetal with surface Fermi arcs in the transition metal monopnictide TaAs class. *Nat. Commun.*, **6**.

Íñiguez, J. 2008. First-principles approach to lattice-mediated magnetoelectric effects. *Phys. Rev. Lett.*, **101**, 117201.

Joannopoulos, J. D., Johnson, S. G., Winn, J. N., and Meade, R. D. 2011. *Photonic Crystals: Molding the Flow of Light*. Princeton University Press.

Jones, R. O. 2015. Density functional theory: its origins, rise to prominence, and future. *Rev. Mod. Phys.*, **87**, 897–923.

Jotzu, G., Messer, M., Desbuquois, R., Lebrat, M., Uehlinger, T., Greif, D., and Esslinger, T. 2014. Experimental realization of the topological Haldane model with ultracold fermions. *Nature*, **515**, 237–40.

Jungwirth, T. 2002. Anomalous Hall effect in ferromagnetic semiconductors. *Phys. Rev. Lett.*, **88**, 207208.

Kane, C. L. 2008. Condensed matter: an insulator with a twist. *Nature Phys.*, **4**, 348–9.

Kane, C. L., and Mele, E. J. 2005a. Quantum spin Hall effect in graphene. *Phys. Rev. Lett.*, **95**, 226801.

Kane, C. L., and Mele, E. J. 2005b. Z_2 topological order and the quantum spin Hall effect. *Phys. Rev. Lett.*, **95**, 146802.

Karplus, R., and Luttinger, J. M. 1954. Hall effect in ferromagnets. *Phys. Rev.*, **95**, 1154–1160.

King-Smith, R. D., and Vanderbilt, D. 1993. Theory of polarization of crystalline solids. *Phys. Rev. B*, **47**, 1651–4.

Kitaev, A., and Preskill, J. 2006. Topological entanglement entropy. *Phys. Rev. Lett.*, **96**, 110404.

Kivelson, S. 1982. Wannier functions in one-dimensional disordered systems: application to fractionally charged solitons. *Phys. Rev. B*, **26**, 4269.

Knez, I., Du, R.-R., and Sullivan, G. 2011. Evidence for helical edge modes in inverted InAs/GaSb quantum wells. *Phys. Rev. Lett.*, **107**, 136603.

Kohmoto, M., Halperin, B. I., and Wu, Y.-S. 1993. Quantized Hall effect in 3D periodic systems. *Physica B*, **184**, 30–33.

Kohn, W. 1996. Density functional and density matrix method scaling linearly with the number of atoms. *Phys. Rev. Lett.*, **76**, 3168–3171.

Kohn, W., and Luttinger, J. M. 1957. Quantum theory of electrical transport phenomena. *Phys. Rev.*, **108**, 590–611.

Kohn, W., and Sham, L. J. 1965. Self-consistent equations including exchange and correlation effects. *Phys. Rev.*, **140**, A1133.

Konig, M., Wiedmann, S., Brune, C., Roth, A., Buhmann, H., Molenkamp, L. W., Qi, X. L., and Zhang, S. C. 2007. Quantum spin hall insulator state in HgTe quantum wells. *Science*, **318**, 766–70.

Lindner, N. H., Refael, G., and Galitski, V. 2011. Floquet topological insulator in semiconductor quantum wells. *Nat. Phys.*, **7**, 490–5.

Littlewood, P. B. 1980. On the calculation of the macroscopic polarisation induced by an optic phonon. *J. Phys. C*, **13**, 4893.

Liu, J., Park, S. Y., Garrity, K. F., and Vanderbilt, D. 2016. Flux states and topological phases from spontaneous time-reversal symmetry breaking in CrSi(Ge)Te$_3$-based systems. *Phys. Rev. Lett.*, **117**, 257201.

Liu, J., and Vanderbilt, D. 2015. Gauge-discontinuity contributions to Chern–Simons orbital magnetoelectric coupling. *Phys. Rev. B*, **92**, 245138.

Lopez, M. G., Vanderbilt, D., Thonhauser, T., and Souza, I. 2012. Wannier-based calculation of the orbital magnetization in crystals. *Phys. Rev. B*, **85**, 014435.

Lu, L., Joannopoulos, J. D., and Soljačić, M. 2014. Topological photonics. *Nat. Photonics*, **8**, 821–829.

Luttinger, J. M. 1958. Theory of the Hall effect in ferromagnetic substances. *Phys. Rev.*, **112**, 739.

Lv, B. Q., Weng, H. M., Fu, B. B., Wang, X. P., Miao, H., Ma, J., Richard, P., Huang, X. C., Zhao, L. X., Chen, G.F., et al. 2015. Experimental discovery of Weyl semimetal TaAs. *Phys. Rev. X*, **5**, 031013.

Malashevich, A., Coh, S., Souza, I., and Vanderbilt, D. 2012. Full magnetoelectric response of Cr$_2$O$_3$ from first principles. *Phys. Rev. B*, **86**, 094430.

Malashevich, A., Souza, I., Coh, S., and Vanderbilt, D. 2010. Theory of orbital magnetoelectric response. *New J. Phys.*, **12**, 053032.

Martin, R. M. 2004. *Electronic Structure: Basic Theory and Practical Methods*. Cambridge University Press.

Marzari, N., Mostofi, A. A., Yates, J. R., Souza, I., and Vanderbilt, D. 2012. Maximally localized Wannier functions: theory and applications. *Rev. Mod. Phys.*, **84**, 1419–75.

Marzari, N., and Vanderbilt, D. 1997. Maximally localized generalized Wannier functions for composite energy bands. *Phys. Rev. B*, **56**, 12847–65.

Mogi, M., Kawamura, M., Tsukazaki, A., Yoshimi, R., Takahashi, K. S., Kawasaki, M., and Tokura, Y. 2017. Tailoring tricolor structure of magnetic topological insulator for robust axion insulator. *Sci. Adv.*, **3**, eaao1669.

Mogi, M., Kawamura, M., Yoshimi, R., Tsukazaki, A., Kozuka, Y., Shirakawa, N., Takahashi, K. S., Kawasaki, M., and Tokura, Y. 2017. A magnetic heterostructure of topological insulators as a candidate for an axion insulator. *Nat. Mater*, **16**, 516.

Mong, R. S. K., Essin, A. M., and Moore, J. E. 2010. Antiferromagnetic topological insulators. *Phys. Rev. B*, **81**, 245209.

Moore, J. E. 2010. The birth of topological insulators. *Nature*, **464**, 194–8.

Moore, J. E., and Balents, L. 2007. Topological invariants of time-reversal-invariant band structures. *Phys. Rev. B*, **75**, 121306.

Morimoto, T., Furusaki, A., and Nagaosa, N. 2015. Topological magnetoelectric effects in thin films of topological insulators. *Phys. Rev. B*, **92**, 085113.

Mostofi, A. A., Yates, J. R., Lee, Y.-S., Souza, I., Vanderbilt, D., and Marzari, N. 2008. Wannier90: a tool for obtaining maximally-localised Wannier functions. *Comput. Phys. Commun.*, **178**, 685–99.

Mostofi, A. A., Yates, J. R., Pizzi, G. Lee, Y.-S., Souza, I., Vanderbilt, D., and Marzari, N. 2014. An updated version of Wannier90: a tool for obtaining maximally-localised Wannier functions. *Comput. Phys. Commun.*, **185**, 2309–2310.

Murakami, S. 2007. Phase transition between the quantum spin Hall and insulator phases in 3D: emergence of a topological gapless phase. *New J. Phys.*, **9**, 356.

Murakami, S., and Kuga, S. 2008. Universal phase diagrams for the quantum spin Hall systems. *Phys. Rev. B*, **78**, 165313.

Nagaosa, N., Sinova, J., Onoda, S., MacDonald, A. H., and Ong, N. P. 2010. Anomalous Hall effect. *Rev. Mod. Phys.*, **82**, 1539–1592.

Nakahara, M. 2003. *Geometry, Topology and Physics*. CRC Press.

Neaton, J. B., Ederer, C., Waghmare, U. V., Spaldin, N. A., and Rabe, K. M. 2005. First-principles study of spontaneous polarization in multiferroic $BiFeO_3$. *Phys. Rev. B*, **71**, 014113.

Nielsen, H. B., and Ninomiya, M. 1983. The Adler-Bell-Jackiw anomaly and Weyl fermions in a crystal. *Phys. Lett. B*, **130**, 389–96.

Nikolaev, S. A., and Solovyev, I. V. 2014. Orbital magnetization of insulating perovskite transition-metal oxides with a net ferromagnetic moment in the ground state. *Phys. Rev. B*, **89**, 064428.

Niu, Q. 1986. Quantum adiabatic particle transport. *Phys. Rev. B*, **34**, 5093.

Niu, Q. 1991. Theory of the quantized adiabatic particle transport. *Mod. Phys. Lett. B*, **5**, 923.

Niu, Q., and Thouless, D. J. 1984. Quantised adiabatic charge transport in the presence of substrate disorder and many-body interaction. *J. Phys. A: Math. Gen.*, **17**, 2453.

Nomura, K., and Nagaosa, N. 2011. Surface-quantized anomalous Hall current and the magnetoelectric effect in magnetically disordered topological insulators. *Phys. Rev. Lett.*, **106**, 166802.

Nourafkan, R., and Kotliar, G. 2013. Electric polarization in correlated insulators. *Phys. Rev. B*, **88**, 155121.

Olsen, T., Taherinejad, M., Vanderbilt, D., and Souza, I. 2017. Surface theorem for the Chern–Simons axion coupling. *Phys. Rev. B*, **95**, 075137.

Onoda, M., and Nagaosa, N. 2002. Topological nature of anomalous Hall effect in ferromagnets. *J. Phys. Soc. Jpn.*, **71**, 19–22.

Ortiz, G., and Martin, R. M. 1994. Macroscopic polarization as a geometric quantum phase: many-body formulation. *Phys. Rev. B*, **49**, 14202.

Pancharatnam, S. 1956. Generalized theory of interference, and its applications. Part I. Coherent pencils. In *Proceedings of the Indian Academy of Sciences, Section A*. Vol. 44. Indian Academy of Sciences, pp. 247–62.

Parr, R. G., and Weitao, Y. 1989. *Density-Functional Theory of Atoms and Molecules*. Vol. 16. Oxford University Press.

Pick, R. M., Cohen, M. H., and Martin, R. M. 1970. Microscopic theory of force constants in the adiabatic approximation. *Phys. Rev. B*, **1**, 910–920.

Po, H. C., Vishwanath, A., and Watanabe, H. 2017. Symmetry-based indicators of band topology in the 230 space groups. *Nat. Commun.*, **8**.

Prodan, E. 2009. Robustness of the spin-Chern number. *Phys. Rev. B*, **80**, 125327.

Prodan, E., and Kohn, W. 2005. Nearsightedness of electronic matter. *Proc. Nat. Acad. Sci. USA*, **102**, 11635–8.

Qi, X. L., Hughes, T. L., and Zhang, S. C. 2008. Topological field theory of time-reversal invariant insulators. *Phys. Rev. B*, **78**, 195424.

Qi, X.-L., Wu, Y.-S., and Zhang, S.-C. 2006. Topological quantization of the spin Hall effect in two-dimensional paramagnetic semiconductors. *Phys. Rev. B*, **74**, 085308.

Qi, X.-L., and Zhang, S.-C. 2011. Topological insulators and superconductors. *Rev. Mod. Phys.*, **83**, 1057–110.

Resta, R. 1992. Theory of the electric polarization in crystals. *Ferroelectrics*, **136**, 51.

Resta, R. 1998. Quantum-mechanical position operator in extended systems. *Phys. Rev. Lett.*, **80**, 1800–3.

Resta, R. 2002. Why are insulators insulating and metals conducting? *J. Phys. Condens. Matter*, **14**, R625.

Resta, R. 2010. Towards a bulk theory of flexoelectricity. *Phys. Rev. Lett.*, **105**, 127601.

Resta, R., and Sorella, S. 1999. Electron localization in the insulating state. *Phys. Rev. Lett.*, **82**, 370–3.

Resta, R., and Vanderbilt, D. 2007. Theory of polarization: a modern approach. In C. Ahn and K. M. Rabe, eds., *Physics of Ferroelectrics: A Modern Perspective*. Springer-Verlag, pp. 31–68.

Rice, M. J., and Mele, E. J. 1982. Elementary excitations of a linearly conjugated diatomic polymer. *Phys. Rev. Lett.*, **49**, 1455.

Rivera, J. P. 1994. On definitions, units, measurements, tensor forms of the linear magnetoelectric effect and on a new dynamic method applied to Cr-Cl boracite. *Ferroelectrics*, **161**, 165–80.

Rivera, J.-P. 2009. A short review of the magnetoelectric effect and related experimental techniques on single phase multiferroics. *Eur. Phys. J. B*, **71**, 299.

Roy, R. 2009. Topological phases and the quantum spin Hall effect in three dimensions. *Phys. Rev. B*, **79**, 195322.

Roy, R., and Harper, F. 2017. Floquet topological phases with symmetry in all dimensions. *Phys. Rev. B*, **95**, 195128.

Rüegg, A., and Fiete, G. A. 2011. Topological insulators from complex orbital order in transition-metal oxides heterostructures. *Phys. Rev. B*, **84**, 201103.

Sakurai, J. J. 1994. *Modern Quantum Mechanics*. 2nd ed. Addison-Wesley.

Sato, M., and Ando, Y. 2017. Topological superconductors: a review. *Rep. Prog. Phys.*, **80**, 076501.

Senthil, T. 2015. Symmetry-protected topological phases of quantum matter. *Annu. Rev. Condens. Matt. Phys.*, **6**, 299–324.

Shi, J., Vignale, G., Xiao, D., and Niu, Q. 2007. Quantum theory of orbital magnetization and its generalization to interacting systems. *Phys. Rev. Lett.*, **99**, 197202.

Smit, J. 1955. The spontaneous Hall effect in ferromagnetics I. *Physica*, **21**, 877–87.

Smit, J. 1958. The spontaneous Hall effect in ferromagnetics II. *Physica*, **24**, 39–51.

Sodemann, I., and Fu, L. 2015. Quantum nonlinear Hall effect induced by Berry curvature dipole in time-reversal invariant materials. *Phys. Rev. Lett.*, **115**, 216806.

Soluyanov, A. A., Gresch, D., Wang, Z., Wu, Q., Troyer, M., Dai, X., and Bernevig, B. A. 2015. Type-II Weyl semimetals. *Nature*, **527**, 495–8.

Soluyanov, A. A., and Vanderbilt, D. 2011a. Computing topological invariants without inversion symmetry. *Phys. Rev. B*, **83**, 235401.

Soluyanov, A. A., and Vanderbilt, D. 2011b. Wannier representation of Z_2 topological insulators. *Phys. Rev. B*, **83**, 035108.

Son, D. T., and Spivak, B. Z. 2013. Chiral anomaly and classical negative magnetoresistance of Weyl metals. *Phys. Rev. B*, **88**, 104412.

Son, D. T., and Yamamoto, N. 2012. Berry curvature, triangle anomalies, and the chiral magnetic effect in Fermi liquids. *Phys. Rev. Lett.*, **109**, 181602.

Souza, I., Íñiguez, J., and Vanderbilt, D. 2002. First-principles approach to insulators in finite electric fields. *Phys. Rev. Lett.*, **89**, 117602.

Souza, I., Íñiguez, J., and Vanderbilt, D. 2004. Dynamics of Berry-phase polarization in time-dependent electric fields. *Phys. Rev. B*, **69**, 085106.

Souza, I., and Vanderbilt, D. 2008. Dichroic f-sum rule and the orbital magnetization of crystals. *Phys. Rev. B*, **77**, 054438.

Souza, I., Wilkens, T., and Martin, R. M. 2000. Polarization and localization in insulators: generating function approach. *Phys. Rev. B*, **62**, 1666–83.

Steinberg, J. A., Young, S. M., Zaheer, S., Kane, C. L., Mele, E. J., and Rappe, A. M. 2014. Bulk Dirac points in distorted spinels. *Phys. Rev. Lett.*, **112**, 036403.

Stengel, M., and Spaldin, N. A. 2006. Accurate polarization within a unified Wannier function formalism. *Phys. Rev. B*, **73**, 075121.

Stengel, M., Spaldin, N. A., and Vanderbilt, D. 2009. Enhancement of ferroelectricity at metal/oxide interfaces. *Nat. Mater.*, **8**, 392–7.

Středa, P. 1982. Quantised Hall effect in a two-dimensional periodic potential. *J. Phys. C*, **15**, L1299–303.

Su, W. P., Schrieffer, J. R., and Heeger, A. J. 1979. Solitons in polyacetylene. *Phys. Rev. Lett.*, **42**, 1698.

Sun, Y., Wu, S.-C., Ali, M. N., Felser, C. and Yan, B. 2015. Prediction of Weyl semimetal in orthorhombic $MoTe_2$. *Phys. Rev. B*, **92**, 161107.

Sundaram, G., and Niu, Q. 1999. Wave-packet dynamics in slowly perturbed crystals: gradient corrections and Berry-phase effects. *Phys. Rev. B*, **59**, 14915–25.

Taguchi, Y., Oohara, Y., Yoshizawa, H., Nagaosa, N., and Tokura, Y. 2001. Spin chirality, Berry phase, and anomalous Hall effect in a frustrated ferromagnet. *Science*, **291**, 2573–6.

Taherinejad, M., Garrity, K. F., and Vanderbilt, D. 2014. Wannier center sheets in topological insulators. *Phys. Rev. B*, **89**, 115102.

Taherinejad, M., and Vanderbilt, D. 2015. Adiabatic pumping of Chern–Simons axion coupling. *Phys. Rev. Lett.*, **114**, 096401.

Tanaka, Y., Ren, Z., Sato, T., Nakayama, K., Souma, S., Takahashi, T., Segawa, K., and Ando, Y. 2012. Experimental realization of a topological crystalline insulator in SnTe. *Nat. Phys.*, **8**, 800–3.

Thonhauser, T. 2011. Theory of orbital magnetization in solids. *Int. J. Mod. Phys. B*, **25**, 1429–1458.

Thonhauser, T., Ceresoli, D., Vanderbilt, D., and Resta, R. 2005. Orbital magnetization in periodic insulators. *Phys. Rev. Lett*, **95**, 137205.

Thouless, D. J. 1983. Quantization of particle transport. *Phys. Rev. B*, **27**, 6083.

Thouless, D. J., Kohmoto, M., Nightingale, M. P., and den Nijs, M. 1982. Quantized Hall conductance in a two-dimensional periodic potential. *Phys. Rev. Lett.*, **49**, 405.

Tsirkin, S. S., Souza, I., and Vanderbilt, D. 2017. Composite Weyl nodes stabilized by screw symmetry with and without time-reversal invariance. *Phys. Rev. B*, **96**, 045102.

Tsui, D., Stormer, H., and Gossard, A. 1982. Two-dimensional magnetotransport in the extreme quantum limit. *Phys. Rev. Lett.*, **48**, 1559–62.

Turner, A. M., Zhang, Y., Mong, R. S. K., and Vishwanath, A. 2012. Quantized response and topology of magnetic insulators with inversion symmetry. *Phys. Rev. B*, **85**, 165120.

Umari, P., and Pasquarello, A. 2002. Ab initio molecular dynamics in a finite homogeneous electric field. *Phys. Rev. Lett.*, **89**, 157602.

Vanderbilt, D. 1997. Nonlocality of Kohn-Sham exchange-correlation fields in dielectrics. *Phys. Rev. Lett.*, **79**, 3966–9.

Vanderbilt, D. 2000. Berry-phase theory of proper piezoelectric response. *J. Phys. Chem. Solids*, **61**, 147–51.

Vanderbilt, D., and King-Smith, R. D. 1993. Electric polarization as a bulk quantity and its relation to surface charge. *Phys. Rev. B*, **48**, 4442–55.

Vanderbilt, D., and Resta, R. 2006. Quantum electrostatics of insulators: polarization, Wannier functions, and electric fields. In S. G. Louie, and M. L. Cohen, eds., *Conceptual Foundations of Materials Properties: A Standard Model for Calculation of Ground- and Excited-State Properties. Contemporary Concepts of Condensed Matter Science*. Vol. 1. Elsevier, pp. 139–63.

Volovik, G. E. 2003. *The Universe in a Helium Droplet*. Oxford University Press.

Wan, X., Turner, A. M., Vishwanath, A., and Savrasov, S. Y. 2011. Topological semimetal and Fermi-arc surface states in the electronic structure of pyrochlore iridates. *Phys. Rev. B*, **83**, 205101.

Wang, J., Lian, B., Qi, X.-L., and Zhang, S.-C. 2015. Quantized topological magnetoelectric effect of the zero-plateau quantum anomalous Hall state. *Phys. Rev. B*, **92**, 081107.

Wang, X. J., Vanderbilt, D., Yates, J. R., and Souza, I. 2007. Fermi-surface calculation of the anomalous Hall conductivity. *Phys. Rev. B*, **76**, 195109.

Wang, X. J., Yates, J. R., Souza, I., and Vanderbilt, D. 2006. Ab initio calculation of the anomalous Hall conductivity by Wannier interpolation. *Phys. Rev. B*, **74**, 195118.

Wang, Z., Weng, H., Wu, Q., Dai, X., and Fang, Z. 2013. Three-dimensional Dirac semimetal and quantum transport in Cd_3As_2. *Phys. Rev. B*, **88**, 125427.

Wang, Z., Sun, Y., Chen, X.-Q., Franchini, C., Xu, G., Weng, H., Dai, X., and Fang, Z. 2012. Dirac semimetal and topological phase transitions in A_3Bi (A = Na, K, Rb). *Phys. Rev. B*, **85**, 195320.

Wen, X. 2007. *Quantum Field Theory of Many-Body Systems: From the Origin of Sound to an Origin of Light and Electrons*. Oxford University Press.

Wen, X.-G. 2017. Colloquium: Zoo of quantum-topological phases of matter. *Rev. Mod. Phys.*, **89**, 041004.

Weng, H., Fang, C., Fang, Z., Bernevig, B. A., and Dai, X. 2015. Weyl semimetal phase in noncentrosymmetric transition-metal monophosphides. *Phys. Rev. X*, **5**.

Weng, H., Yu, R., Hu, X., Dai, X., and Fang, Z. 2015. Quantum anomalous Hall effect and related topological electronic states. *Adv. Phys.*, **64**, 227–82.

Weyl, H. 1929. Gravitation and the electron. *Proc. Nat. Acad. Sci. USA*, **15**, 323–34.

Wilczek, F. 1987. Two applications of axion electrodynamics. *Phys. Rev. Let.*, **58**, 1799.

Wilczek, F. 2012. Quantum time crystals. *Phys. Rev. Lett.*, **109**, 160401.

Wilczek, F., and Shapere, A. 1989. *Geometric Phases in Physics*. Vol. 5. World Scientific.

Wilson, K. G. 1974. Confinement of quarks. *Phys. Rev. D*, **10**, 2445–59.

Witten, E. 1979. Dyons of charge $e\theta/2\pi$. *Phys. Lett. B*, **86**, 283–7.

Wojdeł, J. C., and Íñiguez, J. 2009. Magnetoelectric response of multiferroic $BiFeO_3$ and related materials from first-principles calculations. *Phys. Rev. Lett.*, **103**, 267205.

Wu, L., Salehi, M., Koirala, N., Moon, J., Oh, S., and Armitage, N. P. 2016. Quantized Faraday and Kerr rotation and axion electrodynamics of a 3D topological insulator. *Science*, **354**, 1124–7.

Xia, Y., Qian, D., Hsieh, D., Wray, L., Pal, A., Lin, H., Bansil, A., Grauer, D., Hor, Y. S., Cava, R. J., and Hasan, M. Z. 2009. Observation of a large-gap topological-insulator class with a single Dirac cone on the surface. *Nat. Phys.*, **5**, 398–402.

Xiao, D., Chang, M.-C., and Niu, Q. 2010. Berry phase effects on electronic properties. *Rev. Mod. Phys.*, **82**, 1959–2007.

Xiao, D., Jiang, J., Shin, J.-H., Wang, W., Wang, F., Zhao, Y.-F., Liu, C., Wu, W., Chan, M. H. W., Samarth, N., and Chang, C.-Z. 2018. Realization of the axion insulator state in quantum anomalous Hall sandwich heterostructures. *Phys. Rev. Lett.*, **120**, 056801.

Xiao, D., Shi, J. R., Clougherty, D. P., and Niu, Q. 2009. Polarization and adiabatic pumping in inhomogeneous crystals. *Phys. Rev. Lett.*, **102**, 087602.

Xiao, D., Shi, J., and Niu, Q. 2005. Berry phase correction to electron density of states in solids. *Phys. Rev. Lett.*, **95**, 137204.

Xu, G., Lian, B., and Zhang, S.-C. 2015. Intrinsic quantum anomalous Hall effect in the kagome lattice $Cs_2LiMn_3F_{12}$. *Phys. Rev. Lett*, **115**, 186802.

Xu, S.-Y., Belopolski, I., Alidoust, N., Neupane, M., Bian, G., Zhang, C., Sankar, R., Chang, G., Yuan, Z., Lee, C.-C., et al. 2015b. Discovery of a Weyl fermion semimetal and topological Fermi arcs. *Science*, **349**, 613–7.

Xu, S.-Y., Liu, C., Alidoust, N., Neupane, M., Qian, D., Belopolski, I., Denlinger, J. D., Wang, Y. J., Lin, H., Wray, L. A., et al. 2012. Observation of a topological crystalline insulator phase and topological phase transition in $Pb_{1-x}Sn_xTe$. *Nat. Commun.*, **3**, 1192.

Yan, B., and Felser, C. 2017. Topological materials: Weyl semimetals. *Annu. Rev. Condens. Matt. Phys.*, **8**, 337–54.

Yang, K.-Y., Lu, Y.-M., and Ran, Y. 2011. Quantum Hall effects in a Weyl semimetal: possible application in pyrochlore iridates. *Phys. Rev. B*, **84**, 075129.

Yao, W., Xiao, D., and Niu, Q. 2008. Valley-dependent optoelectronics from inversion symmetry breaking. *Phys. Rev. B*, **77**, 235406.

Yao, Y. G., Kleinman, L., MacDonald, A. H., Sinova, J., Jungwirth, T., Wang, D. S., Wang, E. G., and Niu, Q. 2004. First principles calculation of anomalous Hall conductivity in ferromagnetic bcc Fe. *Phys. Rev. Lett.*, **92**, 037204.

Yaschenko, E., Fu, L., Resca, L., and Resta, R. 1998. Macroscopic polarization as a discrete Berry phase of the Hartree–Fock wave function: the single-point limit. *Phys. Rev. B*, **58**, 1222–9.

Yates, J. R., Wang, X. J., Vanderbilt, D., and Souza, I. 2007. Spectral and Fermi surface properties from Wannier interpolation. *Phys. Rev. B*, **75**, 195121.

Ye, L., Tian, Y., Jin, X., and Xiao, D. 2012. Temperature dependence of the intrinsic anomalous Hall effect in nickel. *Phys. Rev. B*, **85**, 220403.

Young, S. M., Zaheer, S., Teo, J. C. Y, Kane, C. L., Mele, E. J., and Rappe, A. M. 2012. Dirac semimetal in three dimensions. *Phys. Rev. Lett.*, **108**, 140405.

Yu, R., Qi, X. L., Bernevig, A., Fang, Z., and Dai, X. 2011. Equivalent expression of Z_2 topological invariant for band insulators using the non-Abelian Berry connection. *Phys. Rev. B*, **84**, 075119.

Yu, R., Zhang, W., Zhang, H.-J., Zhang, S.-C., Dai, X., and Fang, Z. 2010. Quantized anomalous Hall effect in magnetic topological insulators. *Science*, **329**, 61–4.

Zak, J. 1989. Berry's phase for energy bands in solids. *Phys. Rev. Lett.*, **62**, 2747.

Zhang, H., Haule, K., and Vanderbilt, D. 2017. Metal-insulator transition and topological properties of pyrochlore iridates. *Phys. Rev. Lett.*, **118**, 026404.

Zhong, S., Moore, J. E., and Souza, I. 2016. Gyrotropic magnetic effect and the magnetic moment on the Fermi surface. *Phys. Rev. Lett.*, **116**, 077201.

Zhou, J., Liang, Q.-F., Weng, H., Chen, Y. B., Yao, S.-H., Chen, Y.-F., Dong, J. and Guo, G.-Y. 2016. Predicted quantum topological Hall effect and noncoplanar antiferromagnetism in $K_{0.5}RhO_2$. *Phys. Rev. Lett.*, **116**, 256601.

Index

Printed in the United States
by Baker & Taylor Publisher Services

Printed in the United States
by Baker & Taylor Publisher Services